T0400015

Electronics

Electronics
From Classical to Quantum

Michael Olorunfunmi Kolawole

CRC Press
Taylor & Francis Group
Boca Raton London New York

CRC Press is an imprint of the
Taylor & Francis Group, an **informa** business

First edition published 2021
by CRC Press
6000 Broken Sound Parkway NW, Suite 300, Boca Raton, FL 33487-2742

and by CRC Press
2 Park Square, Milton Park, Abingdon, Oxon, OX14 4RN

© 2021 Taylor & Francis Group, LLC

CRC Press is an imprint of Taylor & Francis Group, LLC

Library of Congress Cataloging-in-Publication Data
Names: Kolawole, Michael O., author.
Title: Electronics : from classical to quantum / Michael Olorunfunmi Kolawole.
Description: First edition. | Boca Raton, FL : CRC Press, 2020. | Includes bibliographical references and index. | Summary: "This book discusses formulation and classification of integrated circuits, develops hierarchical structure of functional logic blocks to build more complex digital logic circuits, outlines the structure of transistors, their processing techniques, their arrangement forming logic gates and digital circuits, optimal pass transistor stages of buffered chain, and performance of designed circuits under noisy conditions. It also outlines the principles of quantum electronics leading to the development of lasers, masers, reversible quantum gates and circuits and applications of quantum cells"— Provided by publisher.
Identifiers: LCCN 2020011028 (print) | LCCN 2020011029 (ebook) | ISBN 9780367512224 (hardback) | ISBN 9781003052913 (ebook)
Subjects: LCSH: Electronics. | Electronic circuits. | Quantum electronics.
Classification: LCC TK7815 .K64 2020 (print) | LCC TK7815 (ebook) | DDC 621.3815—dc23
LC record available at https://lccn.loc.gov/2020011028
LC ebook record available at https://lccn.loc.gov/2020011029

ISBN: 978-0-367-51222-4 (hbk)
ISBN: 978-1-003-05291-3 (ebk)

Typeset in Times
by codeMantra

Contents

Preface

The impetus for writing this book *Electronics: From Classical to Quantum* comes from the realization that there is somehow discontinuity in the understanding of the main ideas and techniques in the field of electronics. The rapid rate of progress in this field has made the new emphasis on electronics focusing on quantum information, thus making it difficult for newcomers to obtain a broad overview of the most important techniques of the field and the results drawn from the developed techniques. As a result, this book brings together the essential learning tools to understanding the basics—classical and integrated circuits arising from Boolean irreversible concepts—to evolving high-speed analytical processes based on the quantum mechanics to developing and designing reversible quantum electronic circuits. While current technology is digital and evolving to quantum, real-life conversations require both analog and digital technology. As such, an understanding of the conversion process involved is also discussed.

The purpose of this book is therefore twofold. First, it introduces the background material in chemistry, physics, and engineering necessary to understanding the formulation of digital integrated and quantum electronic circuits. This is done at a level comprehensible to readers with a background at least the equal of a beginning graduate student in one or more of these three disciplines. Second, it develops in detail the essential digital integrated and quantum electronic circuitries. With thorough study, the reader is enabled to develop a working understanding of the fundamental tools and results of quantum field, either as part of their general education or as a prelude to independent research in quantum electronics and quantum information, leading to enabling substantial improvements in the electronics in terms of innovation, flexibility, and efficiency.

In order to provide a simple but excellent explanation of the technical aspects of electronics engineering from digital to quantum, which is a recent phenomenon, as well as enabling the reader to move on to more complex design and operational concepts without great difficulty, this book is structured into nine chapters covering:

1. The processes involved in producing electronic devices or circuits that comprise many layers of single or compound elements, their fabrication processes, limitations, and classification, as well as the concepts behind *small-scale integration* (SSI), *medium-scale integration* (MSI), *large-scale integration* (LSI), *very large-scale integration* (VLSI), and *ultra-large-scale integration* (ULSI).
2. Analysis and design of logic gates as well as interfacing between various logic families: primitive, combinational, and sequential logic circuits.
3. Principles and design of transistors, as well as their applications and use in logic circuits designs including switching and wave-shaping circuits, regenerative circuits (astable and monostable), and understanding operational amplifier fundamentals including multivibrators, comparators, and Schmitt triggers.
4. Analysis, design, fabrication process, and application of Field-Effect Transistors (FETs) including *metal oxide semiconductor* (MOS) and *complementary metal oxide semiconductor* (CMOS) and variants.
5. The process of data conversion: analog-to-digital (ADC) and digital-to-analog (DAC) conversion processes and application.
6. An overview of the electronic properties of nanomaterials from bulk to nanoscale.
7. The fundamentals of laser and maser, as well as the underlying principles of quantum electronics from basic quantum mechanics to the formulation of quantum gates and circuits, including gates developed based on quantum cells, fabrication methods including self-assembly and tunneling, error correction, and an overview of quantum cryptography and the distribution of cryptographic keys in particular.

The materials in each chapter are arranged so that the reader need not continually refer elsewhere, as far as possible. Each chapter contains worked examples, real-world designs, and problems for further study. While this book is primarily intended for use as an undergraduate or postgraduate textbook, its structure will also appeal to practicing and design engineers.

MATLAB® is a registered trademark of The MathWorks, Inc. For product information, please contact:

The MathWorks, Inc.
3 Apple Hill Drive
Natick, MA 01760–2098 USA
Tel: 508–647–7000
Fax: 508–647–7001
E-mail: info@mathworks.com
Web: www.mathworks.com

Acknowledgment

I thank my sons—Remi Kolawole and Seyi Kolawole—for their love and encouragement throughout their lives. Most importantly, many thanks to my beautiful companion—Dr Helen Kolawole—for her love, care, support, and understanding over many years. You are the loveliest of women.

Author

Michael Olorunfunmi Kolawole is professor emeritus of electrical engineering: a highly respected academic and project leader, who is known for his expertise in satellite, radar and communication engineering, and remote sensing. He earned a BEng (1986) at Victoria University, Melbourne, Australia; an MEnvSt (1988) at the University of Adelaide, Australia; and a PhD (2000) at the University of New South Wales (UNSW), Sydney, Australia. He has held various professional and academic positions in Australia, Italy, and Nigeria, including Telstra Corporation, Melbourne, Australia; Tenix Defence Systems (now called BAE Systems Australia), Williamstown; Australian Bureau of Meteorology Melbourne; Australian Defence Force Academy, Canberra; Victoria University, Melbourne, Australia; and Federal University of Technology Akure, Nigeria. He holds two patents and has overseen three major operational innovations. He has published four books:

1. Satellite Communication Engineering, *2nd Edition, New York: CRC Press, 2013 (*Satellite Communication Engineering, *1st Edition, New York: Marcel Dekker, 2002);*
2. Basic Electrical Engineering, *Akure: Aoge Publishers, 2012;*
3. A Course in Telecommunication Engineering. *New Delhi: S Chand, 2010; and*
4. Radar Systems, Peak Detection and Tracking, *Oxford: Elsevier (Newnes), 2003.*

He has also published over 70 peer-reviewed papers and 25 industry-based technical reports/reviews. Professor Kolawole is the executive director of Jolade Consulting Company, Melbourne, Australia, where he has provided vision and leadership since its establishment in 2000. He has been an invited speaker at professional and academic workshops and conferences. He plays clarinet and saxophone, and also arranges and composes music.

1 Formulation and Classification of Electronic Devices

Circuits have been fundamental to the development of electronic devices: from the basic analog to integrated digital systems and to the evolving quantum computing systems. This chapter discusses the technological processes involved in producing electronic devices or circuits that comprise many layers of formation from basic single to compound elements. These elements include the type of chemical bonding involved (ionic or covalent) as well as the process of creating features in a specialized layer of material through chemical modification and how pattern transfer is carried out multiple times to form the device or circuit, and their limitations. The classification of ICs by integration levels is also discussed. An IC, also called *chip*, is an assembly of electronic components, fabricated as a single unit with miniature devices and their interconnections, built up on a thin substrate of semiconductor material(s).

1.1 INTRODUCTION

Circuits have been fundamental to the development of electronic devices. A circuit that allows continuous, variable signal is called an analog circuit. Analog circuits within electrical equipment have conveyed information through changes in the current, voltage, or frequency. An analog circuit can be used to convert the original continuous, variable signal into binary bits, usually expressed as "0" and "1": synonymous, respectively, to "off" and "on." (More is said about the conversion technique in Chapter 7.) A digital circuit operates on quantized binary bits: "0" or "1." A basic quantum circuit, however, operates on a small number of signals, called *qubits*, which is the quantum version of the classical binary bit physically realized with a two-state device. In the classical logic circuits, behavior is governed implicitly by classical physics: no restrictions on copying or measuring signal, whereas the behavior of quantum circuits is governed by quantum mechanics. (More is said about *classical logical circuits* and *quantum circuits* in Chapters 2–6 and 9, respectively.) Technically, all these circuits attempt to find circuit structures for needed operations that are functionally correct and independent of physical technology, and, wherever possible, that use the minimum number of components, logic gates, or quantum bits (*qubits* for short). Such development of circuit structures has led to miniaturized circuits of modern integrated circuit (IC) electronics. An *IC* is an electronic device that integrates (or gathers) a number of electronic components on a small semiconductor chip. Technically, an IC has a particular functionality. ICs' electronics have evolved and continued to gain from continuous research. The research outcomes have led to improvements in IC characteristics and fabrication processes used.

1.2 BASIC PROPERTIES OF SEMICONDUCTORS

Devices such as modern computer processors and semiconductor memories fall into a class known as *ICs*. They are so named because all of the components in the circuit (and their "wires") are connected together and formed on the same substrate or wafer. An IC, also called *chip*, is an assembly of electronic components, fabricated as a single unit with miniature devices and their interconnections, built up on a thin substrate (or wafer) of semiconductor material(s).

Semiconductors are unique materials, usually a solid chemical element—such as silicon (Si), or compound—such as gallium arsenide (GaAs) or indium antimonide (InSb), which can conduct electricity or not under certain conditions. Semiconductor behavior is not restricted to solids:

there are liquid conductors, for instance, mercury (Hg) and those regarded as "Type II alloys" superconductors. It should be noted that the resistivity of most metals increases with an increase in temperature and vice versa. As such, there are some metals and chemical compounds whose resistivity becomes zero when their temperature is brought near zero-degree Kelvin ($0°K$) (i.e. $-273°C$). {Note that absolute temperature, T, in degree Kelvin ($°K$), is expressed by $T = 273 + t$, where t = measured temperature in degree Celsius.} At this stage such metals or compounds are said to have attained superconductivity. The transition from normal conductivity to super-conductivity takes place almost suddenly over a very narrow range of temperature. Mercury, for example, becomes superconducting at approximately $4.5°K$. Type-II superconductors usually exist in a mixed state of normal and superconducting regions [1]. However, because of atomic diffu-sion, regions with different dopings will mix rapidly and a stable device with an inhomogeneous structure is not possible [2].

Semiconductors can be made to conduct electricity by adding other impurity elements to the semiconductor material. However, areas of semiconductor material that are highly pure and with little or no impurities would impede the free flow of electrons from atom to atom (and molecule to molecule) and would act as insulators. When electrons flow freely, the semiconductor's electrical conductivity lies in magnitude between that of a conductor and an insulator: specifically in the range of 10^{-3} to 10^{-8} S cm^{-1} (Siemens per centimeter).

An atom is a form of matter that consists of a very small nucleus surrounded by orbiting elec-trons, as depicted in Figure 1.1. The nucleus is made up of a core of protons (positively charged particles) and neutrons (particles having no charge). Protons and neutrons are composite particles, each consisting of three *quarks*, and as such, they may not be considered as fundamental particles in the true sense of the term. Nevertheless, they are regarded as being fundamental particles for all chemical purposes.

A proton's charge distribution decays approximately exponentially. Electrons are arranged around the nucleus in energy levels. The total number of electrons in an atom is the same as the number of protons in the nucleus. When the number of protons in a nucleus equals the number of electrons orbiting the nucleus, the atom is electrically neutral. The number of protons in an atom is called its *atomic number* (also called the *proton number*). Attached to an electron is an extra degree of freedom called its *spin*. The operator associated with the measurement of a spin has a vector character of its own. (More is said about *quarks* and the operator associated with the measurement of a spin in Chapter 9, Section 9.2.)

Atoms can lose or gain electrons to form charged particles called *ions*. For example, if an atom loses or gains one or more electrons, it becomes positively or negatively charged ions, respectively.

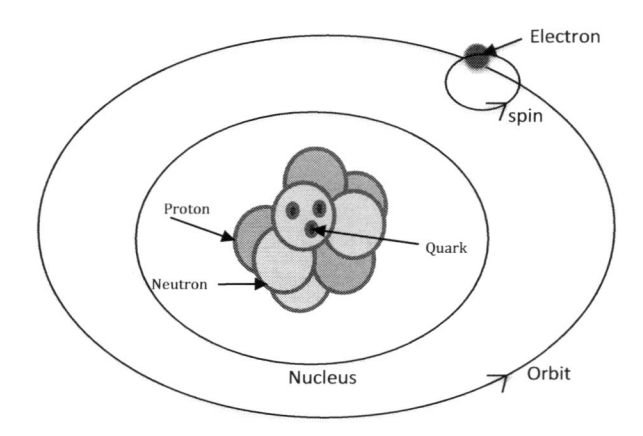

FIGURE 1.1 Fundamental particles of an atom.

Also, nuclei (plural of nucleus) that consist of an even (odd) number of protons and even (odd) number of neutrons are said to be *even-even* (*odd-odd*). However, nuclei that consist of an even (odd) number of protons and odd (even) number of neutrons are said to be *even-odd* (*odd-even*). Atomic energy levels are classified according to angular momentum, and selection rules for radiative transitions between levels are governed by angular-momentum addition rules [3]. Fundamental to the theoretical studies of atomic structure and atomic transitions is knowledge of the quantum mechanics of angular momentum.

Semiconductors are made up of individual atoms bonded together in a regular, periodic structure to form an arrangement whereby each atom is surrounded by a number of electrons. The material properties of a semiconductor are determined by its atomic bond structure. The forces in the atom are repulsion between electrons and attraction between electrons and protons, where the neutrons play no significant role.

The specific properties of a semiconductor depend on the impurities, or *dopants*, added to it. For instance, an *n-type* semiconductor carries current mainly in the form of negatively charged electrons, in a manner similar to the conduction of current in a wire. However, a *p-type* semiconductor carries current predominantly as electron deficiencies called *holes*. A hole has a positive electric charge, equal and opposite to the charge on an electron. In semiconductor physics, the flow of holes occurs in a direction opposite to the flow of electrons. Primary elements or constituents of semiconductors include antimony (Sb), arsenic (As), boron (B), carbon (C), germanium (Ge), selenium (Se), silicon (Si), sulfur (S), and tellurium (Te).

Chemical elements are categorized in the *Periodic Table of elements*—by placing them with their atomic numbers and weights, names, and symbols—in the text as shown in Figure 1.2. Chemical elements in a column share similar properties.

The conventional way of representing an atom of an element (say, of chemical symbol X) and its nuclear properties is

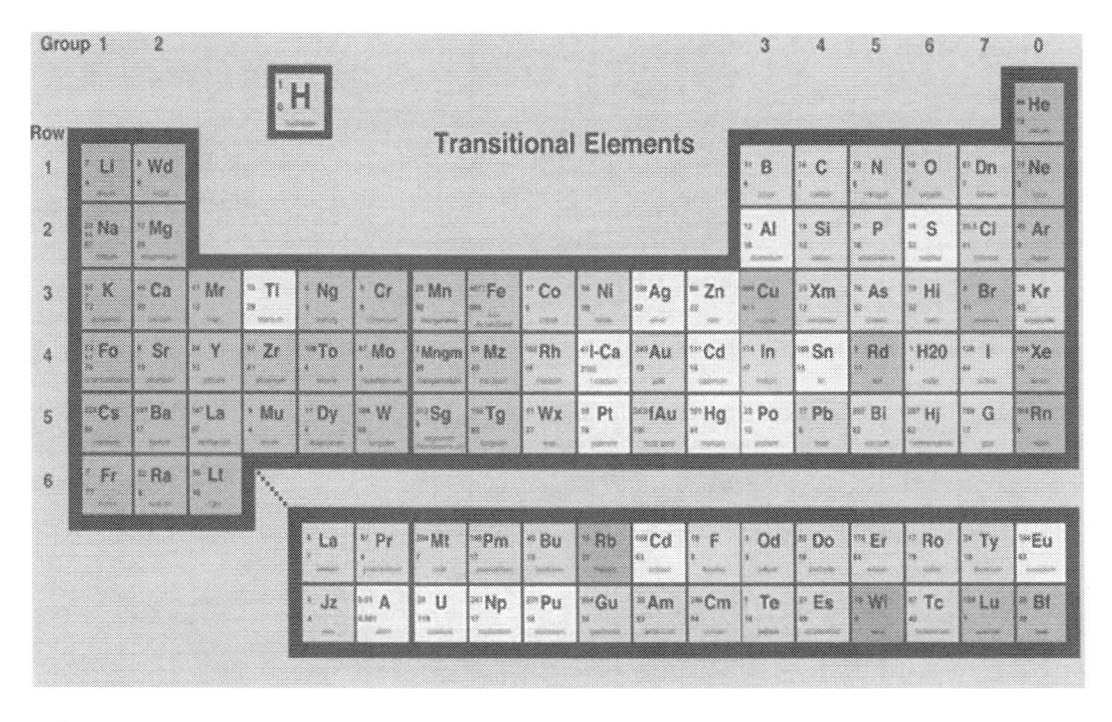

FIGURE 1.2 Periodic table of elements.

$$\ce{^{A}_{Z}X}$$ (1.1a)

where

A is the mass number, placed as the left-hand superscript to the element symbol, which is the whole number closest to the accurate relative mass (equal to the sum of the numbers of protons and neutrons in the nucleus), with the nuclear charge Ze^{+}, where e is the elementary unit of electron charge $(= 1.60217733 \times 10^{-19} \text{ coulombs})$.

Z is the atomic number, placed as the left-hand subscript, which indicates the number of protons in the nucleus. The atomic number equals to the number of electrons in non-ionized (neutral) atom. Note that each atomic number identifies a specific element, but not the isotope—a variant of an atom. Isotopes are different atoms that have the same atomic number but have different numbers of neutrons. Although all isotopes, of a given atom, have the same chemical properties their physical properties are different. This difference in properties helps their usage for different purposes. For example, an isotope of Uranium is used as a fuel in the nuclear reactors, whereas that of Cobalt is used in the treatment of cancer.

The difference between the mass number and atomic number specifies the number of neutrons, N_n; that is,

$$N_n = A - Z$$ (1.1b)

Mass numbers are very close to being whole numbers because the relative masses of nuclei are composed of a number of protons and neutrons whose relative masses are very close to 1 on the relative atomic unit (RAM) scale. The actual mass of an atom, M_a, can be expressed by the equation

$$M_a = Zm_X + N_n m_n - \frac{E_B}{c^2}$$ (1.2)

where

m_X is the mass of element X atom,

m_n is the mass of a neutron,

E_B represents the nuclear binding energy, i.e. the energy released when the atom is formed from the appropriate numbers of X atoms and neutrons. This energy is converted into a mass by Einstein's equation $(E = mc^2)$, where c is the speed of light $(c \approx 3 \times 10^8 \text{ m s}^{-1})$. For practical purposes, the last term in Equation (1.2), i.e. (E_B/c^2) is ignored, as it makes very little difference to the determination of the mass number of the atom. So,

$$M_a = Z(m_X - m_n) + Am_n$$ (1.3)

Intuitively, the mass of an electron moving relative to the speed of light can be resolved using Einstein's theory of relativity. Every known electron at rest has the same mass as every other known electron at rest. By time dilation, the speed of an electron relative to that of light can also be estimated. For relativistic motion, the mass m of a moving object is dependent upon its velocity v according to Einstein's equation:

$$m = \gamma m_o$$ (1.4a)

where m_o is the particle's rest mass; mass of an electron as measured when its speed is zero relative to an observer, which is approximately $9.1013897 \times 10^{-31}$ kg, and γ the *gamma factor* is expressed by the equation

$$\gamma = \frac{1}{\sqrt{1 - \left(\dfrac{v}{c}\right)^2}} \tag{1.4b}$$

Example 1.1

Calculate the wavelength of an electron moving at 95% of the speed of light.

Solution:

From Equations (1.4a) and (1.4b), an electron moving at 95% of its velocity of light (i.e. 0.95c) would have a mass

$$m = \frac{9.1013897 \times 10^{-31}}{\sqrt{\left(1 - 0.95^2\right)}} = 2.91478 \times 10^{-30} \text{ kg}$$

The wavelength λ of the electron can be obtained from Einstein's equation; that is,

$$E = mc^2 = \frac{hc}{\lambda} \tag{1.5}$$

where fundamental constant h is called Planck's constant and is equal to 6.62608×10^{-34} Js. Note that: $1\,\text{J} = 1\,\text{kg} \cdot \text{m}^2\,\text{s}^{-2}$. Solving for wavelength λ

$$\lambda = \frac{h}{mc} = \frac{6.62608 \times 10^{-34} \text{ Js}}{\left(2.91478 \times 10^{-30} \times 3 \times 10^8\right) \text{kg} \cdot \text{ms}^{-1}} = 7.578 \times 10^{-13} \text{ m}$$

1.2.1 ELEMENTS OF SEMICONDUCTOR

Semiconductors are made up of different elements with varying properties between them. Some semiconductors are composed of single elements, and some are composed of two or more elements (classified as *compound semiconductors*).

The elemental semiconductors composed of single elements include silicon (Si), germanium (Ge), tin (Sn), selenium (Se), and tellurium (Te); although they come from different groups in the periodic table, they share certain similarities.

The elemental-compound semiconductors composed of two or more elements include GaAs, gallium nitride (GaN), indium phosphide (InP), zinc sulfide (ZnS), zinc selenide (ZnSe), silicon carbide (SiC), silicon germanium (SiGe), InSb, and the oxides of most metals. Prominent of these oxides is GaAs, which is widely used in low-noise, high-gain, weak-signal amplifying devices. Semiconductor materials are nominally small bandgap insulators because of a low-defect interface between them. (More is said about *bandgap* in Chapters 3, 5, and 8.)

1.2.2 CLASSIFICATION OF ICS

The number of transistors and other electronic components they comprise generally classifies ICs. For instance:

- *SSI (small-scale integration)*: when a chip contains up to 100 electronic components.
- *MSI (medium-scale integration)*: roughly when a chip contains from 100 and up to 3,000 electronic components.

- *LSI (large-scale integration)*: when a chip contains from 3,000 to 100,000 electronic components.
- Circuits constructed of SSI, MSI, and LSI ICs are generally referred to as fixed logic, since the functions they realize cannot be changed without physically rebuilding the circuit. So, design changes are difficult and expensive, often impossible to realize.
- *VLSI (very large-scale integration)*: when a chip contains from 100,000 to 1,000,000 electronic components. VLSI circuits are customizable logic devices due to their cost-effectiveness, versatility, and ease of design change. In fact, the availability of computer-aided design (CAD) tools, or software, has accelerated the VLSI design process. Various electronic technologies have been employed in the design and manufacture of digital ICs: the basis arose from transistor–transistor logic (TTL) and complementary metal oxide semiconductors (CMOS). Normally, TTL logic ICs use *npn*- and *pnp*-type bipolar junction transistors while CMOS logic ICs use *CMOS field-effect transistor* (CMOSFET) or *junction field-effect transistor* (JFET) for both their input and output circuitry. (More is said about the *constituencies of bipolar junction transistors* and *CMOS* in Chapters 3 and 6, respectively.)
- ULSI (ultra large-scale integration) when a chip contains more than 1 million electronic components. ULSI (and beyond) circuitry is a future projection that would depend on practical and analogical limits in terms of material properties; that is, non-dependency on silicon and configuration. Progress is being made. For instance, components of transistors, which form the basis of modern electronics, are at the cusp of being further miniaturized—i.e. nanosized—taking advantage of quantum mechanical phenomena. Zavabeti et al. [4] demonstrated the use of metal powder (hafnium dioxide, HfO_2) dissolved in liquid gallium (Ga) alloy forming a metal-oxide skin, which is few atoms thick, and once extracted and lifted off, these oxide layers can be used in electronic components. Note that, hafnium dioxide, HfO_2, (also known as *hafnia*), is a very efficient insulator with a bandgap of $5.3 \sim 5.7\,eV$ and high dielectric constant. Importantly, *hafnia* remains highly insulating even when it becomes very thin. Because these oxide layers have a material that is at the same time efficiently insulating and atomically thin, which means we can arguably make a transistor with a thickness of only three atomic layers. The novel technique (called multisidewall patterning technique, which is an extension of the current IC technology to the nanoscale length scale) creates smooth metal-oxide layers without boundaries so electrons can move unimpeded. (More is said about *nanotechnology* in Chapter 8.)

Further miniaturization of ICs raises the level of energy dissipated per bit inherently deep within irreversible logic gates and circuitry. The energy dissipated per bit is expressed in Ref. [5]

$$E_d = kT \log 2 = E_{th} \log 2 \qquad (1.6)$$

where E_{th} is the thermal energy, k is Boltzmann's constant ($k = 1.3805 \times 10^{-23}$ JK^{-1}), and T is the absolute temperature (in degree Kelvin, K). (Note that absolute temperature $T = 273 + t$ where t = measured temperature in degree Celsius.) Unfortunately, even if all other energy loss mechanisms were eliminated, the circuit would still dissipate this energy per bit, which is the price that comes with logical irreversibility circuits.

Some logic gates (e.g. OR, AND, XOR, NOR, and NAND) that constitute classical circuitries have irreversible properties: meaning that the number of input bits does not equal the number of output bits. Quantum electronics-based circuitries, which are being formulated from reversible logic gates, would reduce, if not suppress, the unwanted heat produced as a side effect of running irreversible logic gates. The advances in technology would enable chip manufacturers to modify their chip designs to use only reversible logic gates. (More is said about *quantum gates and circuits* in Chapter 9.)

While this extremely rapid development of chip technology will continue, technological as well as physical limitations have to be considered. Presently, some interesting (theoretical) research on quantum computing is producing fascinating results, but when such ideas can be used in hardware and for mass production, it is difficult to predict considering the rising optimization problems, their constraints and objectives, as well as the quantum mechanical effects—the very nature of the electronic motion: as device size reduces its wavelength changes, which in turn necessitates additional energy. This will profoundly influence the operation of devices as their dimensions shrink beyond certain limits. (More is said about *quantum electronics* in Chapter 9.)

ICs have been used for a variety of devices, including microprocessors, communications and mainframe computers, audio and video equipment, and automobiles. IC technology changed from time to time. Of recent, instead of integrated electronic circuits, much new development involves photonic ICs (PICs)—a technique that uses light for the data transport step in ICs [6]. Nevertheless, there are four distinct activities required to producing or manufacturing an IC. These are design, fabrication, assembly, and test. These distinct activities are typically developed separately and are used sequentially to create an end product that has high quality and reliability. There are a variety of major manufacturing process technologies used in design and fabrication of silicon-based ICs including metal oxide semiconductor (MOS), bipolar, and combined bipolar and complementary-MOS (BiCMOS). While silicon-based processing dominates in semiconductor manufacturing, other compounds like GaAs—a mixture of gallium (Ga) and arsenic (As), and *graphene*—a monolayer of carbon with a light atomic number and a hexagonal graphite structure—are used as an alternative to silicon for some applications. Graphene displays a semimetallic character such as unique band structure and strong covalent features of its atomic bonding. While graphene might have technological implications, there are still formidable obstacles to making graphene devices and circuits on the scale of those based on any standard semiconductor-based material, let alone highly evolved silicon devices [7]. Also, the inability to accurately measure how temperature will affect transition metal dichalcogenides (TMD) at the atomic level, of which graphene belongs, is impeding the fabrication of much smaller chips and faster microprocessors. Temperature is a measure of the average kinetic energy of the random motions of the particles or atoms that make up a material. As a material gets hotter, the frequency of the atomic vibration gets higher [8]. Optical temperature measurements, which use a reflected laser light to measure temperature, cannot be used on TMD chips because they do not have enough surface area to accommodate the laser beam. (More is said on the essential elements of a laser device in Chapter 9, Section 9.3.)

The next subsections focus primarily on the fundamental processes (also called unit processes), with limited introduction of quantitative expressions, required for the qualitative understanding of modern IC fabrication. The unit processes are grouped by functionality.

1.3 DESIGN PHASE

Basically, IC design flow starts at the system level, describing the device (IC) behavior and system configurations within which the device needs to function. At this level, system requirements, which may be dictated by the end-customer requirements or the manufacturers wishing a competitive edge, are translated to defining key system specifications, such as functional blocks, memory and cycle estimates, pin definitions, and critical performance constraints. A basic design flowchart is shown in Figure 1.3.

An architectural specification is derived from the system specification using a variety of CAD tools, upon which a high-level architectural model of the system is created. Depending on the complexity of design, the CAD tools employed can be any object-oriented method providing a step-by-step of how the interconnectedness can be visualized and achieved. This tool, or tools, should aid in documenting design specification as per the details of each block.

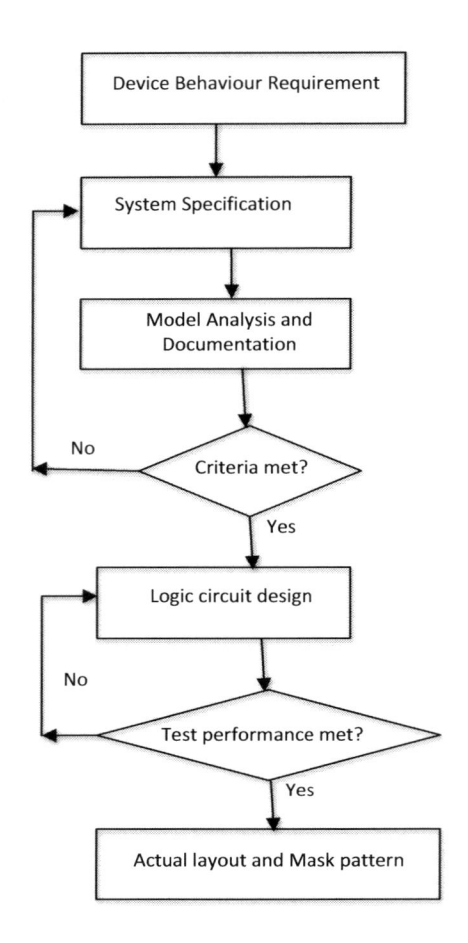

FIGURE 1.3   IC design process.

Analysis of the high-level architectural model can then be performed using a variety of languages and tools such as C/C++, MATLAB®, SIMULINK®, and SPICE (Simulation program for IC engineering), to name a few. Whichever tool, or a combination of tools, a designer uses must be capable of guaranteeing the quality of their design, enabling traceability, easy system testing and verification, and allowing for fast and low-cost design changes. It is important that, when performing a high-level architectural analysis of the model, the designer understands the trade-offs between cost competitiveness and quality of design. In the event the anticipated simulated performance conditions are not met, a closer look at the system specification and the analysis expressions and procedures is mandatory. If however the simulated performance conditions are met, the logic circuit is then designed.

During logic circuit design phase, a logic circuit diagram is drawn to determine the electronic circuit that meets the functional requirements. Figure 1.4 shows a simple circuit design, although in practice the circuit could be a schematic of thousands of discrete, interconnected circuit elements, such as transistors, diodes, resistors, capacitors, and metal wires, forming a set of functional electronic circuit blocks.

From basic physics, these electronic circuit elements perform different functions. For instance, resistors R provide a fixed amount of resistance to current flow; capacitors C store electronic charge until discharged; diodes allow current to flow in one direction but not in the opposite direction; transistors provide two major modes of action: as a switch turning current flow on and off, and act

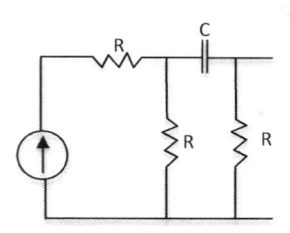

FIGURE 1.4 A simple circuit diagram.

as an amplifier whereby an input current produces a larger output current. (More is said about transistors and their functionalities in Chapter 3.) The "wires" interconnect these elements and provide conducting paths between them.

Upon completion of the logic circuit diagram, a series of simulations is performed to test the circuit's operation. If there is no issue with the operation, the actual layout pattern for the devices and interconnections is designed and a mask pattern is drawn by any CAD tool. It is during this circuit-modeling step and physical verification that engineers design the IC for maximum reliability, often including extra circuits to protect the IC from excessive external currents and voltages. Figure 1.5 is an illustration of a mask pattern.

Mask patterning, or circuit layout, is the set of mask patterns for particular layers (one in this case, as in Figure 1.5): it could be as complex as possible showing planar geometric shapes that correspond to the pattern of discrete layers making up the components of the IC. The layout must pass a series of checks ensuring there are no defects. This process of checks is known as physical verification. Afterwards, a photomask is needed to transcribe the design onto the wafer. The photomask is a copy of the circuit pattern, drawn on a glass plate coated with a metallic film, and a graphic illustration is shown in Figure 1.6.

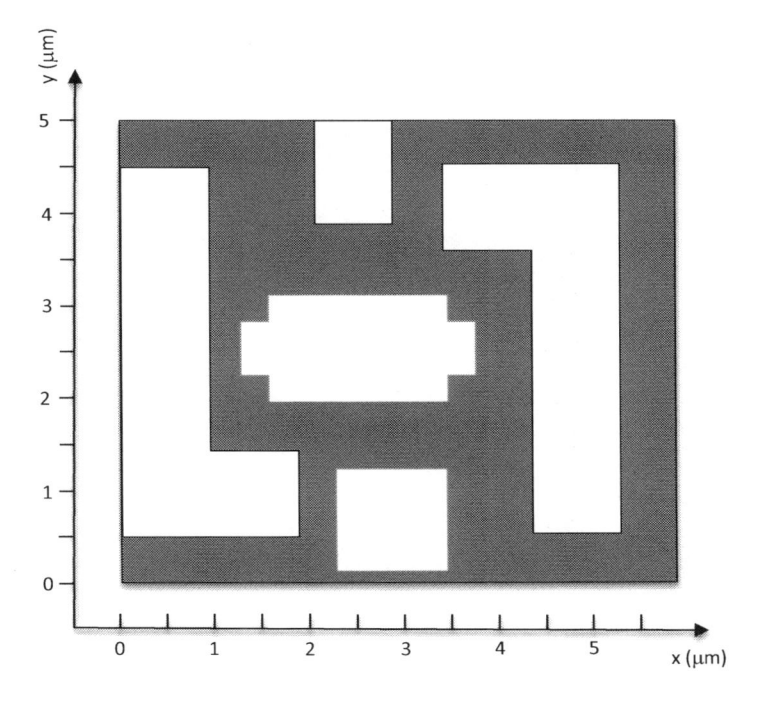

FIGURE 1.5 Mask pattern.

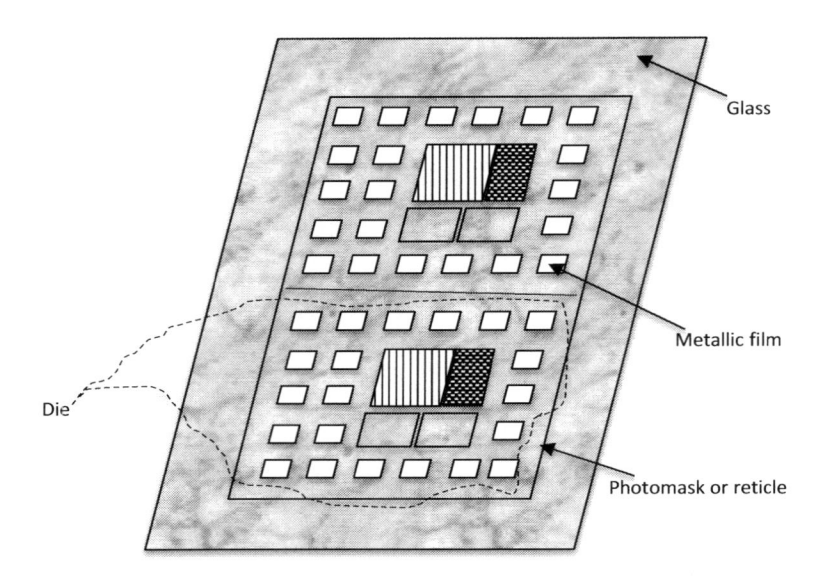

FIGURE 1.6 Mask creation.

The glass plate allows light to pass through, but the metallic film does not. Due to increasingly high integration and miniaturization of the pattern, the size of the photomask (reticle) is usually magnified four to ten times the actual size, depending on the irradiation equipment employed [9]. Typically, a reticle may have the patterns for a few *die* on it and will be stepped across the wafer exposing the pattern after each step to cover the wafer with patterns. Each self-contained area in a reticle is known as a *die* (dotted curve in Figure 1.6). In order to ease the task of reticle fabrication and make the process less defect-sensitive, reticle patterns are either five or four times the size of the desired feature on the wafer, and the reticle pattern is optically shrunk before reaching the wafer.

1.4 MANUFACTURING PROCESS

The process of manufacturing or fabricating integrated semiconductor devices and integrated electronic-microfluidic devices with multilayer configuration is a complex series of sequential steps. Practically all semiconductor devices are fabricated in a planar geometry, with very few exceptions, e.g. large volume coaxial detectors. The materials used in chips fabrication are mostly semiconductors—such as silicon (Si) or germanium (Ge), or compounds—such as GaAs or InSb—whereas that of microfluid chips can be ceramics (e.g. glass), semiconductors (e.g. silicon), or polymer—such as polydimethylsiloxane (PDMS). The main development objective of IC chips is to add the correct dopants to the substrate material (e.g. silicon) in proper amounts in the right places and to connect the transistors thus produced with thin films of metal separated by thin films of insulating materials [10]. While that of microfluidics is to incorporate multiple channels (fluidics), electronic and mechanical components or chemical processes onto a single chip-sized substrate [11]. This process can be achieved by an alignment bonding of top and bottom electrode-patterned substrates fabricated with conventional lithography, sputtering, and lift-off techniques [12]. Consequently, the fabrication processes adopted from the semiconductor-IC can still be used to fabricate some of the integrated-microfluidic devices. Thus, this section briefly describes the established steps and techniques used in a modern IC manufacturing process and the interaction between the various steps. These techniques are characterized by four key steps, namely *wafer fabrication*, *lithography*, *device formulation*, and *assembling* and *testing*; as depicted in Figure 1.7. Each process is discussed under appropriate headings.

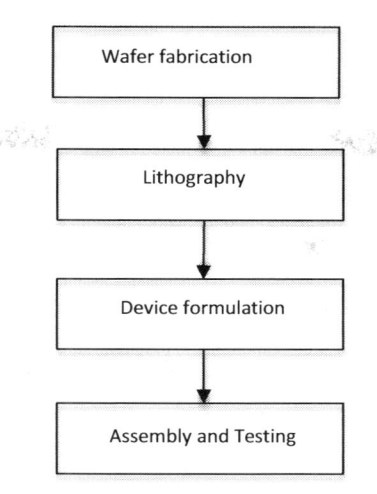

FIGURE 1.7 IC manufacturing process.

1.5 WAFER FABRICATION

The wafer growing process is as follows. A seed crystal of silicon is immersed in a bath of molten silicon. It is slowly pulled out, and because crystal growth occurs uniformly in all directions, the cross section is circular and, as it is pulled, it forms a cylindrical ingot of pure silicon. The ingot pulling process lasts for almost 24 hours and produces a cylinder of diameter larger than is desired. The ingot is ground down to the required diameter, and then is sliced into individual wafers, which, at this stage, would have a rough texture and need to be finely polished to meet the surface flatness and thickness specifications. The polished wafers are then extremely cleaned ensuring particle free at the start of fabrication. Figure 1.8 shows two types of wafers: both (a) and (b) are circular slices of varying diameters (typically available in diameters of 150, 200, or 300 mm) but are distinguished by end grounding. For instance, Figure 1.8a has a flat ground section onto one or more edges to mark how the crystal planes are oriented in the wafer, whereas in Figure 1.8b a small notch in place of a flat. A wafer is approximately 1 mm thick. In essence, the resulting wafer is a highly polished, nearly perfect single crystal with its planes endeavored to precisely orient towards the wafer surface.

There are three major types of silicon wafers currently in use for IC fabrication [13]:

- *Raw wafers*: these are silicon wafers that require no additional processing. This type is mainly used for *dynamic random-access memory* (DRAM): main memory in personal computers.
- *Epitaxial wafers*: these are silicon wafers with a single crystal silicon layer deposited on them. The deposited layer typically has different properties from the underlying raw wafer.

FIGURE 1.8 Wafer orientation indications: (a) wafer with flat edge and (b) wafer with notch edge.

Epitaxial layers allow the properties of the layer in which the devices are formed to be more tailored than in a raw wafer.

- *Silicon on insulator (SOI) wafers*: these are silicon wafers upon which an insulating layer is formed with a thin single crystal silicon layer on top of the insulating layer. SOI wafers reduce the amount of power drawn by an IC when the circuit is switching at a high speed.

1.6 LITHOGRAPHY

In order to build a multicomponent IC, we need to perform different modifications to different areas of the wafer. The method by which we define which areas will be modified is known as *lithography*. Lithography is similar to photographic printing. It creates patterns in a layer of photoresist that coats a prepared wafer (or substrate). Once the patterns are created, the elements of the IC—the electrodes of the transistor and/or the wires connecting them—are formed by etching, depositing material, or ion implantation. These processes are described as follows.

i. Clean wafer and apply on the wafer surface a thin uniform layer of a viscous liquid (photoresist) and heat the surface to drive off moisture and improve adhesion of photoresist. This process is called *priming*. Dispense onto the center of the wafer a small amount (about 1 µm) of photoresist and then spin at high speed to produce a uniform thin film.

ii. The photoresist is hardened by baking and is then selectively removed by projection of light through a reticle containing mask information.

iii. Image the mask from reticle (photomask) on a photosensitive film (photoresist) that coats the wafer and then scan ("step") the wafer to the next position.

iv. Dissolve photoresist in alkaline solutions (called "developer") when it has been exposed to ultraviolet (UV) light. This approach is called *positive resist.*

Some clarification here. Positive resist decomposes UV light. Wherever the underlying material is to be removed, the resist is exposed to UV light. Exposure to the UV light changes the chemical structure of the resist so that it becomes more soluble in the developer. The exposed resist is then washed away by the developer solution, leaving windows of the bare underlying material. The mask, therefore, contains an exact copy of the pattern, which is to remain on the wafer. However, negative resist polymerizes when exposed to the UV light and becomes more difficult to dissolve. Therefore, the negative resist remains on the surface wherever it is exposed, and the developer solution removes only the unexposed portions. Masks used for negative photoresists, therefore, contain the inverse (or photographic "negative") of the pattern to be transferred.

v. Develop the image; bake the resist to higher temperature to toughen it against etching, and stabilize the film prior to subsequent processing. The end result is that for a well-designed process: the postdevelop step ensures that the pattern from the reticle is replicated in the photoresist.

vi. Transfer pattern to an underlying film by selectively etching it.

Etching refers to the physical or chemical removal of unwanted material from the surface of the wafer (e.g. oxide films and metallic films). Etching with liquid chemicals is called "wet etching," and etching with gas is called "dry etching". The oxide films are etched with chemicals, and the metallic films are etched with plasma. An example of "selective etching" is removing an underlying film without etching photoresist. When etching is done at the same rate in all directions, it is called *isotropic* etching, but when it is done in one direction, it is called *anisotropic* etching.

It is possible that, following etching, the photoresist pattern may be stripped off and the wafers sent on for further processing that might include further photolithographic steps.

vii. Remove photoresist using oxygen plasma or organic solvents.

Plasma is a partially ionized gas made up of equal parts of positively and negatively charged particles. Flow of gases through an electric or magnetic field can generate plasma. These electromagnetic fields remove electrons from some of the gas molecules, thereby causing liberated electrons to be accelerated, or energized, by the fields. The energized electrons slam into other gas molecules, liberating more electrons, which are accelerated and liberate more electrons from gas molecules, thus sustaining the plasma.

There are different types of lithographic methods, depending on the radiation used for exposure, namely *photolithography* (also called *optical lithography*), *X-ray lithography*, *electron beam lithography* (EBL), and *ion beam lithography.*

The sequence of photolithography (*optical lithography*) can be modeled, as shown in Figure 1.9, and briefly summarized as follows. Plant or deposit a thin film of some viscous liquid on a substrate of the wafer (Figure 1.9a). The exposed portions (photoresist layers) are dissolved in the developer solution (Figure 1.9b). An UV light is shone through the mask onto the photoresist (Figure 1.9c). The developed resist is transferred onto the pattern mask (Figure 1.9d). Selectively remove unwanted material from the surface of the wafer, particularly where it is exposed through the openings in the resist, as in Figure 1.9e. And finally, remove the resist leaving the patterned oxide, as in Figure 1.9f.

In the age of the lens system architecture, the resolution of the optical lithographic system R_{res} (that is, the smallest feature size that can be patterned or distinguished) is related to the wavelength of the light used, which is determined using the Rayleigh criterion (that calculates the resolution of photolithography equipment) as in Ref. [14]

$$R_{res(pl)} = k_1 \frac{\lambda}{NA} \tag{1.7}$$

where k_1 is a process-dependent (Rayleigh) constant—typically in the range of 0.4–0.9; λ is the wavelength; and NA is the numerical aperture of the lens system. Numerical aperture is defined as

$$NA = \mu \sin \alpha \tag{1.8}$$

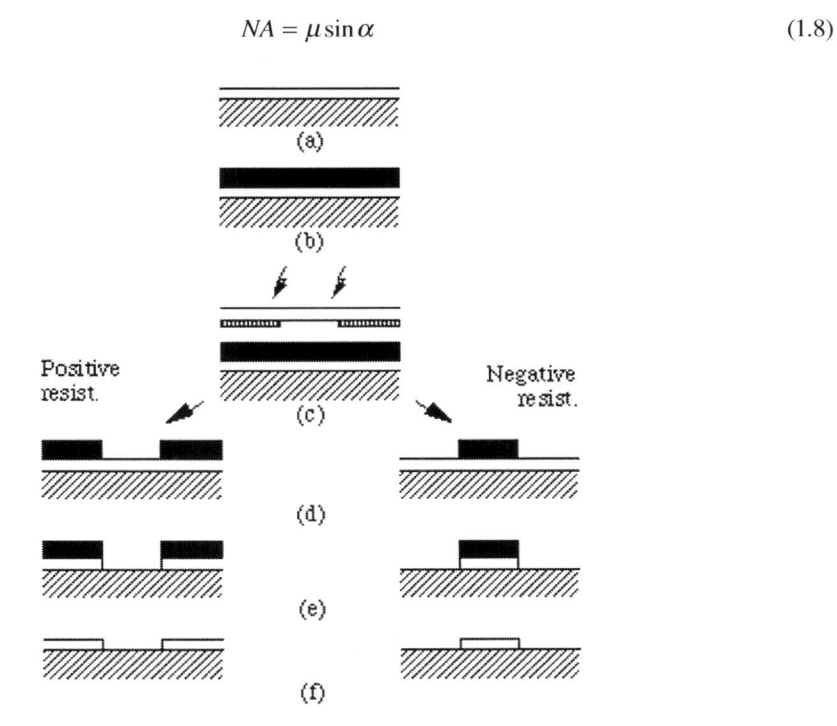

FIGURE 1.9   Photolithography model.

where μ is the refractive index of the medium between the lens and the wafer and α is the semi-angle of the exit lens (or half angle of the cone of light converging to a point image at the wafer). To increase NA, the value of the refractive index, μ, can be increased to that higher than air. This is called *immersion* lithography. Whilst Rayleigh criterion allowed for the realization of improvements to resolution through increasing NA, shortening wavelengths, λ, or reducing the k_1 factor, light diffraction sets a fundamental limit on optical resolution, thereby posing a critical challenge to the downscaling of nanoscale manufacturing [15]. It has been argued that, since light includes broad wavelengths, we could utilize shorter wavelengths (for X-rays or EBL) if resolution improvement based on the Rayleigh criterion is to be achieved [16].

X-Ray lithography, similar to traditional photolithography, is a simple one-to-one shadow-projection process (i.e. step-and-repeat technique) in vacuum and represents a particle density lower than the atmospheric pressure (e.g. ultra-high vacuum, UHV) or low pressure in helium (He) atmosphere. An X-ray lithography schematic diagram is shown in Figure 1.10.

This type of lithography uses X-rays to transfer a pattern from a mask to an X-ray resist on the substrate. Mask is made of an absorber (e.g. typically of gold (Au) or compounds of tungsten (W) on membrane (e.g. silicon substrate, Si_3N_4). The X-ray source can be a plasma radiation (hot ionized gas) or a synchrotron radiation (electromagnetic radiation due to charged acceleration). The main factors limiting the resolution are Fresnel diffraction, fast secondary electrons, the relatively low mask contrast attainable in the soft X-ray range, and, most importantly, the individual radiation characteristics of the X-ray source in question [17]. Resolution limit, $R_{res(X\text{-}ray)}$, due to diffraction can be quantified using

$$R_{res(X\text{-}ray)} = \frac{3}{2}\sqrt{\lambda\left[g + \frac{t_r}{2}\right]} \tag{1.9}$$

where λ is the wavelength, g is the gap, and t_r is the resist thickness. Precise gap must be maintained: too close will damage membrane mask, and too far will reduce resolution [18].

Optical lithography and X-ray lithography methods are basically used to make microdevice structures.

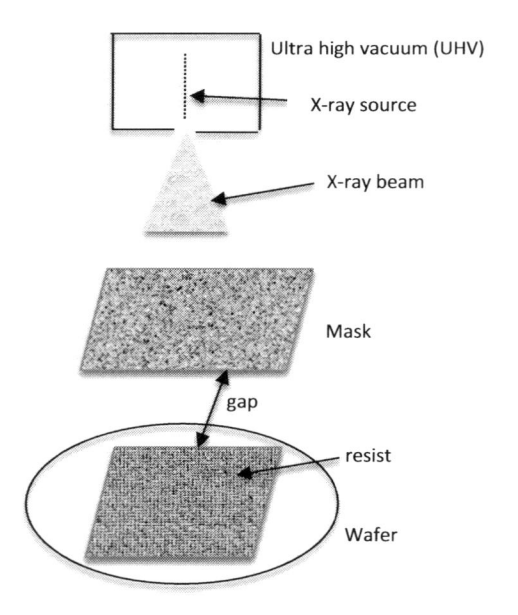

FIGURE 1.10   X-Ray lithographic schematic.

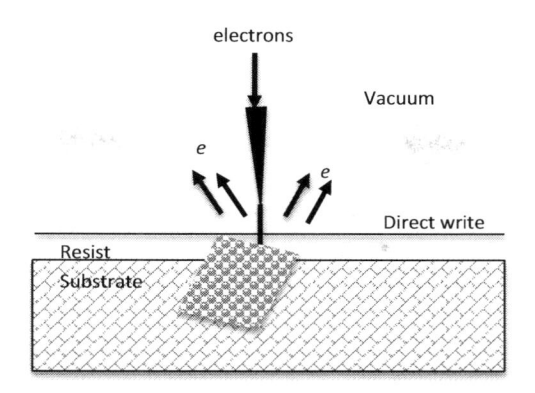

FIGURE 1.11   Electron-beam lithography.

In the EBL, the pattern is recorded in an electron-sensitive film (or resist) deposited on the sample before exposure by spin coating. The resist pattern schematic is shown in Figure 1.11. In the process, the electron beam induces a change in the molecular structure and solubility of the resist film. Expose the pattern with an electron beam. Afterwards, develop the resist in a suitable solvent to selectively dissolve either the exposed or unexposed areas of the resist.

The advantage of using EBL is the use of a direct-write method that focused on electron beam to expose resist. However, EBL suffers from long-range proximity effects at high energies due to backscattered electrons. To enhance lithographic accuracy and resolution, proximity effects correction can be considered.

The ion beam lithography is a variation of the EBL technique using a focused ion beam (FIB) instead of an electron beam. An ion beam scans across the substrate surface and exposes an electron-sensitive coating. The wafer can be exposed with a photomask, or like in EBL without a mask, allowing a direct doping of the wafer without the use of masking layers. As in EBL, electron beams in FIB generate more low-energy secondary electrons to activate the physical or chemical reactions in the resist materials. In FIB lithography, proximity effects are negligible due to the very low number of ions that are backscattered [19].

To further advance mass production of nanoturized device structures, as well as in emerging quantum device fabrication, there are many lithographic approaches currently being implemented in industry and research, where patterning is performed using proximal probe nanolithography, scanning tunneling microscope (STM)-based lithography, or atomic force microscope (AFM)-based lithography, as well as deposition and control of nanoparticles and atoms/molecules. Other lithographic procedures, such as nanoimprint (mold) lithography, plasmonic-assisted lithography, laser interference lithography, nanosphere lithography, chemistry-based nanofabrication, local electro etching, dip pen lithograph with AFM are being studied, tested, or considered [15,20].

1.7   DEVICE FORMULATION

The device formulation stage constructs the transistors on the wafer, and the formulation process is depicted in Figure 1.12, and is summarized as follows for silicon substrate.

i. *Grow oxide films*:

Grow oxide films (insulation films) on the wafers sliced from uniformly doped ingot. (More is said about *doping* in Section 1.7.1.1.) The oxide films protect the wafers but must be thin enough not to block the implantation. An example of the insulation films is silicon oxide (SiO_2). Silicon dioxide is also known as silicon (IV) oxide. Structurally, each silicon atom is bridged to its neighbors by an oxygen atom, as seen in Figure 1.13.

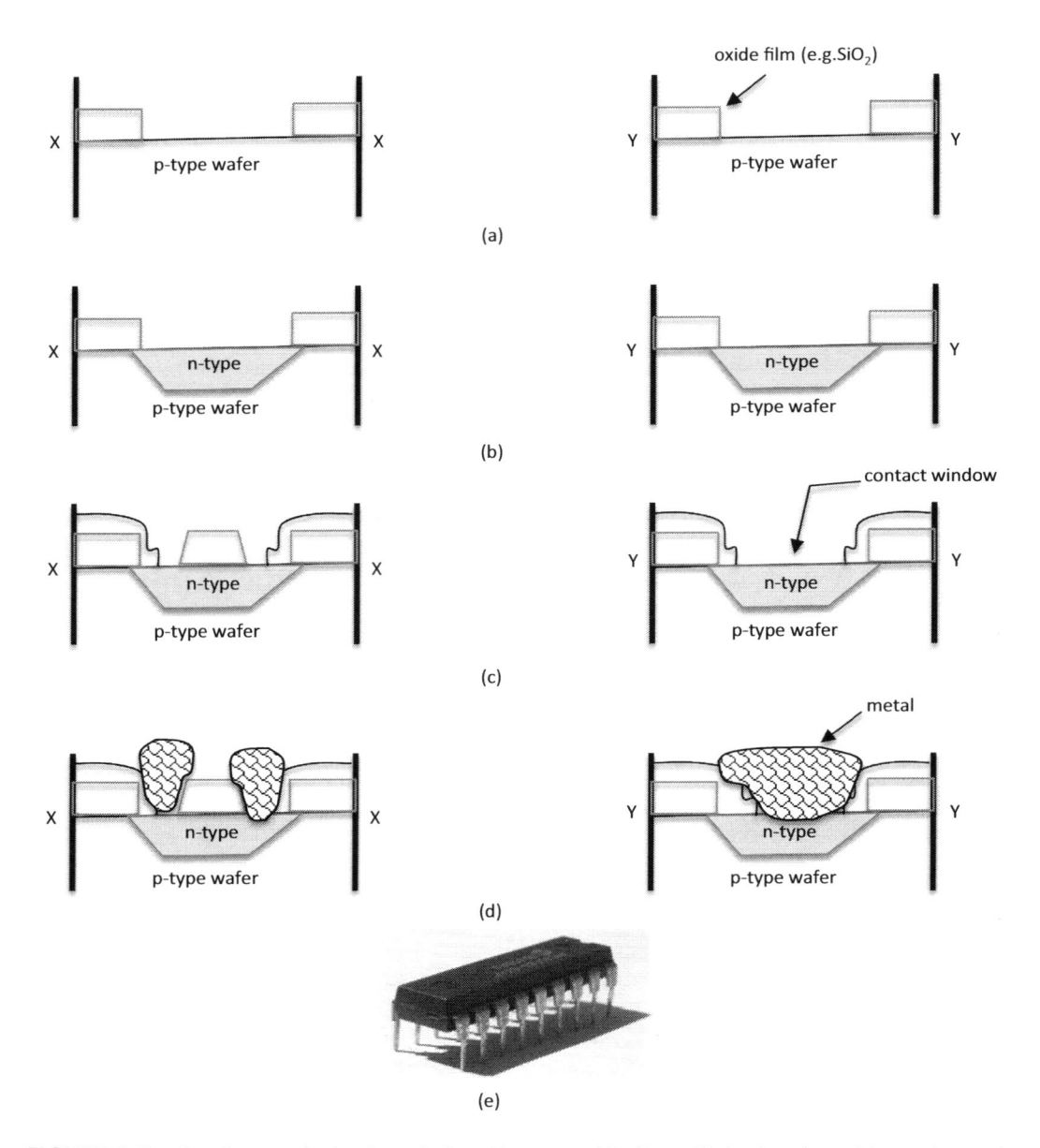

FIGURE 1.12 Step-by-step device formulation: (a) grow oxide films, (b) implant impurities and anneal, (c) deposit dioxide films, (d) create window opening and sputter metal, and (e) test formulated device and seal into packages.

We know from basic chemistry that silicon dioxide (SiO_2), like some ceramic materials, can exist in either crystalline or non-crystalline (amorphous) form. In crystalline materials, a lattice point is occupied either by atoms or ions depending on the bonding mechanism. These atoms (or ions) are arranged in a regularly repeating pattern, as shown in Figure 1.13, in long-range order (i.e. in three dimensions). Each silicon atom corresponds to a volume of $2 \times 10^{-23}\,\text{cm}^3\,(= 0.02\,\text{nm}^3)$. In contrast, in amorphous (non-crystalline) materials, the atoms exhibit only short-range order. The type of bonding (ionic or covalent) and the internal structure (crystalline or amorphous) affect the mechanical, electrical, thermal,

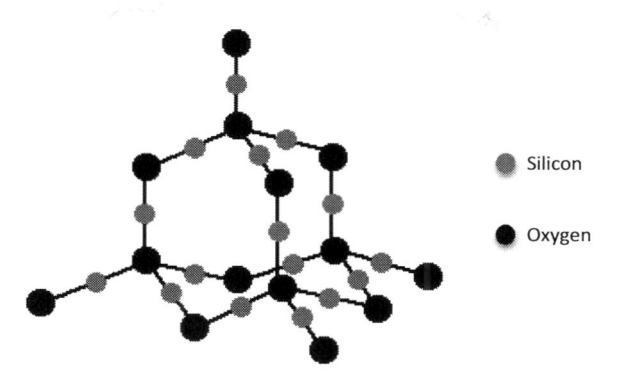

Silicon

Oxygen

FIGURE 1.13 Regular pattern of crystalline silicon dioxide.

and optical properties of ceramic materials. Note that chemical bonding is about the rearrangement of electrons due to the arrangement of the nuclei.

ii. *Implant impurities and anneal*:

Implant the dopant impurities into the wafers. The most widely used technique to introduce dopant impurities into semiconductor is the *ion implantation* technique. The ionized particles are accelerated through an electrical field and targeted at the wafer. These impurities are used to change the electrical properties of silicon. The induced ions create regions (e.g. source and drain) with different properties of the wafer. Following implantation, heating the wafer at a moderate temperature, for a few minutes, repairs lattice damage to the crystal. This process is called *annealing.*

iii. Deposit dioxide films on the wafer via chemical vapor deposition (CVD) process to act as an insulator between the gate and the silicon. There are various dioxide film materials: examples are listed in Table 1.1. The CVD process reactions bring about a chemical decomposition of certain elements or compounds by using heat, light, pressure/vacuum, and/or plasma energy to form a stable solid. CVD-deposited oxides are used at several different places in the manufacturing sequence. For example, to increase the oxide thickness of thermal oxides; capacitor dielectric; dielectric over polysilicon or metal; buffer oxide layer to match mechanical requirements; masking oxide layer; and final passivation (to protect surface from damage). Note that deposited oxide is not normally used as a gate oxide for MOS transistors. Gate oxide for MOS transistors is formed by some thermal oxidation technique.

iv. Create a contact window opening in the oxide where to build the transistor's gate region. Deposit on the interlayer insulation film by sputtering. Sputtering is the process of covering a surface with metal. Aluminum is the standard "wire" for ICs and is usually deposited by

TABLE 1.1

Chemical Vapour Deposition (CVD) Commonly Deposited on IC Films

Film Materials	Chemical Symbols
Silicon dioxide	SiO_2
Silicon nitride	Si_3O_4
Silicon oxynitride	$Si_xO_yN_z$
Polysilicon	Si
Titanium nitride	TiN
Tungsten	W

"sputtering." When a Physical Vapor Deposition (PVD) process deposits polysilicon, the process is often known as "sputtering." Tungsten (grown from a gas reaction) is sometimes used, with increasing interest in copper. Note that the gate is a conductive layer, which is separated from the bulk silicon by a thin gate oxide. The gate oxide needs to be thin since the electrical field must transfer across this insulator. The thinner the metal oxide layer, the faster and more energy efficient the electronic component is. After metallization is complete, the surface is covered with another interlayer insulation film (or the "cover layer") for protection, and the wafer is now complete.

v. Test each chip on the completed wafer, electronically ensuring that individual IC meets the expected direct current (DC) and function characteristics. Then seal ICs into packages. IC packaging is receiving much more attention now than in the past. Today, systems manufacturers, as well as IC manufacturers, realize that an increasing percentage of system performance, or performance limits, is determined by the IC-and-package combination, rather than just the IC.

1.7.1 ESTIMATING IC RESISTANCE AND CAPACITANCE AND UNCERTAINTY OF FABRICATION

1.7.1.1 Layout Resistance

Everything has some level of resistance. Likewise, in an IC, every material used in its fabrication has its resistance value, which varies. We know from our basic science that the resistance of a wire depends on three factors: the length of the wire (L) [in meter, m], the cross-sectional area (a) [in meter-squared, m^2] of the wire, and the resistivity of the material composing the wire (ρ) [in ohm-meter, Ω m]. The resistance can be formalized as

$$R = \rho \frac{L}{a} (\Omega) \tag{1.10}$$

If we consider a rectangular sheet of polysilicon IC of Figure 1.14 having length L (in microns) and width W (in microns) and thickness t_{ox} (in microns), its resistance value is

$$R = \rho \frac{L}{W t_{ox}} = \left(\frac{\rho}{t_{ox}}\right) \frac{L}{W} (\Omega) \tag{1.11}$$

the significance of the (.) term becomes obvious later.

Equation (1.11) assumes uniform doping concentrations of the sheet, which in practice is not, thereby having varying resistivity. As a result, it is convenient to work with a parameter called sheet

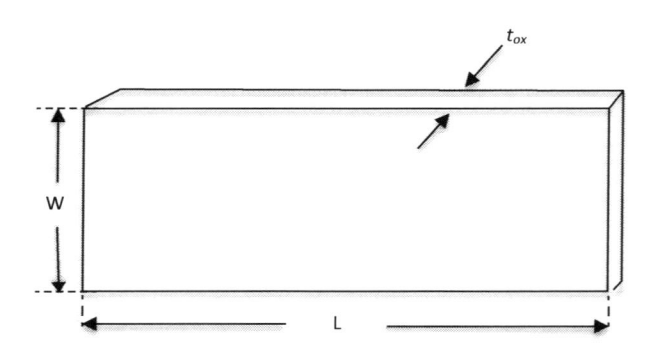

FIGURE 1.14   A rectangular sheet of polysilicon layout.

resistance R_s, which factors in the varying doping concentration and thicknesses. So, for a region of length L and width W, the IC layout resistance is expressed as

$$R = R_s \frac{L}{W} (\Omega) \tag{1.12}$$

Note that when $L = W$, the layout is square and $R = R_s$. It becomes clearer in Chapters 5 and 6 that the sheet resistance R_s is dependent on the doping method and concentrates.

Doping literally means adding impurities to something that is intrinsically pure. In the case of semiconductor, for example, doping is used purposefully for modulating its electrical properties. The impurities used are dependent upon the type of semiconductor. The most common dopants used to dope silicon (Si) (of n-type semiconductor, see Figure 1.15a), are phosphorus (P) or arsenic (As).

As shown in Figure 1.2, the periodic table of the chemical elements, phosphorus (P) or arsenic (As), has five valence electrons; although they behave like Si plus an extra electron. This extra electron contributes to electrical conductivity (σ), and with a sufficiently large number of such dopant atoms, the material can display metallic conductivity. In the case of p-type semiconductor, as in Figure 1.15b, the dopants used are boron (B) or aluminum (Al). B or Al has one electron less than Si. It thereby creates a "hole" around silicon, which can hop from one site to another. The hopping of a hole in one direction corresponds to the hopping of an electron in the opposite direction, again contributing to the doped silicon electrical conductivity (σ). Conductivity by positive holes is called *p-type*. While conductivity by negative electrons is called *n-type*. In the case of *p*-type semiconductor, as in Figure 1.15b, the dopants used are boron (B) or aluminum (Al). B or Al has one electron less than Si. It thereby creates a "hole" around silicon, which can hop from one site to another. The hopping of a hole in one direction corresponds to the hopping of an electron in the opposite direction, again contributing to the doped silicon electrical conductivity (σ). In both doping instances, rather than n and p being equal, the n electrons from the donor usually totally outweigh the intrinsic n- and p-type carriers so that the semiconductor electrical conductivity is dependent on the electronic charge, q, across the material, the "electron" or "hole" mobility, μ, and doping concentration, N.

When considering the microelectronics of the semiconductor, the sheet resistance is expressed thus as

$$R_s = \left(\frac{\rho}{t_{ox}} \right) = \left(\frac{1}{q\mu N t_{ox}} \right) \tag{1.13}$$

where electronic charge $q = 1.602 \times 10^{-19}$ Coulomb.

The IC layout is sort of a maze of rectangular or square metallic layers. Each layer has some level of resistance, calculated using Equation (1.12). Non-rectangular regions can be accounted for by defining an effective number of squares, where L/W is the number of squares.

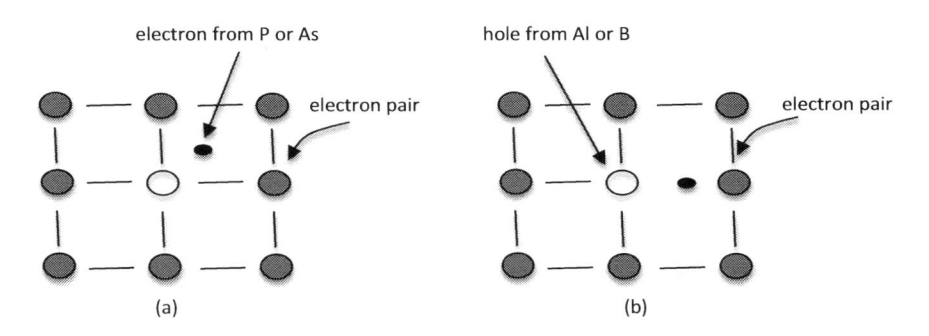

FIGURE 1.15 Doping of semiconductors: (a) n doping and (b) p doping

Example 1.2

For a typical layer thickness of $0.5\,\mu m$ and average doping levels of $10^{15}\,cm^{-3}$ to $10^{19}\,cm^{-3}$, the sheet resistance ranges from $100\,k\Omega\,sq^{-1}$ to $10\,\Omega\,sq^{-1}$. Typical sheet resistances R_s are shown in Table 1.2.

1.7.1.2 Layout Capacitance

The layout capacitance can be modeled as a parallel plate, as shown in Figure 1.16. We know from basic electromagnetic theory that the charge and electric field characteristics in an actual parallel plate capacitor can be approximated using the *ideal parallel plate capacitor* model. In a real parallel plate capacitor, the surface charge densities are not uniform since the charge density grows large at the edges of the plates: an effect called *electric field fringing*. This effect is small for closely spaced large plates, justifying modeling of a parallel plate capacitor. This model becomes more accurate as the capacitor plate area grows larger and the capacitor plate separation grows smaller. The uniform electric field, E, in the parallel plate capacitor means that the electric flux density within the capacitor is also uniform, and no electric field is outside the volume.

So, the capacitance of the parallel plate capacitor is determined by

$$C = \frac{Q}{V} = \frac{\varepsilon a}{d_s} = \frac{\varepsilon}{d_s}(L \cdot W) \tag{1.14}$$

where Q, V, ε, d_s, and a are the surface charge, potential, insulator permittivity, separation distance, and the plate area, respectively. Note that the charge across the capacitor may be written as the product of uniform electric field and separation distance (i.e. $V = E \cdot d_s$).

Given that the electric field in the volume between the plates, as well as the energy density, is uniform, the total energy stored W_E in the capacitor can be found by integrating the energy density associated with the capacitor electric field, thus

$$W_E = \iiint_V \frac{1}{2}\varepsilon E^2 \, dv = \left(\frac{1}{2}\varepsilon E^2\right)(ad_s) = \frac{1}{2}\underbrace{(\varepsilon E a)}_{Q}\underbrace{(E d_s)}_{V}$$

$$= \frac{1}{2}QV = \frac{1}{2}\frac{Q^2}{C} = \frac{1}{2}CV^2 \tag{1.15}$$

This expression is valid for any capacitor on the assumption that the medium is a perfect insulator. However, if the medium between the conductors of a capacitor is not a perfect insulator, there is a finite resistance R_Δ between the conductors. This resistance can be expressed as

$$R_\Delta = \frac{d_s}{\varepsilon a}\frac{\varepsilon}{\sigma} = \frac{d_s}{\sigma LW} \tag{1.16}$$

TABLE 1.2

Materials and Their Ohmic Resistance R_s Values per Area

Description	Ohmic Value (Ω sq^{-1})
Implanted layers in silicon (Si)	10–10^5
n+ polysilicon (using thickness, $t_{ox} = 0.5\,\mu m$)	20
Aluminum (using thickness, $t_{ox} = 1.0\,\mu m$)	0.07
Silicided polysilicon	5
Silicided source/drain diffusion	3

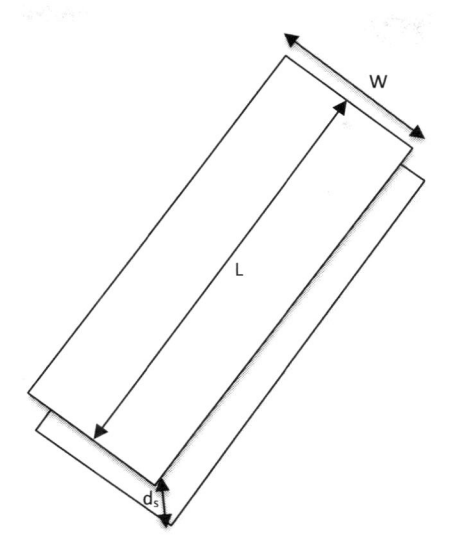

FIGURE 1.16   Capacitance of parallel plate.

where σ is the charge density.

Given the capacitance for a particular capacitance geometry—whether parallel plate, coaxial, or spherical, the corresponding relaxation time T_x can be determined by

$$T_x = CR_\Delta \tag{1.17}$$

The ratio of charge to potential for the capacitor for cylindrical and spherical geometry capacitances is expressed as follows.

Cylindrical capacitor, Figure 1.17, assumes the medium is a perfect insulator. The capacitance of the cylindrical geometry can be found by solving Laplace's equation in cylindrical coordinates. In this space, the E-field falls off as $(1/r)$, where $r = r_2/r_1$. Radii r_1 and r_2 are inner radius and outer radius separated by an insulating medium ε, respectively. Thus, the integral of the field (voltage) varies as the log of the distance. The capacitance is determined by

$$C = \frac{Q}{V} = \frac{2\pi\varepsilon L}{\log\left(r_2/r_1\right)} \tag{1.18}$$

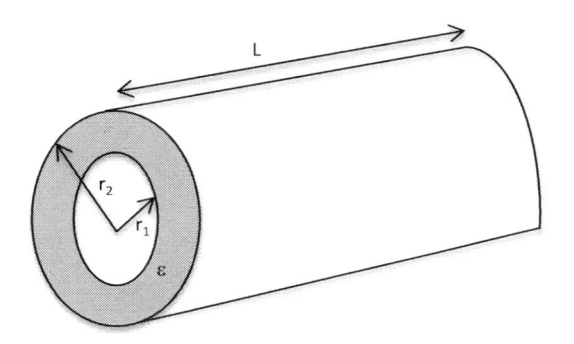

FIGURE 1.17   Geometry of cylindrical capacitor.

And, for an imperfect insulator, the finite resistance $R_{\Delta Co}$ between the conductors is

$$R_{\Delta Co} = \frac{\log(r_2/r_1)}{2\pi\sigma L} \tag{1.19}$$

Spherical capacitor, Figure 1.18, assumes the medium is a perfect insulator. The process used to determine the capacitance of the ideal spherical capacitor is the same as that used for the ideal cylindrical capacitor.

Consequently,

$$C = \frac{Q}{V} = \frac{4\pi\varepsilon}{\left[\dfrac{1}{r_1} - \dfrac{1}{r_2}\right]} \tag{1.20}$$

And, for imperfect insulator, the finite resistance $R_{\Delta Sp}$ between the conductors is

$$R_{\Delta Sp} = \frac{\left[\dfrac{1}{r_1} - \dfrac{1}{r_2}\right]}{4\pi\sigma} \tag{1.21}$$

Each of the above considered capacitor geometries have contained a homogeneous insulating medium between the capacitor conductors. As will be shown in the succeeding chapters, for some geometry with inhomogeneous dielectrics, the capacitance can be represented by a simple series or parallel combination of homogeneous dielectric capacitances.

1.7.1.3 Uncertainty of Fabrication

The precision of transistors and passive components fabricated using IC technology is relatively low just like most devices. Sources of variations include the following

- ion implant dose varies from point to point over the wafer and from wafer to wafer;
- thicknesses of layers after annealing vary due to temperature variations across the wafer;
- irregular linewidths of regions due to imperfect wafer flatness and raggedness in the photoresist edges after development.

This imprecision severely affects the performance of devices because the tolerance ranges of these processes are relatively large. We can explain the effect of linewidth raggedness on the IC resistor uncertainty by considering Figure 1.19. The notations, as seen in Figure 1.19, $<L>$, $<W>$, and δ

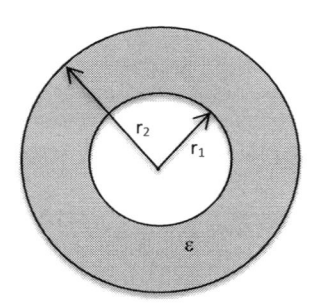

FIGURE 1.18 Spherical capacitor.

denote mean length, mean width, and edge unevenness (raggedness) depth of the rectangular wafer, respectively.

The width W and length L of the IC can be expressed considering the uncertainty in its development as

$$W = \langle W \rangle \pm \frac{\delta}{2} \pm \frac{\delta}{2} = \langle W \rangle \left(1 \pm \frac{\delta}{\langle W \rangle}\right) = \langle W \rangle (1 \pm \varepsilon_W) \tag{1.22}$$

$$L = \langle L \rangle \pm \frac{\delta}{2} \pm \frac{\delta}{2} = \langle L \rangle \left(1 \pm \frac{\delta}{\langle L \rangle}\right) = \langle L \rangle (1 \pm \varepsilon_L) \tag{1.23}$$

where ε_W and ε_L are the width and length normalized uncertainties, respectively. If the width W and length L of the IC are subject to variation, so also are the variables N (dope concentration), t_{ox} (thickness), and μ (mobility).

Similar to Equations (1.22) and (1.23), we can define the normalized uncertainty for these variables:

$$N = \langle N \rangle (1 \pm \varepsilon_N) \tag{1.24}$$

$$\mu = \langle \mu \rangle (1 \pm \varepsilon_\mu) \tag{1.25}$$

$$t_{ox} = \langle t_{ox} \rangle (1 \pm \varepsilon_t) \tag{1.26}$$

By factoring contributions of the overall uncertainty in the IC resistance, we express

$$R = \langle R \rangle (1 + \varepsilon_R) \tag{1.27}$$

where

$$\varepsilon_R = \varepsilon_N \pm \varepsilon_t \pm \varepsilon_\mu \pm \varepsilon_W \pm \varepsilon_L \tag{1.28a}$$

$$\langle R \rangle = \left(\frac{1}{q\langle N \rangle \langle \mu \rangle \langle t_{ox} \rangle}\right)\left(\frac{\langle L \rangle}{\langle W \rangle}\right) \tag{1.28b}$$

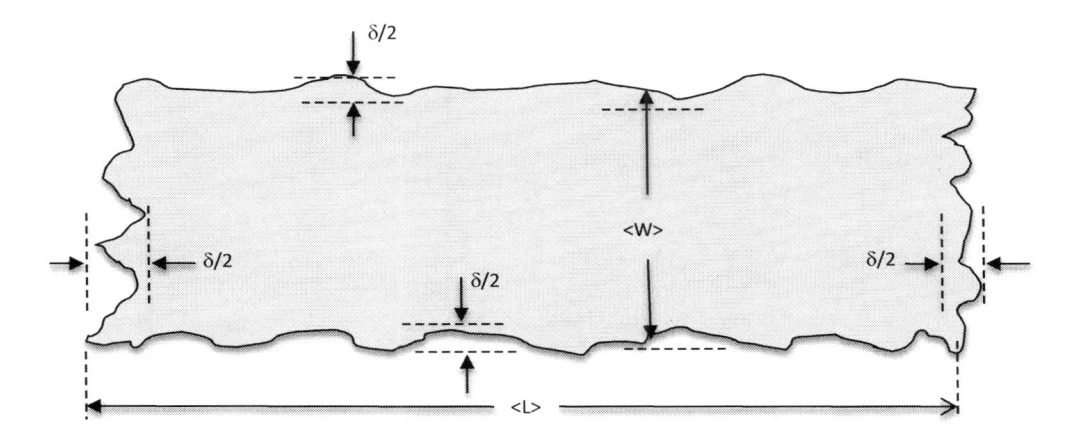

FIGURE 1.19   Ragged edges of a rectangular wafer (greatly exaggerated).

In view of Equations (1.27) and (1.28), it is obvious that there will be minimum and maximum IC resistance $\{(R_{min}), (R_{max})\}$ values due to the fabrication uncertainty, i.e.

$$R_{min} = \langle R \rangle \left[1 - \varepsilon_N - \varepsilon_t - \varepsilon_\mu - \varepsilon_W - \varepsilon_L \right] \tag{1.29a}$$

$$R_{max} = \langle R \rangle \left[1 + \varepsilon_N + \varepsilon_t + \varepsilon_\mu + \varepsilon_W + \varepsilon_L \right] \tag{1.29b}$$

If the variables are independent (meaning that there is no correlation between them), then we can count their contributions to the overall uncertainty by the "root mean square" (rms) variation. In this regard, we count on the "sum of squares" of the normalized variables' uncertainties, i.e.

$$\varepsilon_R|_{rms} = \sqrt{\varepsilon_N^2 + \varepsilon_t^2 + \varepsilon_\mu^2 + \varepsilon_L^2 + \varepsilon_W^2} \tag{1.30}$$

So

$$R_{rms} = \langle R \rangle \left[1 + \varepsilon_R|_{rms} \right] \tag{1.31}$$

In real situation, some variables are correlated, so the resistance value lies between (R_{rms}) and worst-case (R_{min}) estimates. The basic assumption is that the uncertainties are far less than unity, i.e. $\varepsilon_R \ll 1$.

Example 1.3

Given the following normalized uncertainty values, estimate the rms and the worst case: $\varepsilon_N = 0.025$, $\varepsilon_\mu = 0.0125$, $\varepsilon_t = 0.015$, $\varepsilon_W = 0.0285$, and $\varepsilon_L = 0.00002$.

Solution:

Using Equations (1.27a) and (1.29), we have

$$\varepsilon_R|_{rms} = \sqrt{(0.025)^2 + (0.0125)^2 + (0.015)^2 + (0.0285)^2 + (0.00002)^2} = 0.04264$$

$$\varepsilon_R(\text{worse-case}) = 0.025 + 0.0125 + 0.015 + 0.0285 + 0.00002 = 0.08102$$

It is important to know when designing and fabricating IC that each resistor's absolute value is subject to the uncertainties in doping, mobility, and physical dimensions, and the intrinsic uncertainties of fabrication processes can severely affect the performance of devices. Therefore, the analysis of fabrication uncertainties and their outcome on a device performance is a vital task before finalizing the design.

1.8   SUMMARY

This chapter has discussed the technological processes involved in producing electronic devices or circuits that comprise many layers of formation from basic single to compound elements. These elements include the type of chemical bonding involved (ionic or covalent) as well as the process of creating features in a specialized layer of material through chemical modification and how pattern transfer is carried out multiple times to form the device or circuit, and their limitations. The classification of ICs by integration levels is also discussed. An IC, (also called *chip*), is an assembly of electronic components, fabricated as a single unit with miniature devices and their interconnections, built up on a thin substrate of semiconductor material(s).

Small wafer of semiconductor materials embedded with integrated circuitry is shedding more light in the way of manufacturing them leading to increase in electronic components' efficient design and size reduction. Of course, the increase in complexity of chip technology and miniaturization will continue; the technological as well as physical limitations due to heat and leakage have to be considered. More is said about fabrication of very large-scale-integrated and nanosized circuitries and beyond later in Chapters 6, 8, and 9. The next chapters consider what constitute ICs or chips, how their circuitries are formulated from Boolean functions and used in combinatorial and sequential devices we take for granted, and advances in the interconnects.

QUESTIONS

1.1   Explain why silicon is preferred over GaAs in semiconductor development.
1.2   What do mainstream semiconductor IC processes start with?
1.3   If silicon crystals were defective in the development of ICs, what effects would this defect have on the ICs?
1.4   Describe the process of producing an IC.
1.5   What possible errors can you envisage in laying out masks?

REFERENCES

1. Blatt, F.J. (1992). *Modern Physics*. New York: McGraw-Hill.
2. Seeger, K. (2004). *Semiconductor Physics: An Introduction*. London: Springer.
3. Walter, R.J. (2007). *Atomic Structure Theory: Lectures on Atomic Physics*. Berlin: Springer.
4. Zavabeti, A., Ou, J.Z., Carey, B.J., Syed, N., Orrell-Trigg, R., Mayes, E.L.H., Xu, C., Kavehei, O., O'Mullane, A.P., Kaner, R.B., Kalantar-zadeh, K. and Daeneke, T. (2017). A liquid metal reaction environment for the room-temperature synthesis of atomically thin metal oxides. *Science*, 358 (6361), 332–335.
5. Landauer, R. (1961). Dissipation and heat generation in the computing process. *IBM Journal of Research and Development*, 5, 183–191.
6. Kitayama, K-I., Notomi, M., Naruse, M., Inoue, K., Kawakami, S. and Uchida, A. (2019). Novel frontier of photonics for data processing—Photonic accelerator. *APL Photonic*, 4, 090901; doi:10.1063/1.5108912
7. Caine, T.H. (2012). *Graphene and Carbon Nanotube Field Effect Transistors*. New York: Nova Science Publishers.
8. Hu, X., Yasaei, P., Jokisaari, J., Öğüt, S., Salehi-Khojin, A., and Klie, R.F. (2018). Mapping thermal expansion coefficients in freestanding 2D materials at the nanometer scale. *Physical Review Letters*, 120 (5), 055902.
9. REC (Renesas Electronics Corporation). Device type. http://www2.renesas.com/fab/en/line/line3.html. Accessed 6 September 2011.
10. Britannica (2019). The science of Electronics. https://www.britannica.com/technology/electronics/. Accessed 1 November, 2019.
11. Erickson, D. and Li, D. (2004). Integrated microfluidic devices. *Analytica Chimica Acta*, 507 (1), 11–26.
12. Li, M., Li, S., Wu, J., Wen, W., Li, W. and Alici, G. (2012). A simple and cost-effective method for fabrication of integrated electronic-microfluidic devices using a laser-patterned PDMS layer. *Microfluidics and Nanofluidics*, 12 (5), 751–760.
13. Jones, S.W. (2011). Introduction to integrated circuit technology. http://www.ICknowledge.com. Accessed 1 September 2011.
14. Mack, C.A. (1999). Understanding focus effects in submicron optical lithography. *SPI*, 922, 135–148.
15. Nayfeh, M. (Ed.) (2018). *Fundamentals and Applications of Nano Silicon in Plasmonic and Fullerines: Current and Future Trends*. Amsterdam: Elsevier ScienceDirect.
16. Takahashi, N. and Kikuchi, H. (2017). Rayleigh criterion: The paradigm of photolithography equipment. *Annals of Business Administrative Science*, 1–11, doi:10.7880/abas.0170525a
17. Heuberger, A. (1984). Photolithography and X-ray lithography, In *Methods and Materials in Microelectronic Technology*. The IBM Research Symposia Series. Bargon, J. (eds). Boston, MA: Springer, 99–126.
18. Cui, Z. (2008). *Nanofabrication: Principles, Capabilities and Limits*. Boston, MA: Springer.

19. Shi, X. and Boden, S.A. (2016). Materials and processes for next generation lithography. *Frontiers of Nanoscience*, 11, 563–594.
20. Notargiacomo, A., Foglietti, V., Cianci, E., Capellini, G., Adami, M., Faraci, P., Evangelisti, F. and Nicolini, C. (1999). Atomic force microscopy lithography as a nanodevice development technique. *Nanotechnology*, 10 (4), 458

BIBLIOGRAPHY

Cerofolini, G. F. (2009). *Nanoscale Devices*. Berlin: Springer-Verlag.

Hurtarte, J.S., Wolsheimer, E.A. and Tafoya, L.M. (2007). *Understanding Fabless IC Technology*. Oxford: Newnes.

Kolawole, M.O., Adegboyega, G.A. and Temikotan, K.O. (2012). *Basic Electrical Engineering*. Akure: Aoge Publishers.

2 Functional Logics

Many tasks in modern computer, communications, and control systems are performed by logic circuits. Logic circuits are made of *gates*. A *logic gate* is a physical device that implements a Boolean function that performs a logic operation on one or more logical inputs (or terminals) and produces a single logical output. This chapter examines the basic principles of logic gates: the types—from primitive to composite gates—and how they are arranged to perform basic and complex functions.

2.1 THE LOGIC OF A SWITCH

Basic logic circuits with one or more inputs and one output are known as *gates*. Logic gates (or simply *gates*) are used as the building blocks in the design of more complex digital logic circuits. A *logic gate* is a physical device that implements a Boolean function that performs a logic operation on one or more logical inputs (or terminals) and produces a single logical output. Practically, gates function by "opening" or "closing" to allow or reject the flow of digital information. For any given moment, every terminal is in one of the two binary conditions "0" (low) or "1" (high). These binary conditions represent different voltage levels; that is, any voltage v up to the device threshold voltage, V_{th}, (i.e. $0 \leq v \leq V_{th}$); and in the conduction ranges $0 \leq v \leq 2.5\,V$ and $2.5 < v \leq 5\,V$ represent logic states "0" and "1," respectively.

Note that machine arithmetic is accomplished in a two-value (binary) number system, but Boolean algebra—a two-value symbolic logic system—is not a generalization of the binary number system. The symbols 0, 1, +, and • are used in both systems but with totally different meanings in each system. (The meanings of these symbols become obvious during discussions in the next paragraphs.) This symbolic tool allows us to design complex logic systems with the certainty that they will carry out their function exactly as intended. Conversely, Boolean algebra can be used to determine exactly what function an existing logic circuit was intended to perform. So, writing Boolean functional expressions allows us to observe the expected output logic of the simple or complex gates' design or construction.

To have a better understanding of this functional categorization, we refer to our basic circuit theory where we will use simple switches for the purpose of explaining the basic logic functions and combinations. Consider a simple circuit of Figure 2.1a along with its corresponding *truth table*—sometimes called *function table*—Figure 2.1b, and electronic symbolic and logic expression (Figure 2.1c). We describe how the circuit works and how the truth table (i.e. Boolean logical table) is derived.

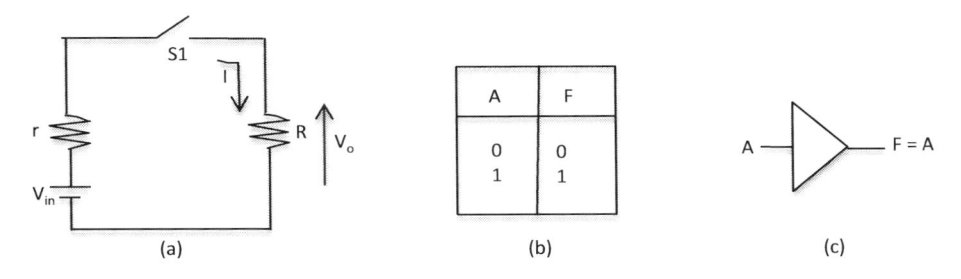

A	F
0	0
1	1

(a)   (b)   (c)

FIGURE 2.1 Simple switch circuit and logic: (a) a switch circuit, (b) truth table of circuit (a), and (c) conventional symbol for Buffer gate.

Suppose a voltage source V_{in} having an internal resistance r is connected in series with a switch $S1$ and a resistance R. When the switch is closed, current I flows. From the basic circuit analysis, the magnitude of current I can be expressed as

$$I = \frac{V_{in}}{r + R} \qquad (2.1)$$

and the amount of voltage drop across resistor R is

$$V_o = IR = \frac{V_{in} R}{r + R} \qquad (2.2)$$

If we assume that $r \ll R$ then the output voltage approximates the source voltage, i.e. $V_o \approx V_{in}$ and the voltage across the switch is zero.

If the switch $S1$ is open, no current flows (i.e. $I = 0$) and no voltage drop is across resistor R (i.e. $V_o = 0$). The two states of output voltage can be defined in terms of a Boolean function F. When $V_o = 0$, $F = 0$, and when $V_o \approx V$, $F = 1$. These two states correspond to the two possible states of the switch, open, and close, which can also be described in terms of a Boolean variable X. When switch is open, $A = 0$ and when switch is closed, $A = 1$. These results are tabulated in the truth table (see Figure 2.1b). It can be seen in the truth table that the input A is passed unchanged to its output: $F = A$. A logic gate that behaves in this way is called *Buffer* (or direct) gate, as shown in Figure 2.1c.

The opposite of buffer gate is a *NOT* gate; i.e. $F = A' = \overline{A}$, as seen in Figure 2.2, which is also called an *inverter*. The logic is denoted as A' or $\overline{A}$ (i.e. a bar over the top of A).

Instead of one switch, suppose two switches are connected in (a) series—as depicted in Figure 2.3a; and (b) in parallel—as depicted in Figure 2.4a.

a. *Series*: both switches $S1$ AND $S2$ (Boolean variables, say, A and B) must be closed for current to flow through resistance R. This operation can be expressed as $F = A \cdot B$, as logically presented in the Truth table in Figure 2.3b. The dot "·" notation is used to represent "AND" operation. The conventional symbol for AND gate is shown in Figure 2.3c.

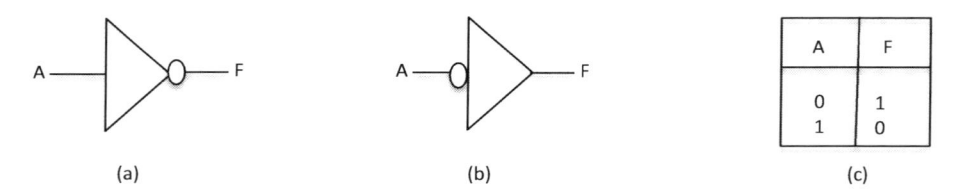

(a)   (b)   (c)

FIGURE 2.2   NOT gate (inverter): (a) most common symbol of NOT gate, (b) alternate symbol of NOT gate, and (c) truth table of gate (a) and (b).

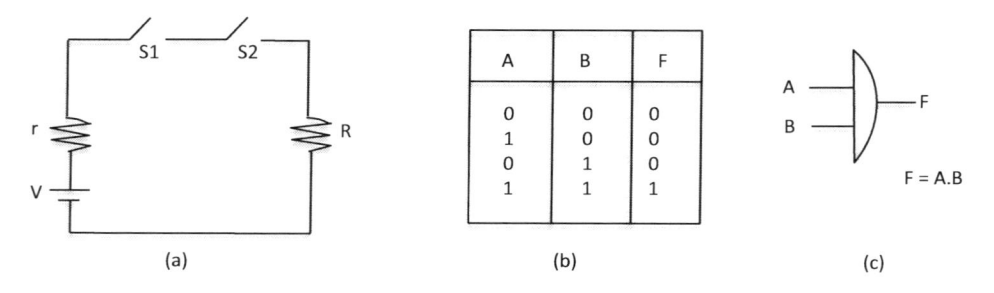

(a)   (b)   (c)

FIGURE 2.3   Series switch logic: (a) two switches in series circuit, (b) truth table of circuit (a), and (c) conventional symbol for AND gate.

An AND gate gives a *high* output "1" only if *all* its inputs are high. The logic expression is written as

$$F = A \cdot B \tag{2.3}$$

The rules of Boolean multiplication are identical to those of binary multiplication; i.e. $0 \cdot 0 = 0$, $0 \cdot 1 = 0$, $1 \cdot 0 = 0$, and $1 \cdot 1 = 1$. In practice, the AND function is implemented by a high-speed electronic gate capable of operating in a few nanoseconds.

b. *Parallel*: Switch $S1$ (Boolean variable A) is connected in parallel to Switch $S2$ (Boolean variable B). Either Switch $S1$ (A) OR Switch $S1$ (B), in Figure 2.4a, must be closed to ensure current flows. This operation is expressed as

$$F = A + B \tag{2.4}$$

reflecting the computation seen in the Truth table in Figure 2.4b. Note that a plus "+" is used to show the "OR" operation. The conventional symbol for OR gate is shown in Figure 2.4c. An OR gate gives a high output "1" if *one* or *more* of its inputs are high.

c. Suppose the switch has a pair of gang contacts, meaning one switch is normally open whilst the other is normally closed. Suppose the upper switch contact is A and the lower contact is $\bar{A}$ (the *inverse* of A), as depicted in Figure 2.5a. We can develop Boolean expressions for both *series* and *parallel* connections for Figures 2.5 and 2.6, respectively.

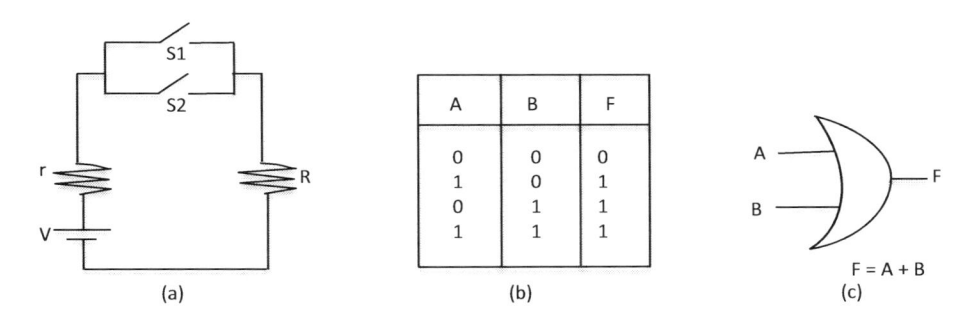

FIGURE 2.4 Parallel switch logic: (a) two switches in parallel circuit, (b) truth table of circuit (a), and (c) conventional symbol for OR gate.

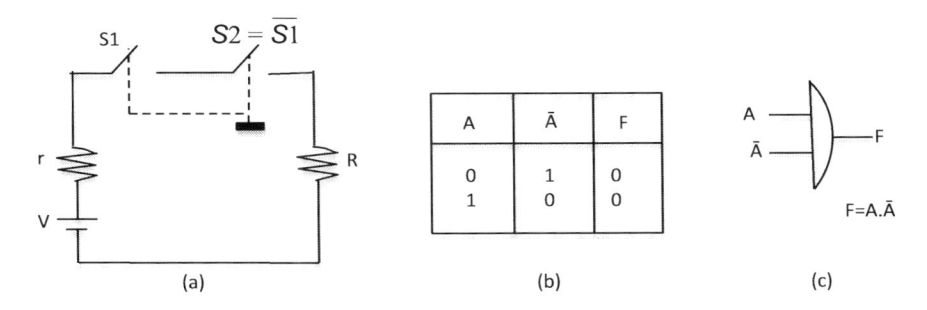

FIGURE 2.5 Series gang-switch logic: (a) two switches in series circuit, (b) truth table of circuit (a), and (c) conventional symbol for AND gate.

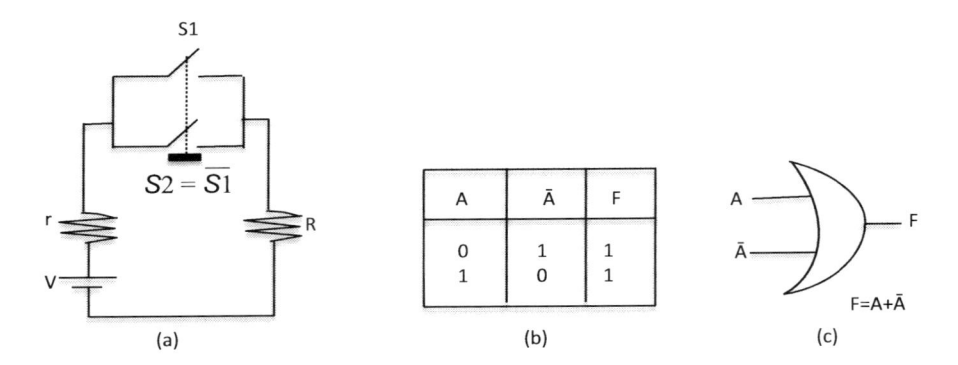

(a)   (b)   (c)

FIGURE 2.6 Parallel gang-switch logic: (a) two switches in parallel circuit, (b) truth table of circuit (a), and (c) conventional symbol for OR gate.

2.2 PRIMITIVE LOGIC AND COMPOSITE GATES

Primitive logic gates (simply basic elementary gates) comprise of *Buffer* (or *direct*), *AND*, *NOT*, and *OR gates*. These gates are also called universal because any Boolean function can be expressed using them. The logic gates are shown in Figure 2.7.

Every digital device—be it a cellular (or mobile) phone, a personal computer, or a complex industrial automation controller—is based on a set of chips designed to store and process information. While these chips come in different shapes, sizes, and forms, they are all made from the same building blocks: primitive logic gates. Of course, there are a few additional gates that are often used in logic design, which are all equivalent to some combination of primitive gates, but they have some interesting properties in their own right. These additional gates include *NAND* (NOT-AND), *NOR* (NOT-OR), *XNOR* or *NXOR* (Exclusive-NOR gate), and *XOR* (eXclusive OR), as shown in Figure 2.8a–d, respectively. Their corresponding *truth tables* and *logic equations* are as follows.

i. *NAND gate* (Figure 2.8a)

A NAND gate is an AND gate with a built-in inverter in the output line (Table 2.1).

$$F = (A \cdot B)' = \overline{AB} \tag{2.5}$$

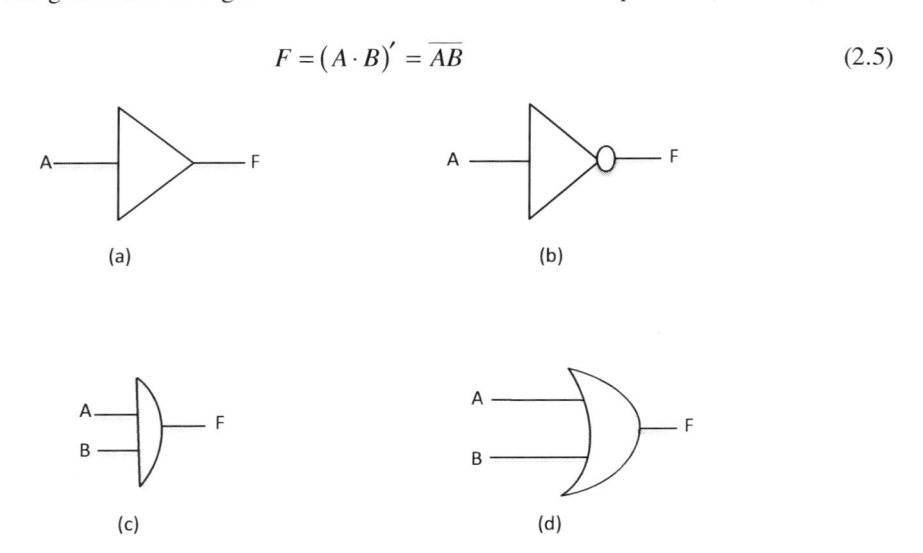

(a)   (b)

(c)   (d)

FIGURE 2.7 Primitive logic gates: (a) buffer (or direct) gate (same as Figure 2.1(c)), (b) NOT gate (inverter) (same as Figure 2.2), (c) AND gate (Same as Figure 2.3(c)), and (d) OR gate (same as Figure 2.4(c)).

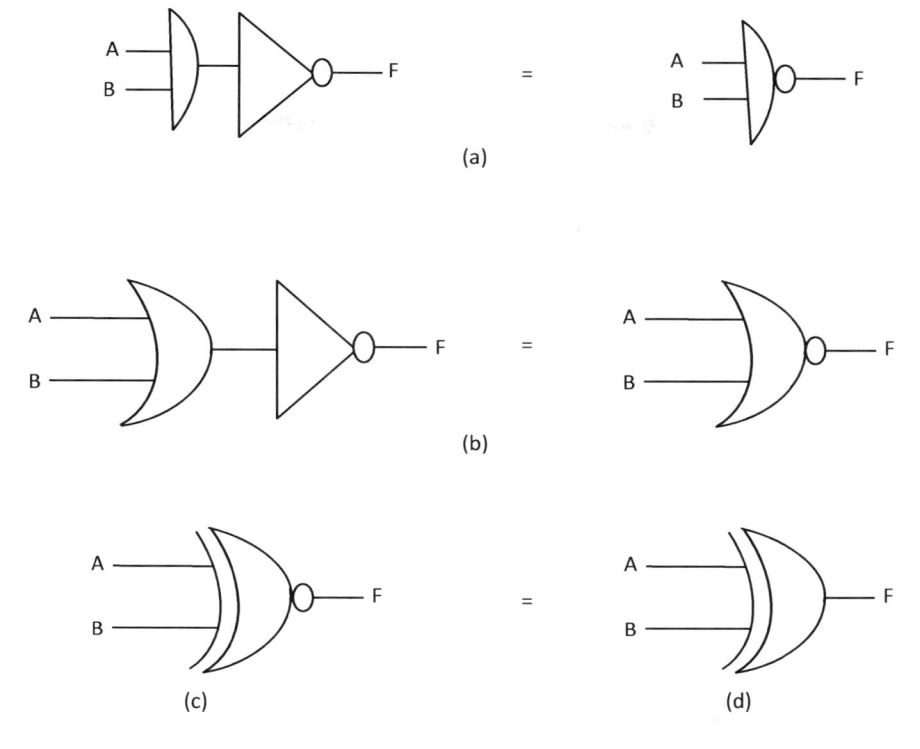

FIGURE 2.8 Other logic gates: (a) NAND gate, (b) NOR gate (NOT OR), (c) XNOR (or NXOR) gate, and (d) XOR gate.

TABLE 2.1
Truth Table of *NAND* Gate

A	B	F
0	0	1
0	1	1
1	0	1
1	1	0

ii. *NOR* gate (Figure 2.8b)

A NOR gate is an OR gate with a built-in inverter in the output line. The NOR operation is the dual of the NAND (Table 2.2).

$$F = \left(A + B\right)' = \overline{A + B} \tag{2.6}$$

iii. Exclusive-NOR (also denoted as XNOR or NXOR) gate (Figure 2.8c)

The complement of the exclusive OR (XOR) function is the XNOR function. The eXclusive-NOR is an Exclusive-OR followed by an inverter: its logic expression can be written as

$$F = \overline{A \oplus B} = \overline{A}\overline{B} + AB \tag{2.7}$$

TABLE 2.2
Truth Table of *NOR* Gate

A	B	F
0	0	1
0	1	0
1	0	0
1	1	0

TABLE 2.3
Truth Table of an *NXOR or NXOR* Gate

A	B	F
0	0	1
0	1	0
1	0	0
1	1	1

Also denoted as

$$F = A \odot B \tag{2.8}$$

A two-input XNOR gate is true when its inputs are equal (Table 2.3).

iv. Exclusive-OR (XOR) gate (Figure 2.8d)

XOR operation can be expressed by either of the following logic expressions:

$$F = A \oplus B = \overline{A}B + A\overline{B} \tag{2.9}$$

$$F = A \oplus B = (A + B)\left(\overline{AB}\right) \tag{2.10}$$

XOR is especially useful for building adders (to be discussed in Section 2.3) and error detection/correction circuits.

We can implement XOR using a combination of three, four, and/or five basic gates as shown in Figure 2.9. For these arrangements, their logic expressions can be written as follows (Table 2.4).

Using a three-gate arrangement, as shown in Figure 2.9a:

$$F = \left((A + B)' + AB\right)'$$

$$= (A + B)\left(\overline{A} + \overline{B}\right)$$

$$= AA' + AB' + A'B + BB'$$

$$= 0 + AB' + A'B + 0 = AB' + A'B \tag{2.11}$$

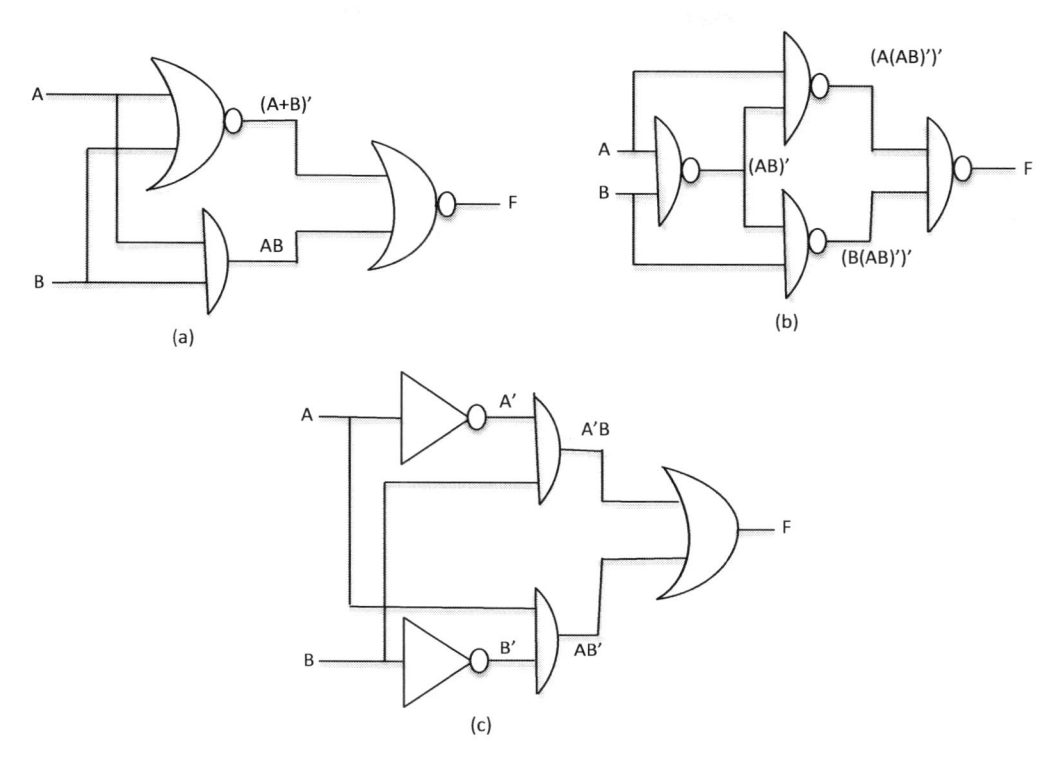

FIGURE 2.9 Implementation of XOR using combinational basic logic gates: (a) XOR derived using three gates, (b) XOR derived using four gates, and (c) XOR derived using five gates.

TABLE 2.4
Truth Table of *XOR* Gate

A	B	F
0	0	0
0	1	1
1	0	1
1	1	0

Using a four-gate arrangement, as shown in Figure 2.9b:

$$F = \left[\left(A(AB)' \right)' \left(B(AB)' \right)' \right]'$$

$$= A\left(\bar{A} + \bar{B} \right) + B\left(\bar{A} + \bar{B} \right)$$

$$= A\bar{A} + A\bar{B} + \bar{A}B + B\bar{B}$$

$$= AB' + A'B = A\bar{B} + \bar{A}B \tag{2.12}$$

The five-gate arrangement, shown in Figure 2.9c, has the same logic expression as for the three-gate arrangement, i.e. Equation (2.11).

Many circuit designers do not consider XOR and XNOR as elementary logic gates in practice.

Majority circuit: A majority circuit is a logic circuit whose output is equal to 1 if the input variables have more 1's than 0's. Otherwise the output is 0. Consider Table 2.5 for functions of three variables x, y, and z. From Table 2.5, a Boolean expression *MAJ(x, y, z)* for implementing the majority circuit can be expressed thus as

$$MAJ\left(x, y, z\right) = \bar{x}yz + x\bar{y}z + xy\bar{z} + xyz \tag{2.13}$$

Figure 2.10 shows the *majority* circuit implemented using Equation (2.13).

As seen in this section, functions can be represented with expressions, truth tables, or circuits. These are all equivalent, and we can arbitrarily transform between them.

Karnaugh map: This is another sort of truth table, which is constructed so that all adjacent entries on the table represent reducible minterm pairs. A minterm is a *product* (AND) term that contains each of the n variables (literals) as factors in either complemented form (if the value assigned to it is 0), or uncomplemented form (if the value assigned to it is 1). For example, Equation (2.13) is composed of four terms; each group of three is a minterm. In it, each variable x, y, or z appears once, either as the variable itself or as the inverse. The first term of Equation (2.13) is $\bar{x}yz$ which is a minterm, whereas $\bar{x}y$ alone is not. Although the minterm form is the most common, another term— called *maxterm*—is also used, so it is necessary to be able to translate from one form into the other.

TABLE 2.5
Truth Table for Functions of Three Variables

X	y	z	MAJ
0	0	0	0
0	0	1	0
0	1	0	0
0	1	1	1
1	0	0	0
1	0	1	1
1	1	0	1
1	1	1	1

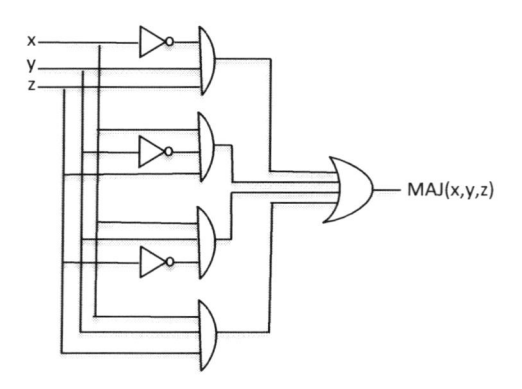

FIGURE 2.10   A majority circuit.

A *sum* (OR) term that contains each of the n variables as factors in either complemented or uncomplemented form is called a *maxterm*. For example, the sum of three variables: $\overline{x} + y + \overline{z}$ is a maxterm, whereas $\overline{x} + y$ is not.

In fact, it can be seen that, by DeMorgan's laws, the maxterms are effectively the minterms with each variable complemented.

By DeMorgan's laws, the inverse of, or complement of, Boolean expressions can be concisely generalized to n variables, thus

a. the complement of the product is the sum of the complements; i.e.

$$\overline{(x_1 \cdot x_2 \ldots x_n)} = \overline{x}_1 + \overline{x}_2 + \cdots + \overline{x}_n$$

b. the complement of the sum is the product of the complements; i.e.

$$\overline{(x_1 + x_2 + \cdots + x_n)} = \overline{x}_1 \cdot \overline{x}_2 \ldots \overline{x}_n$$

These laws can apply to arbitrarily complex expressions.

Although there is more than one way that a Karnaugh map can be labeled and arranged, all versions are used and interpreted in exactly the same way.

2.2.1 COMBINATIONAL AND SEQUENTIAL LOGIC CIRCUITS

There are two kinds of logic: combinational (also called *asynchronous, combinatorial*, or *direct*) logic and sequential (or *synchronous*) logic. Asynchronous logic responds as the input conditions change. No clock input is provided and the circuit is not synchronized with the system clock. Whilst no clock is provided, it is assumed that enough time would elapse for the logic gates constituting the combinational circuit to settle. Unlike the asynchronous logic, synchronous logic responds to the input conditions only at specific times controlled by a clock. As such, its circuit operation is synchronized to the clock.

Combinational logic circuits: Combinational circuits are stateless, whose outputs are functions only of the inputs. Implicitly, a digital circuit is *combinational* if it produces an output that is a Boolean function (combination) of the input values or variables only. Their circuits are made up of gates, which primarily do not have any feedback; that is, outputs are not connected to inputs. A block diagram of combinational circuits is shown in Figure 2.11. The circuits shown in Figures 2.9 and 2.10 can be described as combinational. Combinational circuits have no capacity for storing information.

In digital communication systems, bits are occasionally flipped in transmission. Parity bits are used as the simplest form of error detecting code. A combinational circuit that helps in the check code is called *parity* circuit, which can allow us to detect, and possibly correct, some errors by adding redundancy to the input data string.

Parity circuit: The parity of a bit string is simply whether the number of 1's in that string is odd or even. The parity problem is to determine the parity of an input bit string. A bit string has odd parity if the number of 1's in the string is odd. A bit string has even parity if the number of 1's in the string is even. Suppose input strings are x_2, x_1, x_0 and the condition for odd/even parity is set such that the output z is 1 if *odd* number of inputs is 1; otherwise, z is 0, as shown in truth table (Table 2.6).

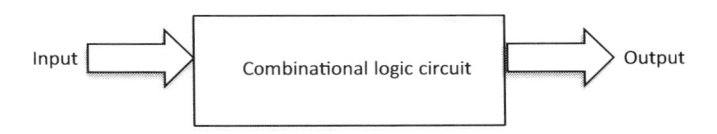

FIGURE 2.11   Structure of combinational circuits.

TABLE 2.6
Odd Parity Circuit Truth Table

Input Strings			Output
x_2	x_1	x_0	Z
0	0	0	0
0	0	1	1
0	1	0	1
0	1	1	0
1	0	0	1
1	0	1	0
1	1	0	0
1	1	1	1

From the truth table (Table 2.6), the functional expression for *odd* parity (z_{odd}) can be expressed as

$$z_{odd} = \bar{x}_2\bar{x}_1x_0 + \bar{x}_2x_1\bar{x}_0 + x_2\bar{x}_1\bar{x}_0 + x_2x_1x_0 \tag{2.14}$$

Likewise, the functional expression for *even* parity (z_{even}) can be expressed as

$$z_{even} = \bar{x}_2x_1x_0 + x_2\bar{x}_1x_0 + x_2x_1\bar{x}_0 \tag{2.15}$$

From these expressions (i.e. z_{odd} and z_{even}), the odd and even parity circuits are drawn as shown in Figure 2.12a and b, respectively.

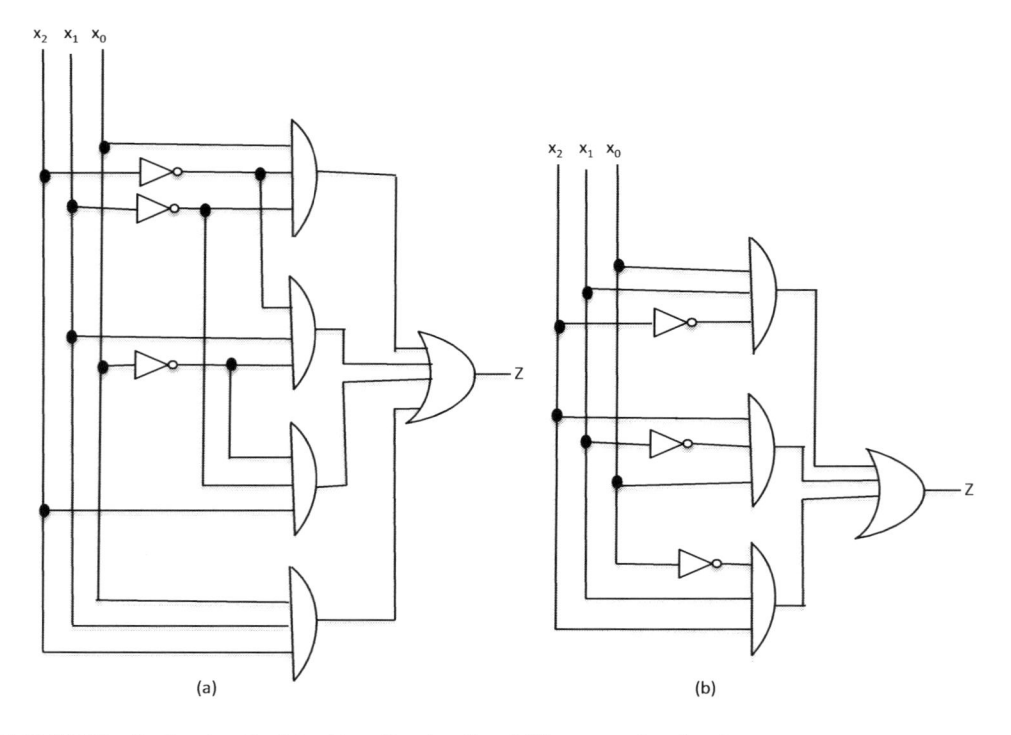

(a)   (b)

FIGURE 2.12   Parity circuits: (a) odd parity circuit and (b) even parity circuit.

Many tasks in communications systems—like coding and decoding, computer systems—like addition, and control systems—like multiplexing—can be performed by combinational logic circuits. How these examples are manifested is discussed in Section 2.3.

Sequential logic circuits: It is possible to make a circuit out of gates that is not combinatorial. This can be achieved by creating feedbacks (or *loops*) in the circuit diagram so that the output values depend indirectly on themselves. If the feedback is *positive* then the circuit tends to have stable states. However, if the feedback is *negative* the circuit tends to have unstable states, and the circuit will tend to oscillate. Figure 2.13 depicts a theoretical viewpoint of how combinational logic and some memory (storage) elements can be used to develop sequential circuits. In it, the combinational logic generates the next state and output(s). State machine inputs/outputs are called primary inputs/outputs.

While feedback allows for a more complex behavior than would otherwise be by just stringing along a lot of logic gates with outputs going directly to inputs of other gates, it can be dangerous if not properly handled. A logic circuit whose output depends not only on the present value of its input signals but also on the sequence of past inputs is called a *sequential logic circuit* (also called *sequential circuit*). Sequential circuits use current input variables and previous input variables by storing the information and putting back into the circuit on the next clock (activation) cycle. Consequentially, the output variables are used to describe the state of a sequential circuit either directly or by deriving state variables from them. This implies that *sequential logic circuits* exist in one of a defined number of states at any one time as well as contain *memory* elements. Talking of *memory* elements, a memory stores data usually one bit per element. A snapshot of the memory is called the *state*. A 1-bit memory is often called a *bistable*; that is, it has two stable internal states. A bistable element can be either static or dynamic and is an essential library element called a register. In bistable circuits, we do observe both "memory state" and "invalid state." Sequential machines occur in nearly every chip design. Basic examples of *sequential logic circuits* are *sequential inverters, latches,* and *flip-flops*.

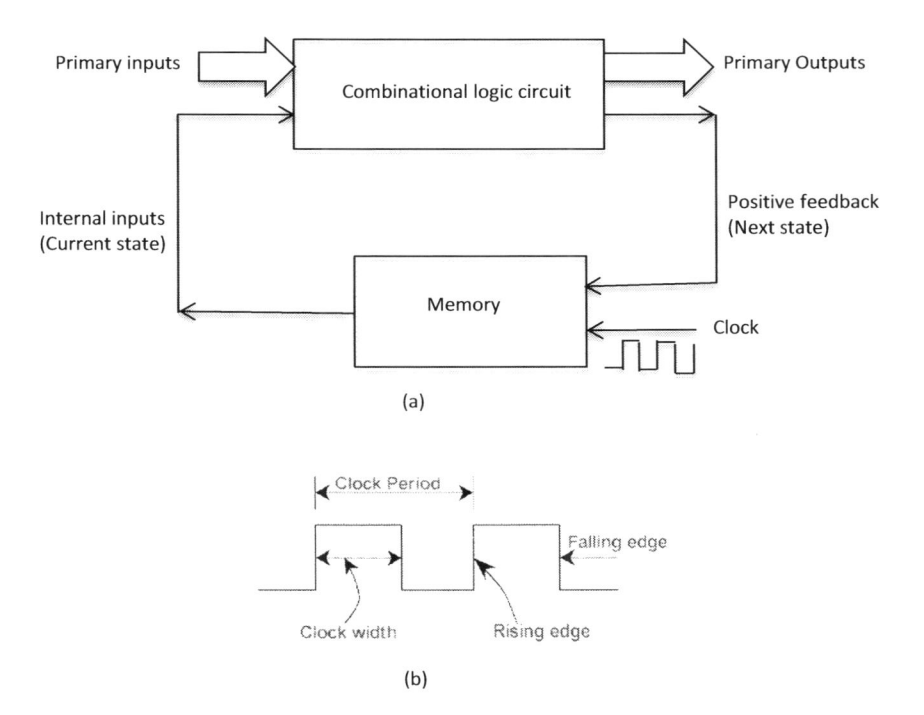

FIGURE 2.13 Theoretical development of sequential logic circuit from combinational and memory (storage) elements: (a) sequential logic: combinational with feedback loop and (b) clock signal.

Sequential inverter: Figure 2.14 depicts two arrangements of a sequential inverter circuit (Figure 2.14a and b). Each simple sequential logic circuit is constructed with two inverters connected sequentially in a loop with no inputs but two outputs (Q and $\bar{Q}$). Given that the circuit has no inputs, we cannot change the values of Q and $\bar{Q}$ since there are no inputs to control the state. Suppose $Q = 0$, when we switch power on, $\bar{Q} = 1$ and $Q = 0$. Suppose $Q = 1$: then $\bar{Q} = 0$ and $Q = 1$. If we use the signal Q as the state variable to describe the state of the circuit, we can say that this circuit (Figure 2.14) has two stable states: $Q = 0$ and $Q = 1$; hence the name *bistable*.

Latches and *flip-flops* are the basic single-bit memory elements used to build sequential circuit with one or two inputs/outputs, designed using individual logic gates and feedback loops. Latches are *asynchronous*, implying that the output changes very soon after the input changes, whereas, flip-flops are *synchronous* version of the latches. The primary difference between a latch and a flip-flop is that a latch watches all of its inputs continuously and changes its outputs at any time, independent of a clocking signal; whereas, a flip-flop samples its inputs and changes its inputs only at times determined by a clocking signal. Asynchronous inputs are usually available for both latches and flip-flops: they are used to either set or clear the storage element's contents independent of the clock. Although *flip-flops* and *latches* are made of *bistable* elements, *monostable* and *astable* circuits can also be used. Basic functions of *latches* and *flip-flops* are described as follows.

Latch: The function of a latch circuit is to "latch" the value created by the input signal to the device and hold that value until another signal changes it. As an example, consider a set/reset latch depicted in Figure 2.15. Figure 2.15a has two inputs: set input (S) and a reset (or rest) input (R). There are four cases to consider in Figure 2.15a, namely, when

a. $R = 1, S = 0$: then $Q = 1$ and $\bar{Q} = 0$;
b. $R = 0, S = 1$: then $Q = 0$ and $\bar{Q} = 1$;

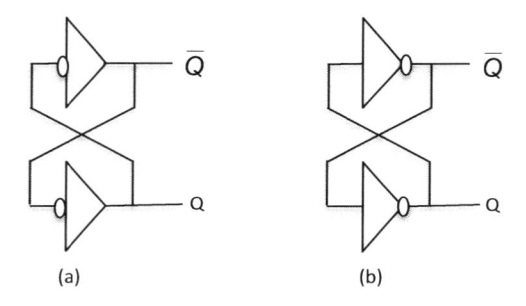

(a)   (b)

FIGURE 2.14   Two arrangements of a sequential inverter circuit: (a) and (b).

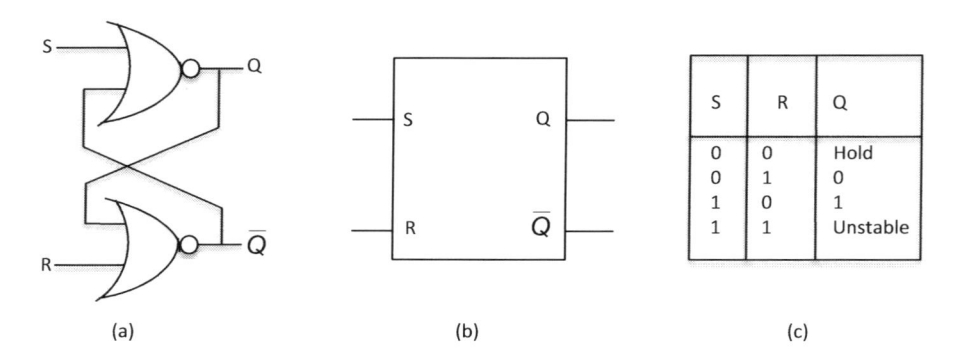

(a)   (b)   (c)

FIGURE 2.15   Set/Reset (SR) latch using NOR gates: (a) SR latch circuit, (b) SR latch symbol, and (c) latch truth table.

c. $R = 0$, $S = 0$: then $Q = Q_{prev}$ and $\overline{Q} = \overline{Q}_{prev}$. This is the memory state (hold behavior).

d. $R = 1$, $S = 1$: then the outputs are forced to zero, $Q = 0$ and $\overline{Q} = 0$. This is an invalid (indeterminate) state because $\overline{Q}$ is **not** Q: the constraint is that Q and $\overline{Q}$ must be complementary. An additional problem with this case is that when the input triggers return to their zero levels, the resulting state of the latch is unpredictable and depends on whatever input is last to go low. Often this state is termed a *metastable* state. We must do something to avoid this state.

Since the enable input on a gated S-R latch provides a way to latch the Q and $\overline{Q}$ outputs without regard to the status of S or R, we can eliminate one of those inputs to create a modified latch circuit with no "invalid" input states. We achieve this by modifying the basic SR latch with two additional inputs, D and clock (*Clk*), placed in front of the basic XOR-based SR latch, as shown in Figure 2.16a. Such a circuit is called a D latch, and its internal logic circuit, as in Figure 2.16a. The clock is sometimes called enable signal or control, and D as data.

The D latch will not respond to a signal input if the clock input is low "0," it simply stays latched in its last state. When the clock input is high "1," however, the Q output follows the D input. Since the reset R input of the SR circuitry has been done away with, this modified latch has no "invalid" state. In essence, the latch "samples" D on the rising edge of clock (*Clk*). When *Clk* rises from 0 to 1, D passes through to Q; otherwise, Q holds its previous value. Q changes only on the rising edge of *Clk*, and Q and $\overline{Q}$ remain *always* opposite of one another. Note that *data* D can be several bits, but 1 bit is used in this case. The D signal should not be associated with the clock's edge (rising or falling), but instead "1" should be considered as a pulse, and "0" as no pulse.

An SR latch can also be implemented using a cross-coupled NAND structure, as shown in Figure 2.17a. When both S and R are low (i.e. $S = 0$ and $R = 0$) at almost the same time, the output is "invalid"; that is, unpredictable because the latch is being set and reset at the same time. On the other hand, when both S and R are high (i.e. $S = 1$ and $R = 1$) at almost the same time, the output remains in state prior to input and does not change; implying that the latch is in normal resting state. This state is considered metastable. When $S = 0$ and $R = 1$, the output Q goes high (i.e. $Q = 1$) and remains high even if the S input goes high. In the final state, when $S = 1$ and $R = 0$, the output Q goes low (i.e. $Q = 0$) and remains low even if the S input goes high.

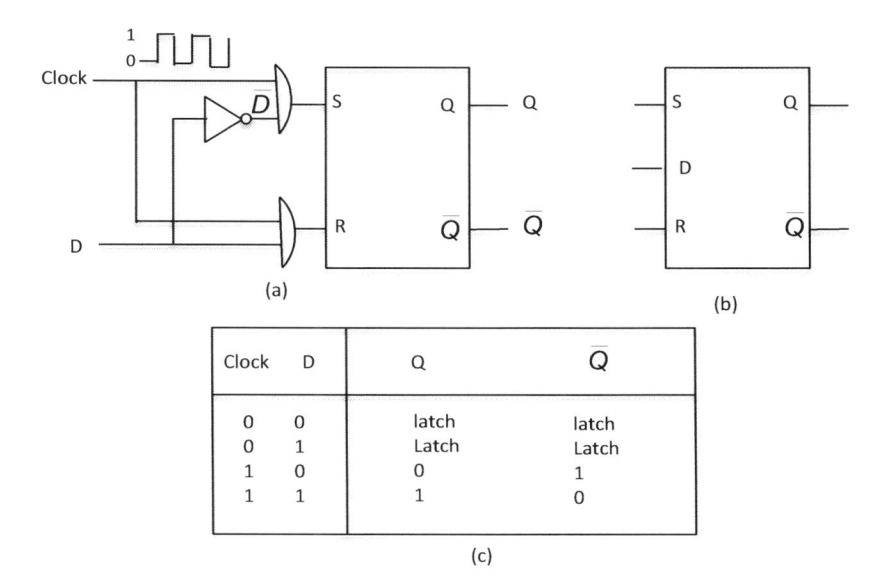

Clock	D	Q	$\overline{Q}$
0	0	latch	latch
0	1	Latch	Latch
1	0	0	1
1	1	1	0

(c)

FIGURE 2.16 D Latch internal circuit: (a) D latch circuit, (b) D latch symbol, and (c) D latch truth table.

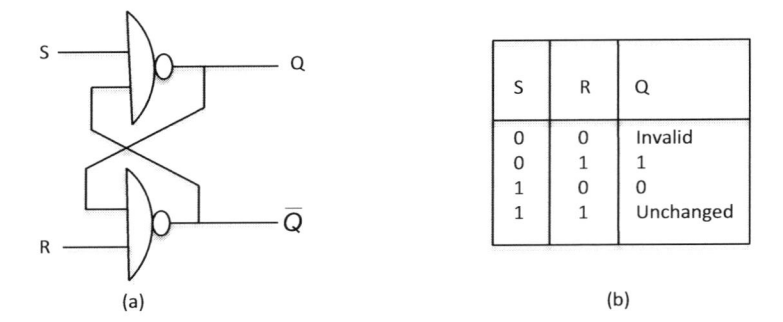

(a)   (b)

FIGURE 2.17 NAND-gate SR latch: (a) NAND-gate latch and (b) latch truth table.

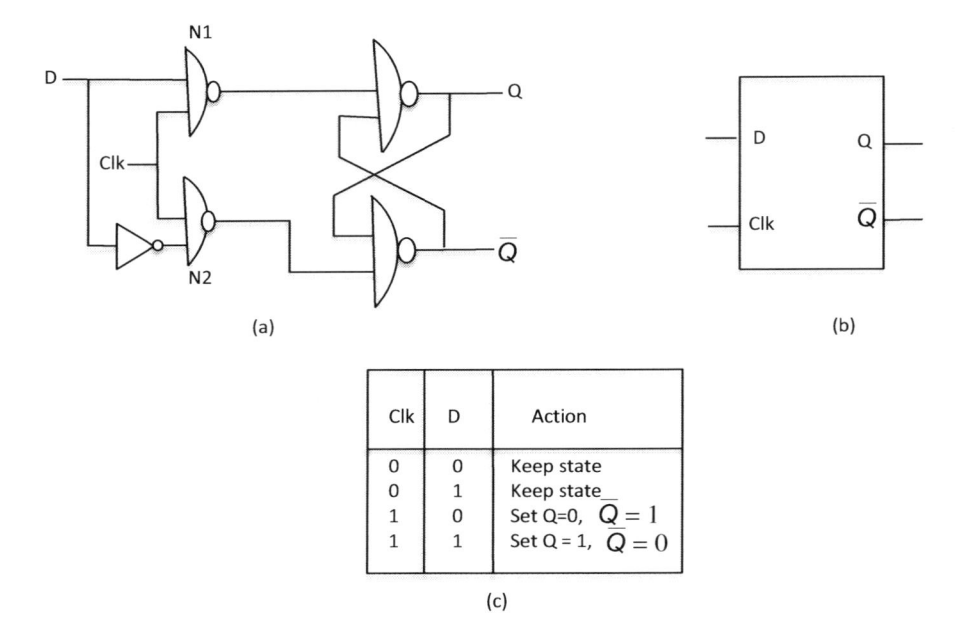

(a)   (b)

(c)

FIGURE 2.18 NAND-based, SR D-latch: (a) gated NAND-based circuit, (b) logic symbol, and (c) truth table.

An SR latch can be derived using gated NAND-based, as shown in Figure 2.18a. Since the SR latch is responsive to its inputs at all times, the clock line (*Clk*) can be used to disable or enable state transitions. So, when the device is disabled (*falling edge*), it holds its current value, and when enabled (or rising edge), it can be set or reset. For instance, since the clock signal (*Clk*) is connected to both $N1$ and $N2$ NAND gates, as in Figure 2.18a, it follows that when the clock signal is zero, the outputs of $N1$ and $N2$ gates are both 1, implying that the latch stores the previous value.

If the clock signal is high (i.e. $Clk = 1$), the output of $N1 = \overline{D}$ (the inverse of the input signal, D), and the output of $N2 = D$. This leads to the input values $S = 1$, $R = 0$; or $S = 0$, $R = 1$ on the latch, which, in turn, enters the corresponding state. Therefore, for high clock signal (i.e. $Clk = 1$), the latch output value follows the value on its D input: this can be said that the latch is "transparent." As we can see, the transparency is controlled by clock (*Clk*).

It is worth noting that there are multiple ways to make a circuit using primitive gates—whether gated or not—that does a particular function. The number of possibilities is enormous especially when designing highly complex circuits. We can cascade latches by connecting the output of one latch to the input of another, as an example, Figure 2.19.

There is a potential problem when operating this cascade: how do we stop changes from racing through chain? To arrest this, we need to (i) be able to control flow of data from one latch to the next, (ii) move one latch per clock period, and (iii) not have to worry about logics between latches that are too fast. We can solve problems (i) and (ii) by breaking data flow using alternating clocks; that is, use positive clock to latch inputs into one SR latch (for the master stage), and use negative clock to change outputs with another SR latch (for the slave stage), as shown in Figure 2.20. By doing this time swap, output changes a few gate delays after the falling edge of clock but does not affect any cascaded flip-flops. To eliminate the 1's catching problem (of (iii)) is by making S and R complements of each other, making input data D. The arrangement depicted in Figure 2.20 is called *D flip-flop* (also known as *Master-slave flip-flop*). The slave-stage output is also the main circuit overall output.

The ambiguity caused by trigger pulses going active simultaneously still remains. This ambiguity can be avoided by adding two feedback lines J and K, as shown in Figure 2.21a, leading to a new configuration called *JK flip-flop*. D flip-flops are most common for integrated circuits (ICs).

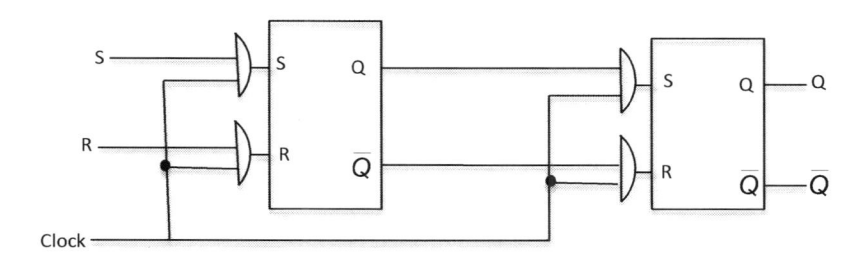

FIGURE 2.19   Cascaded latch circuit.

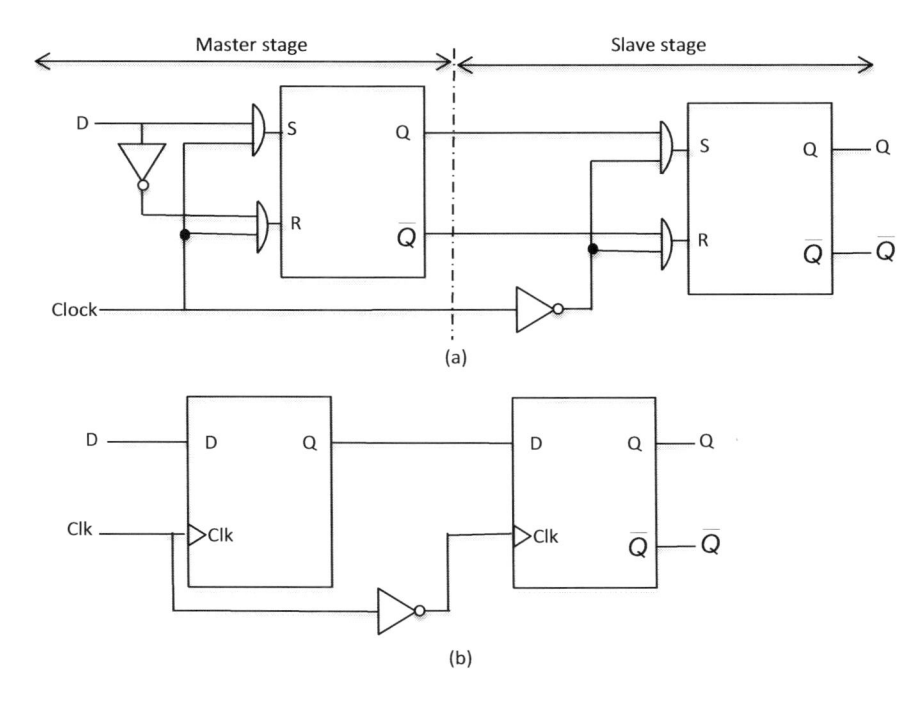

FIGURE 2.20   Master/slave D flip-flop: (a) D-type flip-flop circuit and (b) logic symbol.

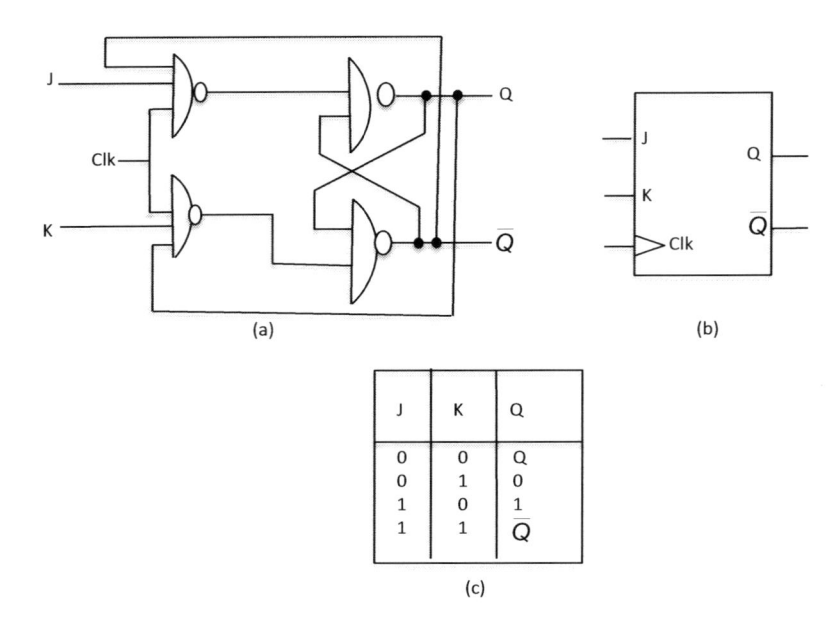

(a)   (b)

J	K	Q
0	0	Q
0	1	0
1	0	1
1	1	$\overline{Q}$

(c)

FIGURE 2.21   JK flip-flop: (a) JK sequential circuit, (b) logic symbol, and (c) truth table.

As seen in Figure 2.21a, if both *J* and *K* are high, and the clock pulses, the output of the *JK flip-flop* is complemented. This implies that the JK flip-flop toggles its state when both inputs are asserted. However, doing so empowers the other input and the flip-flop oscillates. This places some stringent constraints on the clock pulse width; for example, where pulse width is less than the propagation delay through the flip-flop.

Another type of flip-flop is toggle flip-flop (simply T flip-flop), which is an *edge-triggered* device, which we discuss next.

Toggle (T) flip-flop: The toggle T flip-flop is a single input device, which can be made from a JK flip-flop by tying both of its inputs together, as shown in Figure 2.22. As a result, the T-type flip-flop is called a single input JK flip-flop. The T or "toggle" flip-flop changes its output on each clock edge, giving an output, which is half the frequency of the signal to the T input. Here also the restriction on the pulse width can be eliminated with a master-slave or edge-triggered construction. A flip-flop is called an *edge-triggered* device because it is activated on the clock's rising or falling edge. T flip-flop is useful for constructing binary counters, frequency dividers, and general binary addition devices (adders).

In summary, this section has focused on Boolean functions and the *Small-Scale Integration* (SSI) circuits/chips used to implement those functions. The next section shall define some higher-level components that contain the equivalent of a number of SSI components and function at a higher level that may be closer to the logical description of problems we design in communication, computing, and electronics systems. These components may fall under the *Medium-Scale Integration* (MSI) chips classification. Specifically,

- Multiplexers/demultiplexers
- Binary counter and frequency divider
- Encoder/decoder
- Arithmetic circuits (Adder/Subtractor/Multiplier)

which are standard combinational circuits that form standard cells in complex *Very Large-Scale Integration* (VLSI) circuits.

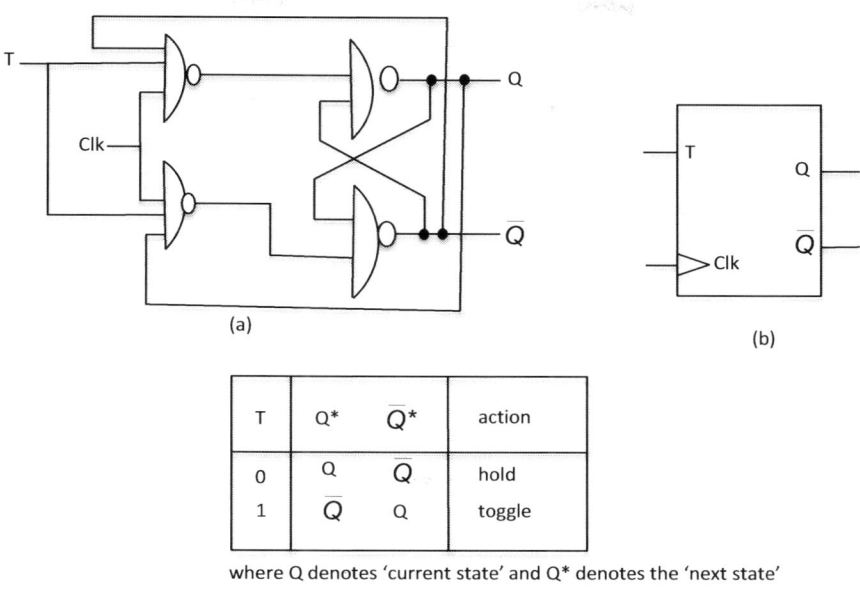

(a)   (b)

T	Q*	$\overline{Q}$*	action
0	Q	$\overline{Q}$	hold
1	$\overline{Q}$	Q	toggle

where Q denotes 'current state' and Q* denotes the 'next state'

(c)

FIGURE 2.22   Toggle (T) flip-flop: (a) T flip-flop circuit, (b) logic symbol, and (c) truth table.

2.3   LOGIC CIRCUITS DESIGN

The goal of circuit design is to build a hardware that computes some given function(s). We can build circuits of any complexity out of these simple gates following these steps. First, by finding (simplified) Boolean expression(s) for the function(s) you want to compute. Second, by calculating the corresponding truth table(s) that completely specifies (specify) its (their) behavior. Afterwards, convert the expression into a circuit. Then perform circuit optimization; i.e. wanting smallest number of product terms that produce fastest circuit. VLSI design is probably the most fascinating application area of combinatorial optimization.

2.3.1   DESIGN EXAMPLES

Multiplexer: Suppose a switch is to be designed that chooses between two data lines X and Y. As such, the circuit must have a control input line (say, C), and, of course, an output, Z. We can interpret the specification as: if the control input is "0," the output is the same as that given by X; and the output is the same as Y when the control input "1." We can draw the circuit's functional truth table as (Table 2.7)

We can see from the truth table that when $C = 0$, $Z = X$, and when $C = 1$, $Z = Y$. We can use two AND gates for the data input lines X and Y. To have the desired objective, we need to "invert" the control input so that when $C = 0$, X is unblocked but Y is blocked, and vice versa. We can write the functional expression as

$$Z = \left(\overline{C} \cdot X\right) + \left(C \cdot Y\right) \tag{2.16}$$

From Equation (2.16), we need additional OR gate to collect the 2-AND outputs as its inputs so that the circuit output Z gives a high output "1" if *one* or *more* of its inputs are high; the logic circuit is as shown in Figure 2.23. This circuit is a *multiplexer* (abbreviated as MUX) because it allows us to select, at different instants, the signal (or binary information) we wish to place on the common line.

TABLE 2.7
Circuit Function Truth Table

Control Line	Input		Output
C	X	Y	Z
0	0	0	0
0	0	1	0
0	1	0	1
0	1	1	1
1	0	0	0
1	0	1	1
1	1	0	0
1	1	1	1

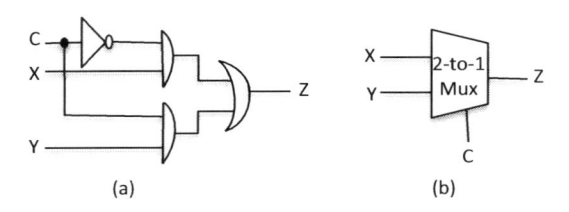

(a)   (b)

FIGURE 2.23   A multiplexer with selection line C: (a) a 2-by-1 multiplexer circuit and (b) multiplexer symbol.

A multiplexer is also called *data selector* because, being a combinational circuit, it selects binary information from one of the input lines and directs the information to a single output line.

In general, the selection of a particular input line is controlled by a set of selection lines. For 2^n data line, we will have n selection lines. For example, for 4-input lines (2^2) and 1-output multiplexer, there must be 2-selection lines. A 4-to-1-line multiplexer is shown in Figure 2.24. In Figure 2.24, 2-selection lines C_0 and C_1 are used to select one AND gate and directs its data to output, as depicted in Figure 2.24c.

It is possible to build parallel multiplexers where multiplexer blocks are combined in parallel with common selection and enable lines to perform selection on multibit quantities. In this situation, the developed MUX must have an enable (E) input line to control the flow of data. As a consequence, when MUX is enabled, it will behave as normal MUX, if not all outputs are zero, as shown in Figure 2.25.

Demultiplexer: A demultiplexer performs the inverse operation of a multiplexer. In this case, it is a combinational circuit that receives input from a single line and transmits it to one of 2^n possible output lines. The selection of the specific output is controlled by the bit combination of n selection lines. Figure 2.26 is an example of a 1-by-4 line demultiplexer.

2.3.2   BINARY COUNTER AND FREQUENCY DIVIDER

We can construct a binary counter using T flip-flops by taking the output of one flip-flop to the clock input of the next, for instance, Q_0 to C_1, Q_1 to C_2, and Q_2 to C_3, as in this case of four T-flip-flops, as shown in Figure 2.27a. The clock generator is connected to the clock input, C_0, of the first flip-flop. The inputs of each flip-flop are set to "1." At each clock's cycle, a toggle is produced. So, for each two toggles of the first T-flip-flop, a toggle is produced in the second T-flip-flop, and so on down to the fourth T-flip-flop. This process produces a binary number equal to the number of cycles of the

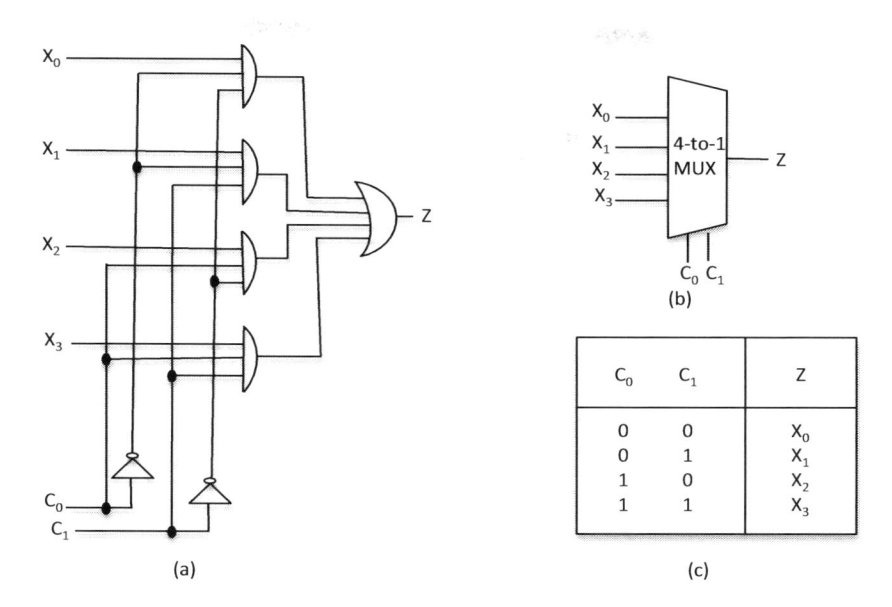

FIGURE 2.24 A 4-to-1 line multiplexer: (a) a 4-by-1 line multiplexer logic circuit, (b) multiplexer symbol, and (c) function table.

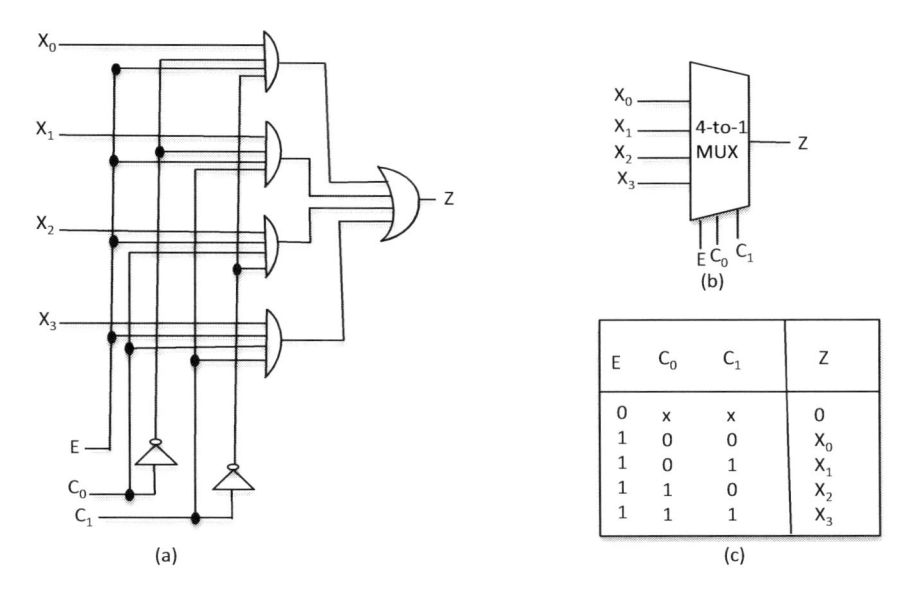

FIGURE 2.25 A 4-to-1 line multiplexer with enable (E) line: (a) a 4-by-1 line multiplexer logic circuit, (b) multiplexer symbol, and (c) function table.

input clock signal, as seen in Figure 2.27b. This construction is also a *ripple through* counter and is sometimes useful as a *frequency divider.*

Shift register: A shift register can be implemented using a chain of D-type flip-flops, as shown in Figure 2.28. It has a serial input D_{in} and parallel outputs Q_0, Q_1, Q_2, and Q_3. The data moves one position to the right on application of the clock rising edge.

We can use this serial input, parallel outputs concept to develop a serial data link. For example, we can utilize the S and R inputs on the flip-flops to provide a parallel data input feature so that data

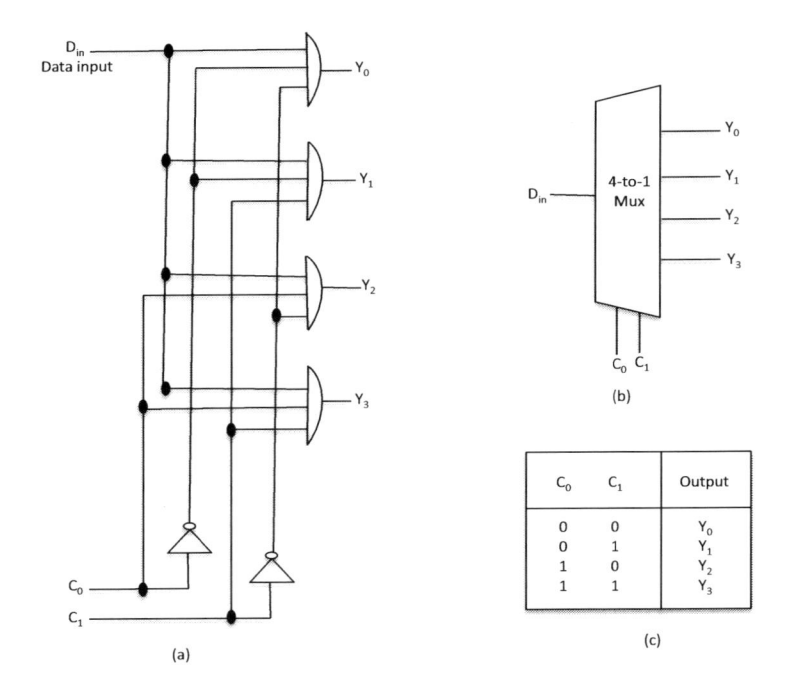

FIGURE 2.26 A demultiplexer: (a) 1-by-4 Line demultiplexer logic circuit, (b) multiplexer symbol, and (c) function table.

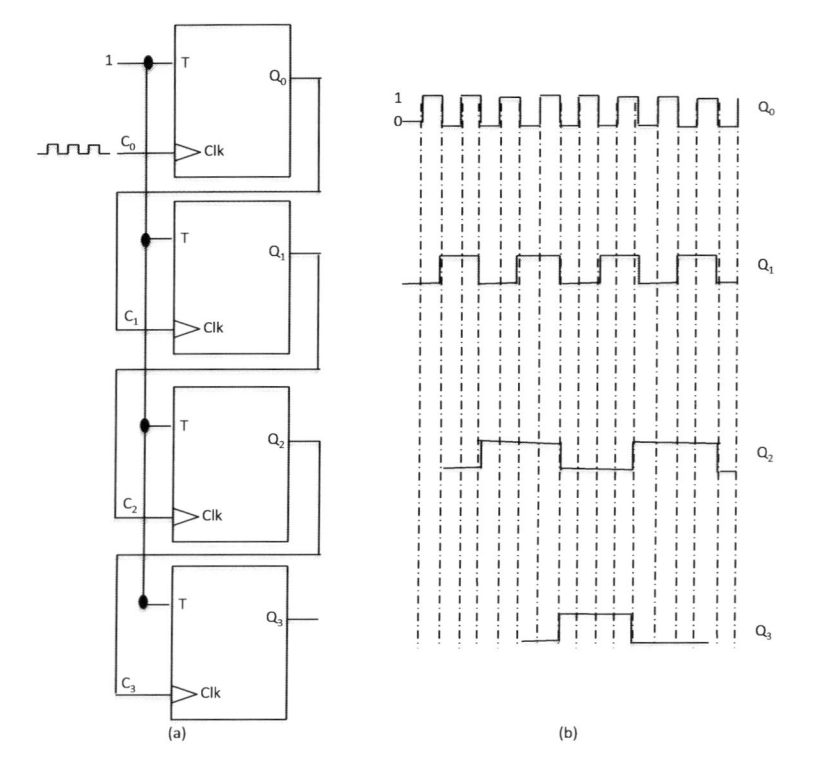

FIGURE 2.27 Binary counter (same as frequency counter): (a) counter circuit and (b) binary number equivalent to number of cycles.

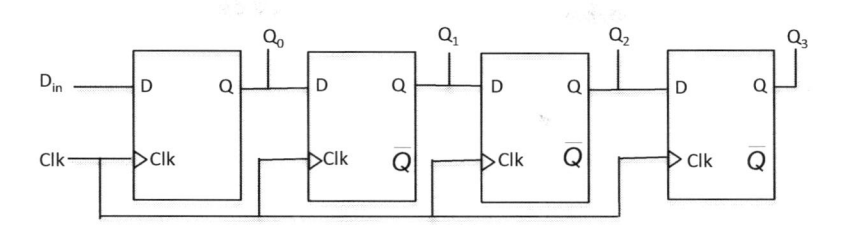

FIGURE 2.28   Shift register using D-type flip-flop.

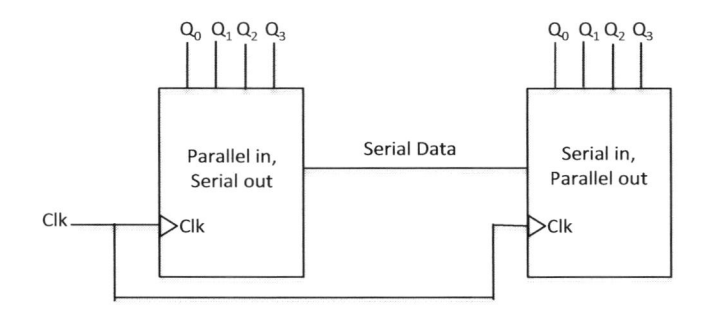

FIGURE 2.29   Serial data link.

can then be clocked out through Q_3 in a serial fashion; i.e. to have a parallel in, serial out arrangement. This, along with the previous serial in, parallel out shift register arrangement can be used as the basis for a serial data link, for instance, as shown in Figure 2.29. One data bit, at a time, is sent across the serial data link.

In essence, it is useful to be able to guarantee certain characteristics—like large and minimum transition density—about the information that gets sent across the serial link. To guarantee these characteristics, an encoder/decoder is used. In this regard, the encoder will take in incoming data along with some metadata—a signal that indicates whether the incoming data represents actual data or control characters—and produce an encoded value that is then transmitted on the link. Likewise, the decoder will take the values from the link and produce the original value, along with some information; for example, whether the incoming character contains any errors, and/or represents a data or control character. Encoders and decoders are combinational logic circuits: their basic designs are discussed next.

2.3.3   ENCODER AND DECODER

Encoder: An encoder converts information from one code format to another with the aim of shrinking information size either for speed, secrecy, or security purpose. Encoders are combinational logic circuits, which perform the reverse operation to decoders. An encoder has 2^n (or fewer) input lines and n output lines, which generate the binary code corresponding to the input values, as shown in Figure 2.30.

Typically, an encoder converts a code containing exactly one bit that is 1 to a binary code corresponding to the position in which the 1 appears. Suppose we have an octal to binary encoder: that is, eight-input lines ($D_0 \ldots D_7$) and three-output lines (Y_0, Y_1, Y_2), as shown in the truth table (Table 2.8). As shown in Table 2.8, the leftmost bit, D_7, is called the *most significant bit* (MSB) while the rightmost bit, D_0, is the *least significant bit* (LSB).

From Table 2.8, the outputs' functional equations can be derived, as follows, allowing the combinational logic circuit of Figure 2.31 to be drawn.

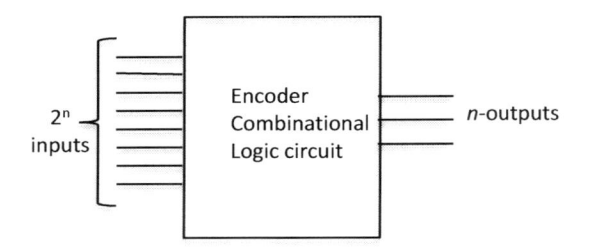

FIGURE 2.30   An encoder symbolic representation.

TABLE 2.8
Truth Table for Octal-to-Binary Encoder

Inputs								Outputs		
D_7	D_6	D_5	D_4	D_3	D_2	D_1	D_0	Y_2	Y_1	Y_0
0	0	0	0	0	0	0	1	0	0	0
0	0	0	0	0	0	1	0	0	0	1
0	0	0	0	0	1	0	0	0	1	0
0	0	0	0	1	0	0	0	0	1	1
0	0	0	1	0	0	0	0	1	0	0
0	0	1	0	0	0	0	0	1	0	1
0	1	0	0	0	0	0	0	1	1	0
1	0	0	0	0	0	0	0	1	1	1

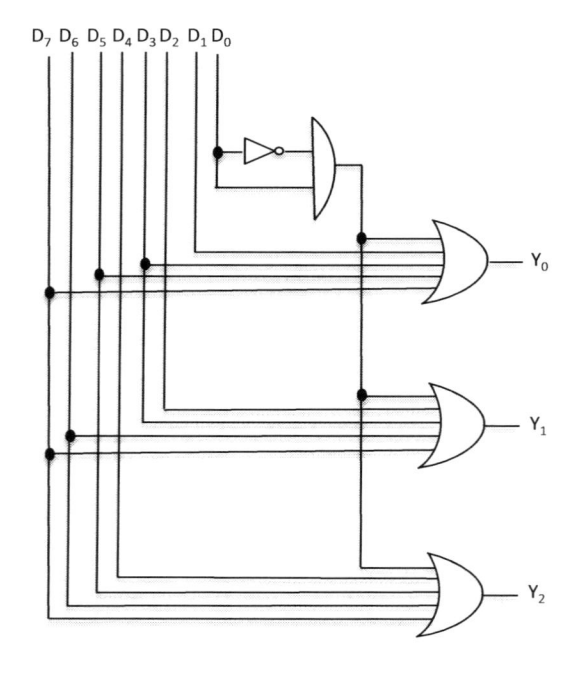

FIGURE 2.31   Designed encoder.

$$Y_0 = D_1 + D_3 + D_5 + D_7 \qquad (2.17a)$$

$$Y_1 = D_2 + D_3 + D_6 + D_7 \qquad (2.17b)$$

$$Y_2 = D_4 + D_5 + D_6 + D_7 \qquad (2.17c)$$

The limitation of just designed encoder is that only one of the inputs is allowed to be "1"; and when all inputs are zeros, the output is zero. But this situation, that is, when the outputs equal zero is the same as when input $D_0 = 1$, as seen in Table 2.8. However, if more than one input value is 1, the encoder just designed in Figure 2.31 will not produce a meaningful result. An encoder that can accept all possible combination of input values and produce a meaningful result is a *priority encoder*.

Priority Encoder generates an encoded result amongst the 1's that appear. The output is an encoded value of the location of the most significant input position (or the least significant input position) containing a 1 and responds with the corresponding binary code for that position. This means that if two or more inputs are equal to 1 at the same time, the input having the highest priority will take precedence. In other words, if D_i and D_j are both "1" simultaneously, and $i > j$, then D_i has priority over D_j. For example, if D_1, D_3, and D_5 are 1 simultaneously, D_5 has the highest priority. As seen in Table 2.9, D_7 is assigned the highest priority and D_0 the lowest, and "x" means "don't care" either "1" or "0." Y_2, Y_1, and Y_0 encode the value of the highest priority 1 and the other coded output V is high if any bit in $D_7 \dots D_0$ is logic 1. If all inputs are zero, there is no valid input and V is zero. From the truth table (Table 2.9), we can write the outputs' functional equations, which will lead to the construction of the priority encoder circuit.

The functional equations are

$$Y_0 = D_1 \bar{D}_2 \bar{D}_4 \bar{D}_6 + D_3 \bar{D}_4 \bar{D}_6 + D_5 \bar{D}_6 + D_7 \qquad (2.18a)$$

$$Y_1 = D_2 \bar{D}_4 \bar{D}_5 + D_3 \bar{D}_4 \bar{D}_5 + D_6 + D_7 \qquad (2.18b)$$

$$Y_2 = D_4 + D_5 + D_6 + D_7 \qquad (2.18c)$$

$$V = D_0 + D_1 + D_2 + D_3 + D_4 + D_5 + D_6 + D_7 \qquad (2.18d)$$

The constructed priority encoder circuit is shown in Figure 2.32. Priority encoders allow us to encode values based on a *priority scheme*.

TABLE 2.9
Truth Table for Octal-to-Binary Priority Encoder

Input								Output			
D_7	D_6	D_5	D_4	D_3	D_2	D_1	D_0	V	Y_2	Y_1	Y_0
0	0	0	0	0	0	0	1	1	0	0	0
0	0	0	0	0	0	1	x	1	0	0	1
0	0	0	0	0	1	x	x	1	0	1	0
0	0	0	0	1	x	x	x	1	0	1	1
0	0	0	1	x	x	x	x	1	1	0	0
0	0	1	x	x	x	x	x	1	1	0	1
0	1	x	x	x	x	x	x	1	1	1	0
1	x	x	x	x	x	x	x	1	1	1	1
0	0	0	0	0	0	0	0	0	0	0	0

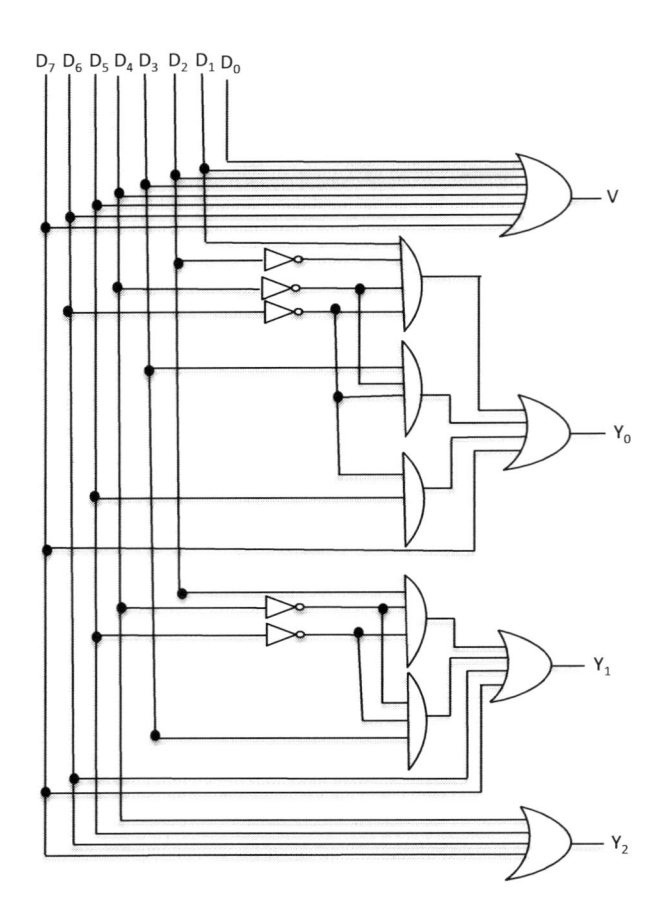

FIGURE 2.32   Octal-to-binary priority encoder circuit.

Decoder: A decoder is an inverse operation of an encoder. It is a combinational logic circuit that converts binary information from n input lines to a maximum of 2^n unique outputs. It is also called the n-to-m line decoders, for example, where decoding involves the conversion of an n-bit input code to an m-bit output code with $n \leq m \leq 2^n$ such that each valid code word produces a unique output code.

Suppose we consider a 3-to-8 decoder that is based on the association of binary numbers to decimal numbers, as is shown in Table 2.10.

When the decoder is not active; that is, when the *enable* input is 0 (i.e. $E = 0$), all outputs are 0. This is true without regard to the inputs. When the enable input is 1 (i.e. $E = 1$), the decoder is active and the outputs correspond to the inputs. From the truth table (Table 2.10), we can write the outputs' functional equations, which will lead to the construction of the enable decoder circuit. The decoder functional equations are

$$
\begin{aligned}
D_0 &= \bar{x}_2 \bar{x}_1 \bar{x}_0 & D_1 &= \bar{x}_2 \bar{x}_1 x_0 \\
D_2 &= \bar{x}_2 x_1 \bar{x}_0 & D_3 &= \bar{x}_2 x_1 x_0 \\
\\
D_4 &= x_2 \bar{x}_1 \bar{x}_0 & D_5 &= x_2 \bar{x}_1 x_0 \\
D_6 &= x_2 x_1 \bar{x}_0 & D_7 &= x_2 x_1 x_0
\end{aligned}
\tag{2.19}
$$

The constructed priority decoder circuit is shown in Figure 2.33.

TABLE 2.10

Decoder

	Input Lines			Output
Enable	x_2	x_1	x_0	
1/0	0	0	0	D_0
1	0	0	1	D_1
1	0	1	0	D_2
1	0	1	1	D_3
1	1	0	0	D_4
1	1	0	1	D_5
1	1	1	0	D_6
1	1	1	1	D_7

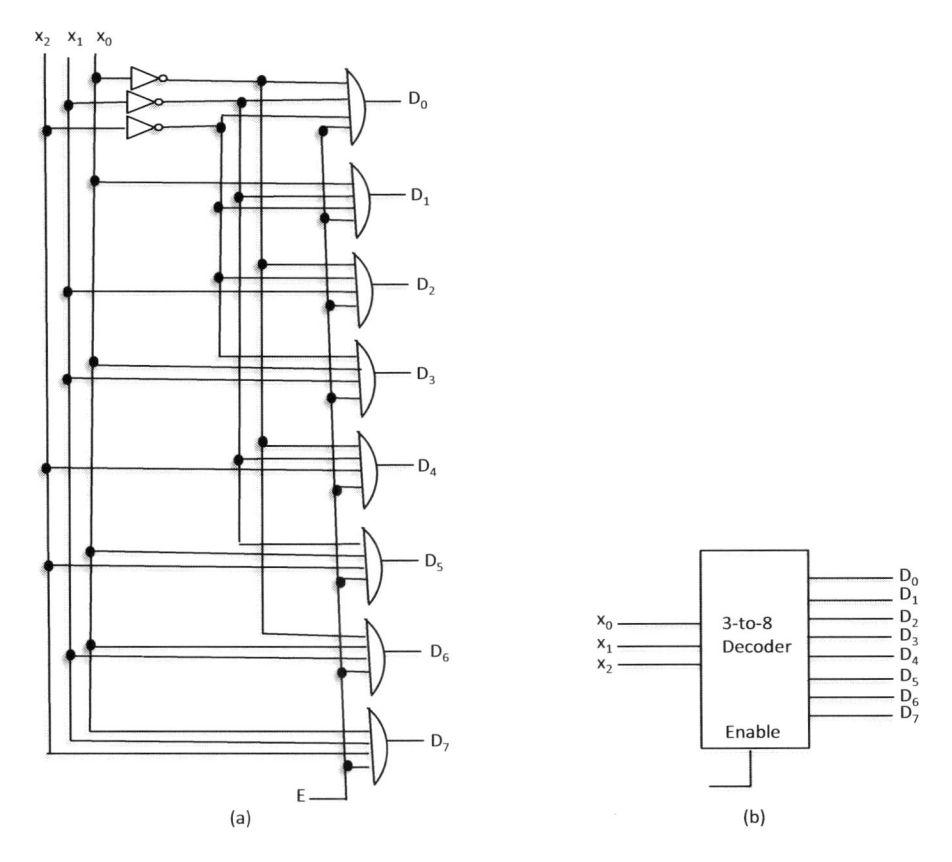

FIGURE 2.33　3-to-8 Line decoder with enable: (a) 3-to-8 line decoder circuit with enable and (b) decoder symbolic representation.

A *decoder* can take the form of a multiple-input, multiple-output logic circuit that converts coded inputs into coded outputs, where the input and output codes are different e.g. n-to-2^n, binary-coded decimal decoders. Decoding is necessary in applications such as data multiplexing, seven-segment display, and memory address decoding. Decoder with enable input can function as a demultiplexer.

A curious reader would observe that there is equivalence between decoders and demultiplexers. The direct correspondence between demultiplexers and decoders with enable is interpretational; that is, what is a control bit and what is a data bit? For instance, the data inputs of a decoder correspond to the control bits of a demultiplexer, and the enable input of a decoder corresponds to the data bit of a demultiplexer. In essence, decoders are essentially demultiplexers. But, encoders are **not** essentially multiplexers.

2.3.4   ARITHMETIC CIRCUITS

Before we develop circuits that perform addition, subtraction, multiplication, or division, it is important we review what binary arithmetic is about. While ordinarily, algebra is a generalization of our human system of arithmetic, machine arithmetic is accomplished in a binary-number system. This generalization does not apply to Boolean algebra because it is not a generalization of the binary number system. For example, the symbols 0, 1, +, and • are used in both systems but with totally different meanings in each system. In Boolean algebra, "0" and "1" denote memory states while "+" and "•" denote "OR" and "AND," respectively. In the machine arithmetic, "0" and "1" denote values (0, 1) while "+" and (*) denote simply, "add" and "multiplication," respectively. As such, Boolean and algebra must be treated as entirely different entities.

One of the most important tasks performed by a digital computer is the operation of addition; i.e. adding two binary numbers. It should be noted that subtraction of two numbers is included in the meaning of addition, since subtraction is performed first by carrying out some operation on the subtrahend and then adding the result. (What operation is first performed depends on the type of computer—either inverting the subtrahend or taking its 2's complement.) Subtrahend is the number that is used to subtract the first number from. For example; say, "X" minus "Y" equals "Z" (i.e. $X - Y = Z$): "X" is the minuend; "Y" is the subtrahend; and "Z" is the difference. This example assumes decimal (base 10) notation, which is the notation most commonly used by humans for representing integers.

2.3.5   REVIEW OF BINARY ARITHMETIC

Binary is base 2. Each digit (known as a *bit*) is either 0 or 1. Decimal to binary conversion can be performed successive division by 2. For example, suppose we want to convert (a) 106_{10} and (b) 25_{10} to binary.

a. Conversion of 106_{10} by successive division by 2 to have

106/2 = 53	remainder = 0
53/2 = 26	remainder = 1
26/2 = 13	remainder = 0
13/2 = 6	remainder = 1
6/2 = 3	remainder = 0
3/2 =1	remainder = 1
1/2 =0	remainder = 1

By reading the remainders from bottom upwards, the binary equivalent of 106_{10} equals 1101010_2.

b. Similarly, conversion of 25_{10} by successive division by 2 is found thus:

25/2 = 12	remainder = 1
12/2 = 6	remainder = 0
6/2 = 3	remainder = 0
3/2 = 1	remainder = 1
1/2 = 0	remainder = 1

Again, reading the remainders from bottom upwards, the binary equivalent of 25_{10} equals 11001_2.

We can rearrange the unsigned numbers 106_{10} and 25_{10} and their binary representation, as well as their associated binary coefficients as follows.

1 1 0 1 0 1 0 = 106_{10}	Binary coefficients
64 32 16 8 4 2 1	
2^6 2^5 2^4 2^3 2^2 2^1 2^0	
MSB *LSB*	

0 0 1 1 0 0 1 = 25_{10}	Binary coefficients
64 32 16 8 4 2 1	
2^6 2^5 2^4 2^3 2^2 2^1 2^0	
MSB *LSB*	

As earlier noted, the leftmost bit is called the MSB while the rightmost bit is called LSB. From these tabulated examples, we can generalize that for n-bit binary number—$a_{n-1}a_{n-2}a_{n-3} \rightleftharpoons a_2a_1a_0$—has the decimal value, B:

$$B = \sum_{i=0}^{n-1} a_i \times 2^i \tag{2.20}$$

where $a = 0$ or 1. We can represent positive integers from 0 to $2^n - 1$. In computing system, an 8-bit-long binary number is known as a *byte*. So, a byte can represent unsigned values from 0 to 255; i.e. 0 to $\{2^8 - 1 = 256 - 1\}$, $2^8 = 256$ different combinations of bits in a byte. What we have done so far represent positive numbers.

Negative numbers: If we want to represent negative numbers, we have to give up some of the range of positive numbers we had before: a popular approach to do this is called 2's complement. The rule for changing a positive 2's complement number into a negative 2's complement number (or vice versa) is to complement all the bits and add 1. So, for n-bit binary number $a_{n-1}a_{n-2}a_{n-3} \rightleftharpoons a_2a_1a_0$ the decimal equivalent of a 2's complement number B is

$$B = -a_{n-1} \times 2^{n-1} + \sum_{i=0}^{n-2} a_i \times 2^i \tag{2.21}$$

For example,

$$1\,1\,0\,1\,0\,0\,1\,0 \ = -a_7 \times 2^7 + \sum_{i=0}^{6} a_i \times 2^i$$

$$= -1 \times 2^7 + \ 1 \times 2^6 + \ 0 \times 2^5 + 1 \times 2^4 + \ 0 \times 2^3 + \ 0 \times 2^2 + \ 1 \times 2^1 + \ 0 \times 2^0$$

$$= -128 \ + \ 64 \ + \ 0 + 16 \ + \ 0 \ + \ 0 \ + 2 \ + \ 0 \ = \ -46$$

This example being *signed* number; the complement number rule deliberately ignores any carry from the MSB. To accommodate for unsigned numbers, we need to modify this rule to detect when a result is out of range. For instance, when working with 8-bit unsigned numbers, we can use the "carry" from the 8th bit (MSB) to indicate that the number has got too big. Therefore, the rule for detecting 2's complement overflow is the carry into the MSB does not equal the carry out from the MSB. Some examples of 8-bit, 2's *Complement Overflow*, are as follows:

```
    00001111    15
    00001111   +15
(0) 00011110    30 (no overflow)

    11110001   −15
    11110001   −15
(1) 11100010   −30 (there is a carry out, no overflow, sum is correct)

    01111111    127
    00000001   +1
(0) 10000000    128 (no carry out, there is an overflow, sum is correct)

    10000001   −127
    11111110   −2
(1) 01111111   −128 (there is an overflow)

    11011001   −39
    11111110   +92
(1) 00110101   +53 (there is a carry out with an overflow; sum is correct)
```

From the preceding examples, we can deduce the rules for detecting overflow in a 2's complement sum, thus

- If the sum of two positive numbers yields a negative result, the sum has overflowed.
- If the sum of two negative numbers yields a positive result, the sum has overflowed.
- Otherwise, the sum has not overflowed.

When there is carry out, the *sum* is too big to be represented with the same number of bits as the two addends. A negative and positive added together cannot overflow because the *sum* is between the addends. Note that the *overflow* and *carry out* can each occur without the other. In unsigned numbers, *carry out* is equivalent to *overflow*. However, if two 2's complement numbers are added, and they both have the same sign (both positive or both negative), then *overflow* occurs if and only if the result has an opposite sign. In reality, the overflow in 2's complement occurs, not when a bit is *carried-out* out of the left column, but when one is *carried in* to it.

2.3.6 BINARY ADDER CIRCUITS

A schematic diagram of a binary full adder (FA) is shown in Figure 2.34, where x_i and y_i are the inputs of multiple variables where each line shown really represents i-digit lines, and the *carry in* is C_i. The outputs are the sum S_i and the *carry out* C_{i+1}.

The truth table for the FA for the two outputs is shown in Table 2.11

We can write the expressions for the two outputs as

1. *Sum*:

$$S_i = \bar{x}_i y_i \bar{C}_i + x_i \bar{y}_i \bar{C}_i + \bar{x}_i \bar{y}_i C_i + x_i y_i C_i \tag{2.22}$$

Removing some redundancies, the sum is concisely written as

$$S_i = x_i \oplus y_i \oplus C_i \tag{2.23}$$

The representation of Equation (2.22) is known as *sum-of-products* (SOP) form.

2. *Carry out*:

$$C_{i+1} = x_i y_i + C_i \left(\bar{x}_i y_i + x_i \bar{y}_i \right)$$

$$= x_i y_i + C_i \left(x_i \oplus y_i \right) \tag{2.24}$$

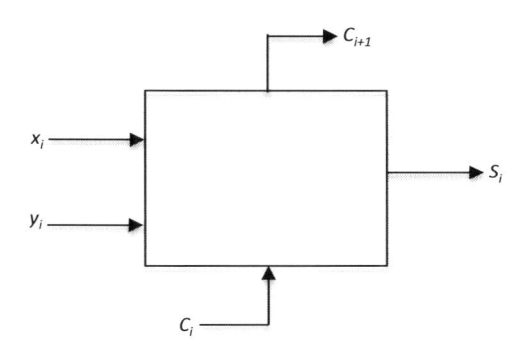

FIGURE 2.34   A schematic diagram of a binary addition.

TABLE 2.11
Truth Table of Binary Adder

Input Lines			Output	
C_i	x_i	y_i	S_i	C_{i+1}
0	0	0	0	0
0	0	1	1	0
0	1	0	0	1
0	1	1	1	0
1	0	0	1	0
1	0	1	0	1
1	1	0	1	1
1	1	1	0	1

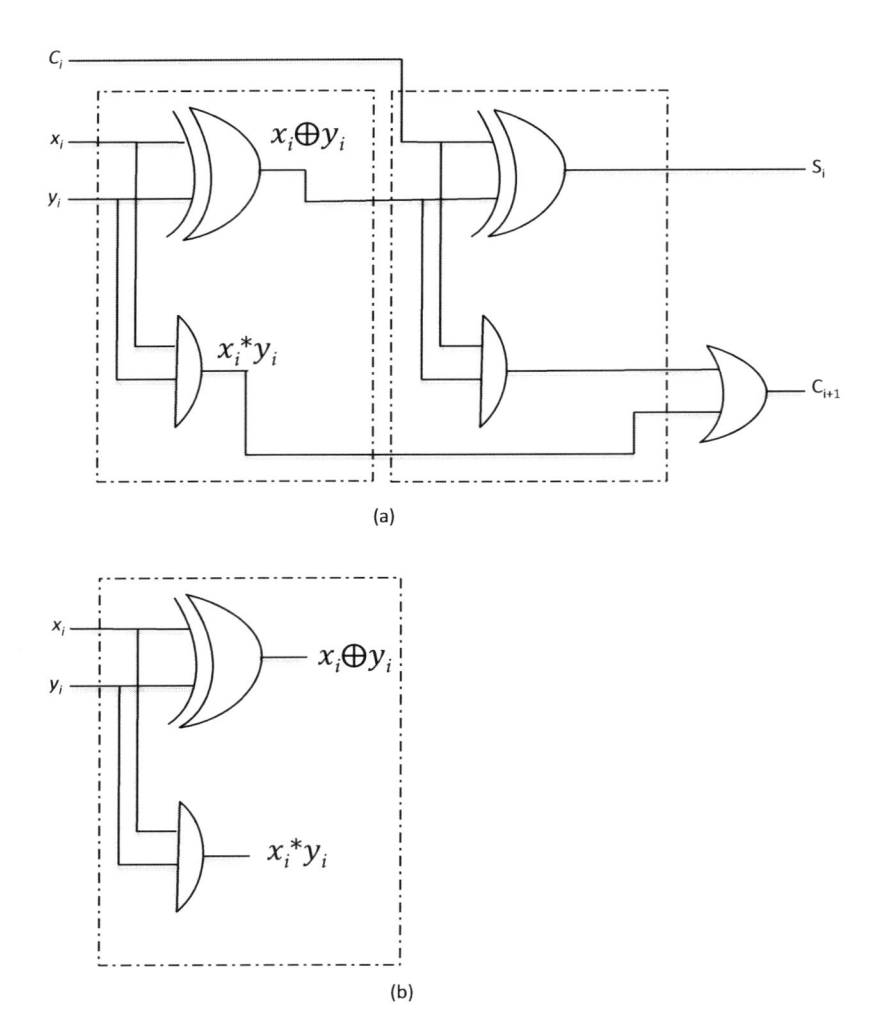
(a)

(b)

FIGURE 2.35   Full binary adder circuit implemented with two half-adders: (a) FA and (b) half adder.

Using Equations (2.23) and (2.24), the full binary adder is drawn, as shown in Figure 2.35a. As observed in Figure 2.35a, there are two identical circuits plus an OR gate comprising the FA circuit. These two identical circuits—denoted by rectangular dashed boxes and reproduced as Figure 2.35b—are called *half-adders* comprising only two inputs (x_i, y_i) to be added without a carry in C_i.

2.3.7   RIPPLE-CARRY ADDER

Suppose we have two multidigit binary numbers, say, two n-bits (A_i, B_i) as input streams. In this case, all digits will be presented in parallel and using an FA to add each corresponding pair of digits, one from each number, to perform the addition process. As a result, multiple FAs will be connected in tandem so that the carry out C_{i+1} from one stage becomes the carry in C_i to the next stage, as illustrated in Figure 2.36 for example for the case of 5-digit numbers. Being connected in tandem, the *carry ripples* through each stage (i.e. from C_0 to C_1 until it propagates through to C_5). For adders, the carry into the first (least significant) stage, C_0, is 0, while the last carry out (the overflow carry) becomes the MSB of the ($n+1$)-bit sum.

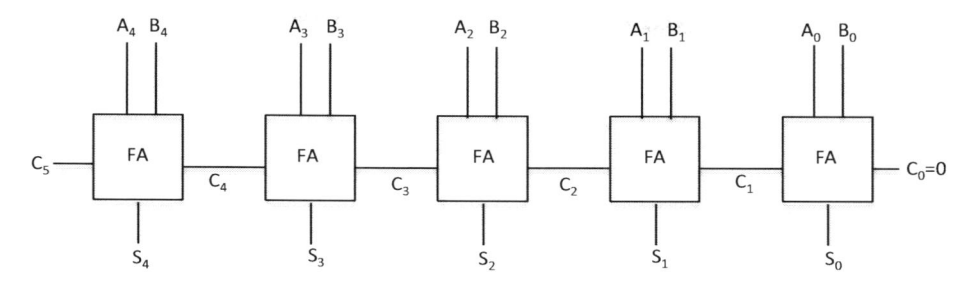

FIGURE 2.36   Five-bit ripple-carry FA.

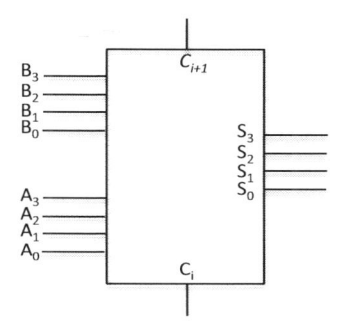

FIGURE 2.37   An MSI package semi-block diagram of 4-bit adder.

The FA and ripple-carry adder circuits described here are available in IC packages. A single FA, for example, is available as a unit. A ripple-carry adder, as illustrated in Figure 2.36 is available as MSI packages. An MSI package, semi-block diagram of 4-bit words carry FA, is shown in Figure 2.37. The greater the number of FA stages included in a unit, the greater the improvement in speed—besides, greater the complexity of the carry out C_{i+1} circuit.

2.3.8   Binary Subtractor Circuits

In digital arithmetic, binary subtraction can be implemented using the same FA circuits with an additional input circuit that negates the subtrahend with a control signal (CS) that decides which operation to perform. A conditional inverter (XOR gate) is used to perform the subtrahend negation. For instance, if one input is 0, then the output is identical to the second input, and if one input is 1, then the output is the complement of the second input. Including the additional input circuit to Figure 2.36, a 2's complement adder/subtractor circuit is constructed as in Figure 2.38. If the CS is 0, then the circuit performs arithmetic addition $(A_i + B_i)$ function; however, if CS is 1, the circuit performs arithmetic subtraction $(A_i - B_i)$ function.

Ensuring overflow is detected, an XOR gate is included sampling the *carry in* C_i and the *carry out* C_{i+1} to the MSBs, as shown in Figure 2.39.

Two's complement addition is the most common method implemented in modern classical computers due to its reduced circuit complexity compared with 1's complement. As such, the 1's complement arrangement is not considered. Unlike classical binary bit that exist at state 0 or 1, but quantum bits (called qubits), which form the basis of quantum electronics, can exist in a continuum of states between "ground" state $|0\rangle$ and "excited" state $|1\rangle$ until they are observed. As such, the configuration of logic gates would differ from those described in this chapter. (More is said of quantum bits (*qubits* for short) and quantum logic gates in Chapter 9.)

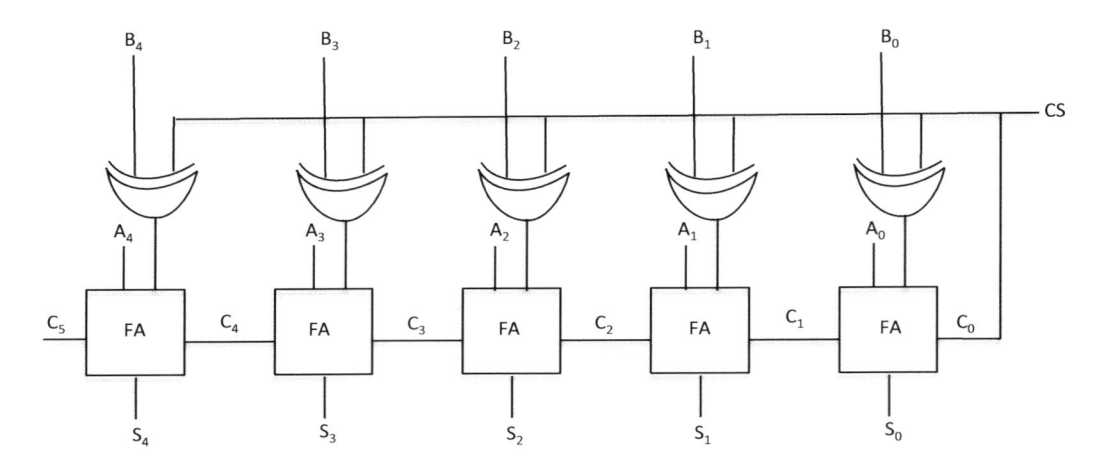

FIGURE 2.38   Two's complement adder/subtractor.

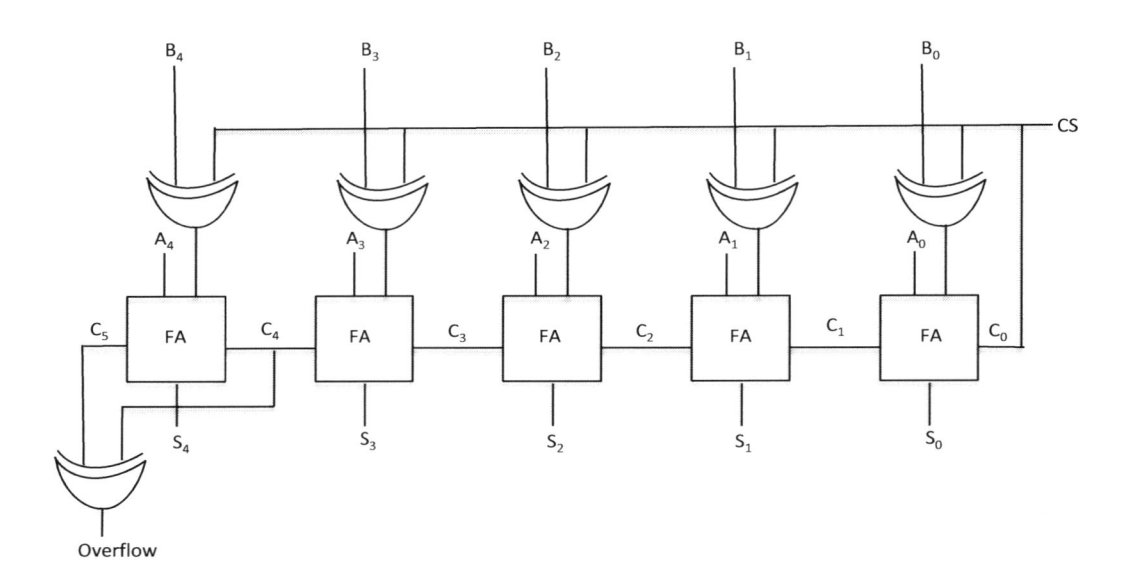

FIGURE 2.39   Two's complement adder/subtractor with overflow detection.

2.3.9   BINARY MULTIPLIER CIRCUITS

Multiplication can be carried out in two steps:

   a. Generate partial products, one for each digit in the multiplier.
   b. Sum the partial products (along with the appropriate carries) to produce the final product.
      Before summing operation, each successive partial product is *shifted* one position to the
      left relative to the preceding partial product.

Let us consider the process of multiplying two 2-bit numbers, $a_1 a_0$ and $b_1 b_0$. The resulting 4-bit
product $P_3 P_2 P_1 P_0$ is obtained, as described by Figure 2.40.

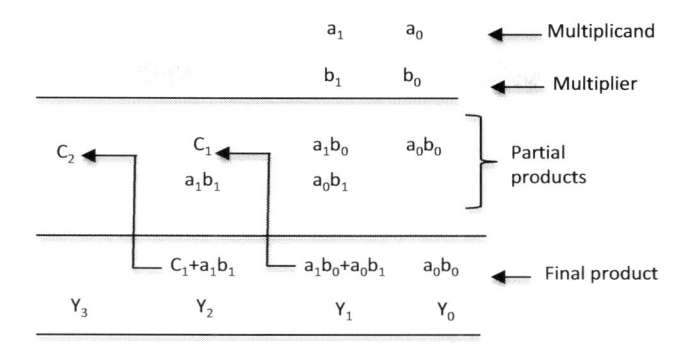

FIGURE 2.40   Two 2-bit multiplication process.

Using this process in Figure 2.40, we can design the multiplier circuit consisting of two modular parts:

- Four AND gates to generate the partial products; i.e. $Y_{00} = a_0b_0$, $Y_{01} = a_0b_1$, $Y_{10} = a_1b_0$, and $Y_{11} = a_1b_1$.
- Two *FA*s to generate the sums of the appropriate partial products. The first *FA₁* determines the sum Y_1 $(= a_1b_0 + a_0b_1)$, where the second bit (Y_1) of the product carries C_1 to the second *FA*. The second *FA₂* determines the sum Y_2 $(= C_1 + a_1b_1)$ to produce Y_2 and Y_3—where carry C_2 results from the addition.

Figure 2.41 depicts the designed 2-bit binary multiplier circuit.

Many different circuits exist for multiplication. Each one has a different balance between speed and amount of logics utilized.

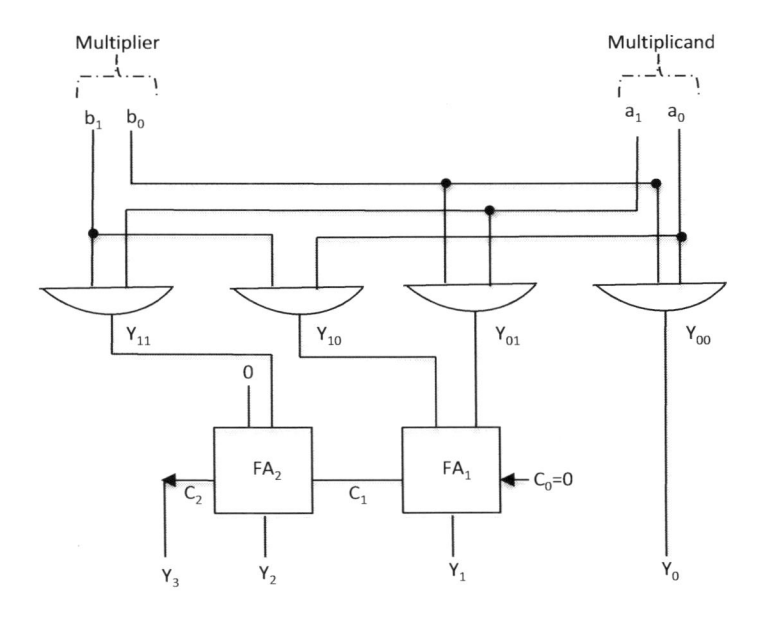

FIGURE 2.41   2-Bit binary multiplier circuit.

Comparator: Another combinational circuit that compares two binary numbers (A_i, B_i) and determines their relationship. For example, determining whether two binary words are equal $(A_i = B_i)$ or interpreting them as signed or unsigned numbers, tells their arithmetic relationship {i.e. greater or less than, $(A_i > B_i)$ or $(A_i < B_i)$, respectively}. Suppose $i = 0, 1, 2, 3$, and using comparison truth table (Table 2.11), we can write Boolean expressions representing the three compared outputs: "equal," "greater than," and "less than," respectively, in terms of the input variables $A_i = A_3 A_2 A_1 A_0$ and $B_i = B_3 B_2 B_1 B_0$, as follows.

a. Case $A_i = B_i$:

$$A_0 = B_0, A_1 = B_1, A_2 = B_2, A_3 = B_3$$

In view of Equation (2.7), the Boolean expression signifying the basic 2-bit equality is the same as ANDing the Exclusive-NORs; that is,

$$y_i = \overline{A_i \oplus B_i} = \overline{A_i}\,\overline{B_i} + A_i B_i \tag{2.25}$$

So, the comparative equality case's output is

$$Y_{A=B} = y_3 \cdot y_2 \cdot y_1 \cdot y_0 \tag{2.26}$$

b. Case $(A_i > B_i)$:

$$\text{Output} = Y_{(A_i > B_i)} = A_3 \overline{B_3} + y_3 A_2 \overline{B_2} + y_3 y_2 A_1 \overline{B_1} + y_3 y_2 y_1 A_0 \overline{B_0} \tag{2.27}$$

c. Case $(A_i > B_i)$ (Table 2.12):

$$\text{Output} = Y_{(A_i < B_i)} = \overline{A_3} B_3 + y_3 \overline{A_2} B_2 + y_3 y_2 \overline{A_1} B_1 + y_3 y_2 y_1 \overline{A_0} B_0 \tag{2.28}$$

Consequently, the comparative cases logic circuits can be drawn as follows.

TABLE 2.12

4-bit Comparator

Compared Input				Output		
A_3, B_3	A_2, B_2	A_1, B_1	A_0, B_0	$A = B$	$A > B$	$A < B$
$A_3 > B_3$	x	x	x	1	1	0
$A_3 < B_3$	x	x	x	0	0	1
$A_3 = B_3$	$A_2 > B_2$	x	x	0	1	0
$A_3 = B_3$	$A_2 < B_2$	x	x	0	0	1
$A_3 = B_3$	$A_2 = B_2$	$A_1 > B_1$	x	0	1	0
$A_3 = B_3$	$A_2 = B_2$	$A_1 < B_1$	x	0	0	1
$A_3 = B_3$	$A_2 = B_2$	$A_1 = B_1$	$A_0 > B_0$	0	1	0
$A_3 = B_3$	$A_2 = B_2$	$A_1 = B_1$	$A_0 < B_0$	0	0	1
$A_3 = B_3$	$A_2 = B_2$	$A_1 = B_1$	$A_0 = B_0$	0	1	0
$A_3 = B_3$	$A_2 = B_2$	$A_1 = B_1$	$A_0 = B_0$	0	0	1
$A_3 = B_3$	$A_2 = B_2$	$A_1 = B_1$	$A_0 = B_0$	1	0	0

'x' denotes "don't care".

Logic circuit for equality: Using Equations (2.25) and (2.26), we draw Figure 2.42.

Logic circuit for the inequalities: Using Equations (2.27) and (2.28), we can draw the logic circuit for the inequalities case, $[(A_i < B_i)$ and $(A_i > B_i)]$, as shown in Figure 2.43.

By combining Figures 2.42 and 2.43, a complete logic comparator circuit is drawn as shown in Figure 2.44 with its logic symbol.

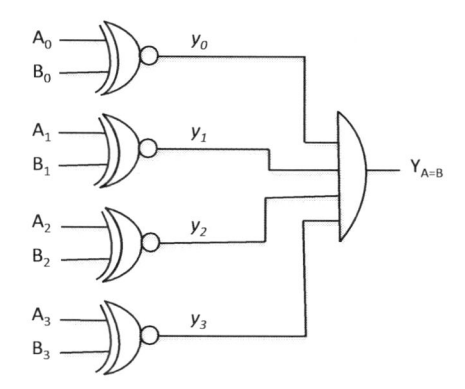

FIGURE 2.42 Comparative equality circuit.

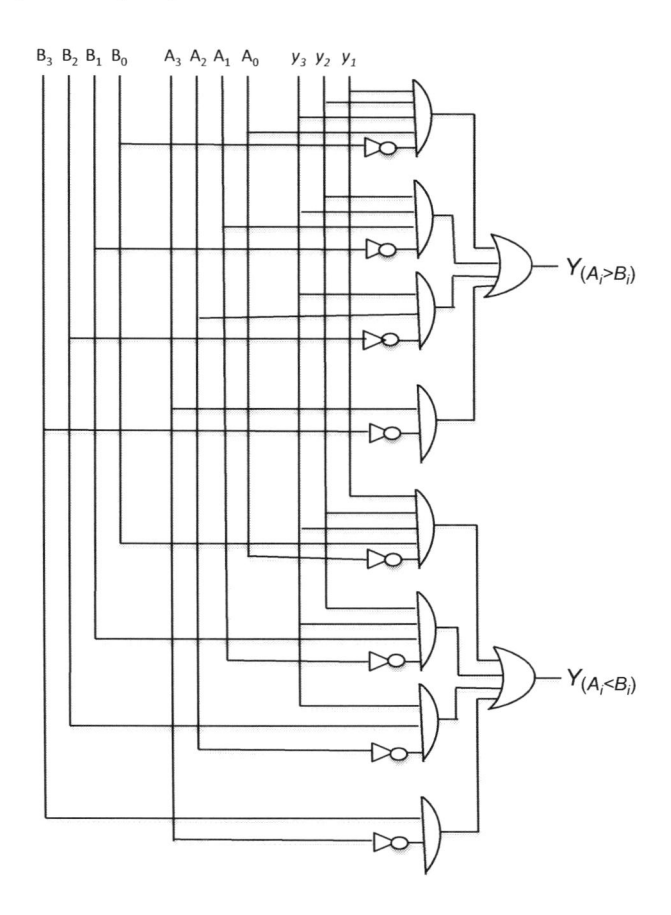

FIGURE 2.43 Logic circuit for inequality comparison.

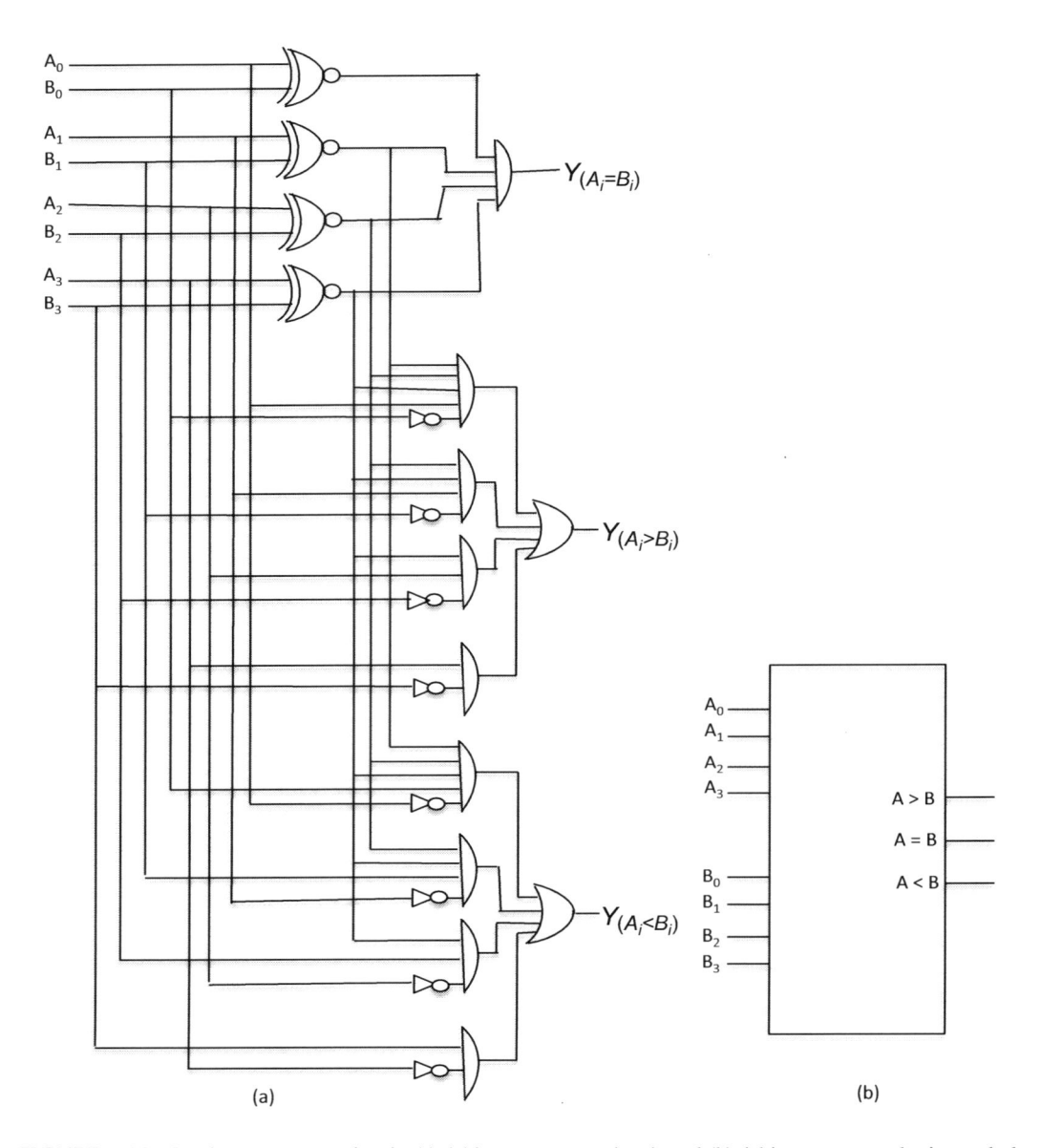

FIGURE 2.44   Logic comparator circuit: (a) 4-bit comparator circuit and (b) 4-bit comparator logic symbol.

2.3.10   ARITHMETIC LOGIC UNIT

We now consider the Arithmetic Logic Unit (ALU)—a collection of logic components—that constitutes the "figuring" unit of the computer providing the calculation capacity. ALU is basically a combinational logic, built from circuits already discussed, where the clock governs its register behavior which times the process. We begin with a very simple 1-bit ALU that performs multiple functions, e.g. AND, OR, and addition, as shown in Figure 2.45. The *select line, Sl*, controls the output of the multiplexer, MUX; i.e. chooses which output to use, as the result and the control lines (k_0, k_1) select which operational function is to be performed. The operational functions are performed internally, but only one result Y_1 is chosen for the ALU's output. The carry out C_{i+1} line from the *FA* is active during the addition operation.

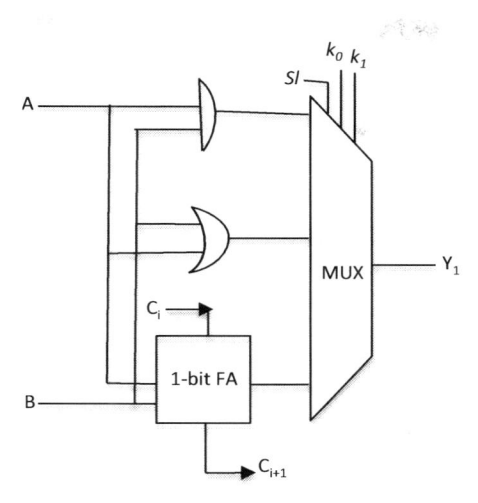

FIGURE 2.45 1-Bit ALU.

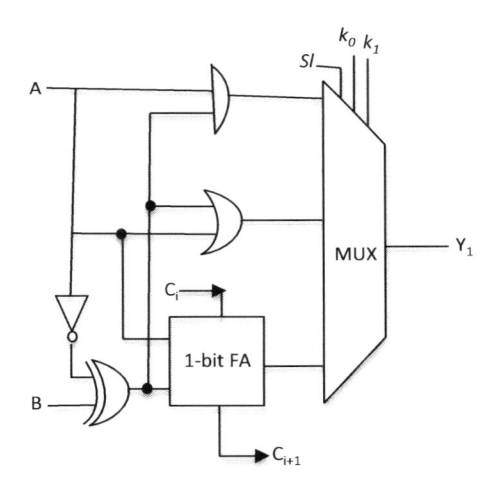

FIGURE 2.46 1-Bit ALU (additional operation—subtraction).

A simple modification to the FA circuit allows addition conversion to a subtraction, as shown in Figure 2.46. Subtraction is performed by forming 2's complement and taking the inverse of every bit and add 1. That is,

$$A - B = A + (-B) \qquad \text{(by arithmetic subtraction)}$$

$$= A + 2\text{'s complement of } B = A + \bar{B} + 1$$

$$= A + 1\text{'s complement of } B + 1 \text{ (given that we only considered 2's complement earlier,}$$

we only indicate this for completeness).

In fact, a large part of ALU design is captured by the design of a 1-bit ALU. When multiple ALUs are connected in tandem, the carry out C_{i+1} from one stage becomes the *carry in* C_i to the next stage.

In essence, ALUs perform many more operations where data can flow in parallel to multiple units and several units producing output internally use many performance optimizations.

2.4 SUMMARY

This chapter has explained what logic gates (or simply *gates*) are, how their functions can be represented with Boolean expressions or truth tables, how digital circuits are designed as a hierarchical structure of functional blocks where each functional block is decomposed into sub-functional blocks until the individual parts are reached, and how simple designs are formulated and used to build more complex digital logic circuits. The use of SOPs to implement circuits shown in this chapter is not necessarily optimal in time (depth of circuit) and space (number of switches). The use of truth tables (Karnaugh maps), which is another method of minimizing Boolean equation as well as the issue of speed (performance), is not exploited in this chapter. Instead the chapter explained in great depth combinational and sequential logics laying the foundations for the design of digital logic circuits, knowing fully the occurrence of sequential machines in nearly every chip design.

Given that logic gates are used to compute binary numbers and store digital information, and each logic gate is itself composed of a number of transistors that do the actual work inside the transistor, it is deemed necessary to consider what constitutes a transistor and transistor circuits, which the next chapter (Chapter 3) deals with. The difference in the ways digital-logic gates and quantum-logic gates are configured is obvious in Chapter 9.

PROBLEMS

2.1 Quizzes to refresh your knowledge of what you have learnt regarding logics:
   a. Gates implement Boolean functions.   True/False
   b. Combinational circuits have feedback lines.   True/False
   c. Sequential circuits can be synchronous and asynchronous.   True/False
   d. A synchronous sequential circuit changes its states at discrete instants of time.
                                                                                 True/False
   e. Asynchronous sequential circuits can have state transitions at discrete instants of time.
                                                                                 True/False
   f. A transition of a clock from 0 to 1 is called the falling edge.   True/False
   g. A clock period is the time between two rising edges or two falling edges. True/False
   h. A toggle flip-flop can be made from a JK flip-flop.   True/False
   i. Flip-flops are the memory used in synchronous sequential circuits.   True/False
   j. Asynchronous sequential circuits can have state transitions at discrete instants of time.
                                                                                 True/False
   k. Synchronous sequential circuits are also known as clocked sequential circuits.
                                                                                 True/False
   l. Encoders are essentially multiplexers.   True/False
   m. Decoders and demultiplexers are equivalent.   True/False
   n. Overflow always occurs when two numbers are of the same sign.   True/False
2.2 Sketch schematics for the following logic functions:
   a. $Y = \overline{x_o x_1 + x_2 \cdot (x_o + x_1)}$;
   b. $Y = x_o x_1 + x_1 x_2 + x_o x_2$ (majority);
   c. 3:2 priority encoder defined by $Y = \overline{x_o} \cdot (x_1 + \overline{x_2})$.
2.3 Design a logic circuit whose *product-of-sum* (POS) function is written as

$$F = (A + BC)(B + \overline{C}A)$$

Can this function be simplified further? If yes, simplify it.
2.4 Obtain the simplified Boolean expressions for the outputs of Figure 2Q.1 and Figure 2Q.2 in terms of their input variables.

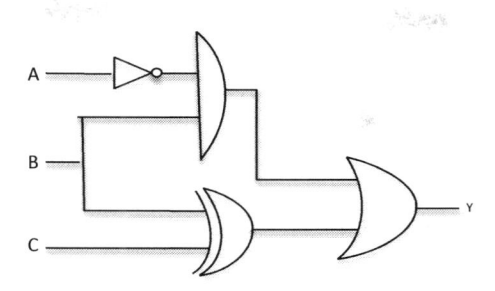

FIGURE 2Q.1

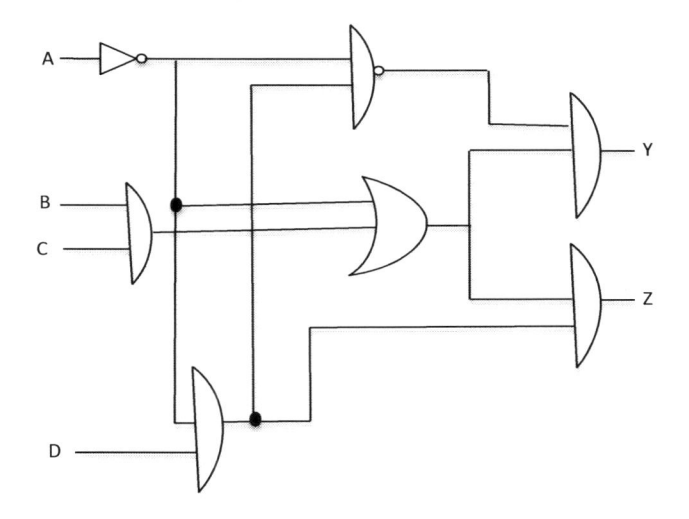

FIGURE 2Q.2

2.5 Design a combinational circuit with three inputs and one output. The output is 1 when the binary value of the inputs is less than 3. The output is 0 otherwise.

2.6 Design a guaranteed timing characteristics and glitch-free traffic light controller using clocked basic flip-flops.

BIBLIOGRAPHY

Holdsworth, B. and Woods, C. (2002). *Digital Logic Design*, 4th edition. Oxford: Newnes.

Mano, M. and Kime, C. (2008). *Logic and Computer Design Fundamentals*, 4th edition. Upper Saddle River, NJ: Prentice Hall.

Roth, C. (1999). *Fundamentals of Logic Design*, 4th edition. New York: PWS Publishing Company.

3 Structure of Bipolar Junction Transistor

In this chapter we present the structure of the bipolar transistor and show how a three-layer structure with alternating *n*-type and *p*-type regions can provide current and voltage amplification. We then present the ideal transistor model and derive an expression for the current gain in the forward-active mode of operation, as well as quantifying the operating point stability. We discuss how to model for small and large signals leading to simpler analysis of the circuits.

Transistors are the most crucial elements in modern electronics. Transistors are used in a great variety of circuits and remain important devices for ultra-high-speed discrete logic circuits such as emitter coupled logic, power-switching applications (including modern electronic digital computers), and in microwave power amplifiers. Transistors, as amplifiers, are used to amplify an electrical signal by allowing a small current or voltage to control the flow of a much larger current from a direct current (dc) power source. An example of transistor usage is in audio systems. There are two general types of transistors: bipolar and field effect. Very roughly, the difference between these two types is that for bipolar devices an input current controls the large current flow through the device, while for field-effect transistors (FET) an input voltage provides the control. In most practical applications, an operational amplifier (abbreviated as *op-amp*) is often used as a source of gain or amplification rather than to build an amplifier from discrete transistors. But there is a compelling case to having a good understanding of transistor fundamentals. For instance, the *integrated circuits* (ICs or chips) used in most computers are made from transistors (as well as diodes and other passive devices—transistors and capacitors), and so the behavior of logic devices depends upon the behavior of transistors. Also, given that op-amps are built from transistors, a detailed understanding of op-amp behavior, particularly input and output characteristics, must be based on an understanding of transistors. Op-amps are discussed in Chapter 4. In addition to the importance of transistors as components of op-amps, logic circuits, and an enormous variety of other ICs, single transistors are still important in many applications.

3.1 CONSTITUENTS OF A TRANSISTOR

A transistor consists of three regions of doped semiconductors. These regions or parts are called a *base*, a *collector*, and an *emitter*. As discussed in Chapter 1, doped semiconductors contain impurities or foreign atoms, which are incorporated into their crystal structure. These impurities can either be unintentional due to lack of control during the growth of the semiconductor or they can be added on purpose to provide free carriers in the semiconductor. When a semiconductor is doped with impurities that are ionized (meaning that the impurity atoms either have donated or have accepted an electron), it will therefore contain free carriers. When the ionized donors provide free electrons in a semiconductor, it is called *n-type* (meaning associated with negative carriers) while ionized acceptors that provide free holes in a semiconductor are called *p-type* (meaning associated with positive carriers) semiconductor. Generally, the ionization of the impurities is dependent on the thermal energy and the position of the impurity level within the energy bandgap. (In the gnarly world of quantum physics, if you move molecules closer or further apart, the "energy gap" changes. The gap affects what wavelengths of light are absorbed and reflected or transmitted—and changes the color we see.) By sandwiching a layer of *n*-type semiconductor between two segments of *p*-type material, a *pnp* transistor is constructed, as shown in Figure 3.1. Alternately, when a layer of *p*-type semiconductor is sandwiched between two segments of *n*-type material, an *npn* transistor

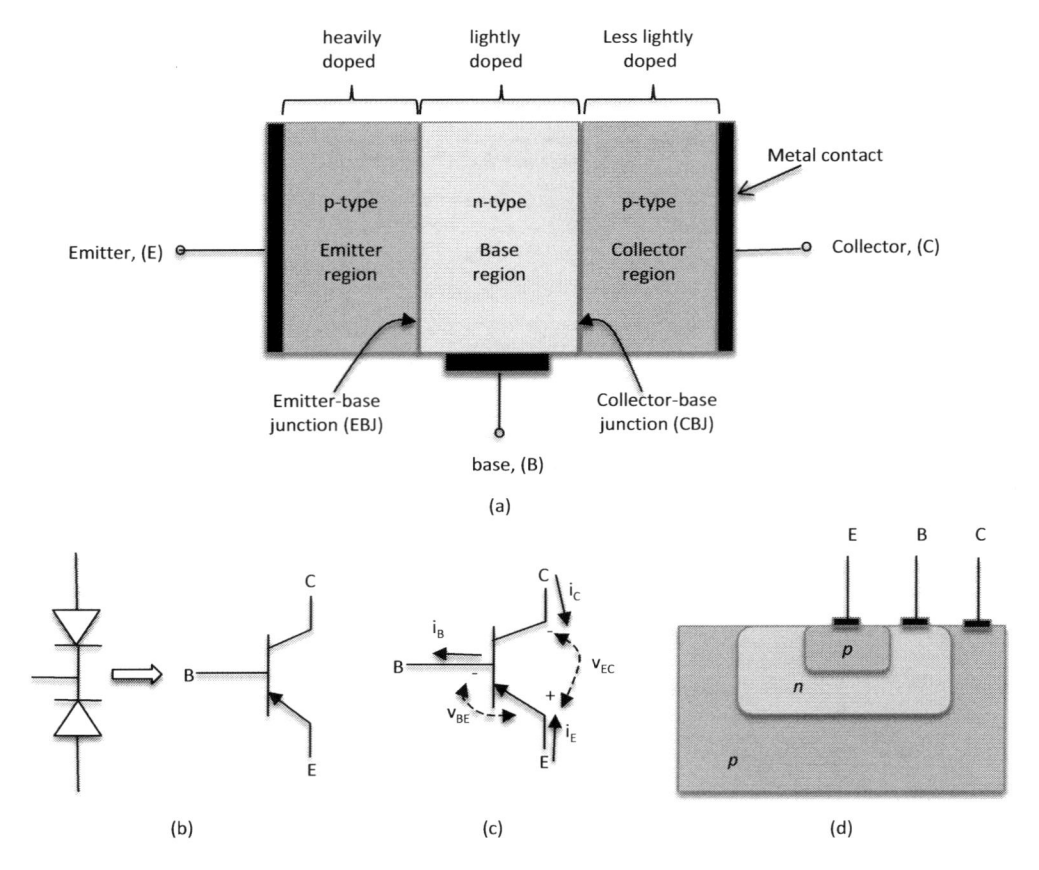

FIGURE 3.1   A *pnp* BJT structure: (a) *pnp* semiconductor, (b) *pnp* transistor symbol, (c) voltage and current notations, and (d) fabrication geometry.

is constructed, as shown in Figure 3.2. These types of transistors are collectively called *bipolar transistors* (or *bipolar junction transistors*, BJTs).

In the three regions, the emitter is the source of majority carriers that result in the gain mechanism of the BJT. These carriers, which are "emitted" into the base, are electrons for the *npn* transistor and holes for the *pnp* transistor. The base, on the other hand, is a region that physically separates the emitter and collector and has an opposite doping: *holes* for the *npn* and *electrons* for the *pnp* BJTs. The word "base" comes from the way that the first transistors were constructed. The base supported the whole structure. Finally, the collector serves to "collect" those carriers injected from the emitter into the base and which reach the collector without recombination.

The structure of *pnp* or *npn* transistor looks like two back-to-back diodes, as seen in Figures 3.1b and 3.2b, but each behaves much differently.

When forward biased, the *pn* junction of a diode has a large recombination rate and thus supports a large current. When forward biased, in the BJT, the *pn* base to emitter junction has a low recombination rate (the base is thin and lightly doped), so the electrons proceed to the collector where they are again the majority carrier. It is worth noting that the arrow is on the emitter terminal (as seen in Figures 3.1b and 3.2b symbol notations) and indicates the direction of emitter current (*into* emitter terminal for the *pnp* and *out of* emitter terminal for the *npn* device). The arrow thus points in the direction of the conventional current flow when the device is in forward-active mode.

As shown in Figure 3.1c, the voltage notation v_{EB}, with the dual subscript, denotes the voltage between the B (base) and E (emitter) terminals. Implicit in the notation is that the first subscript

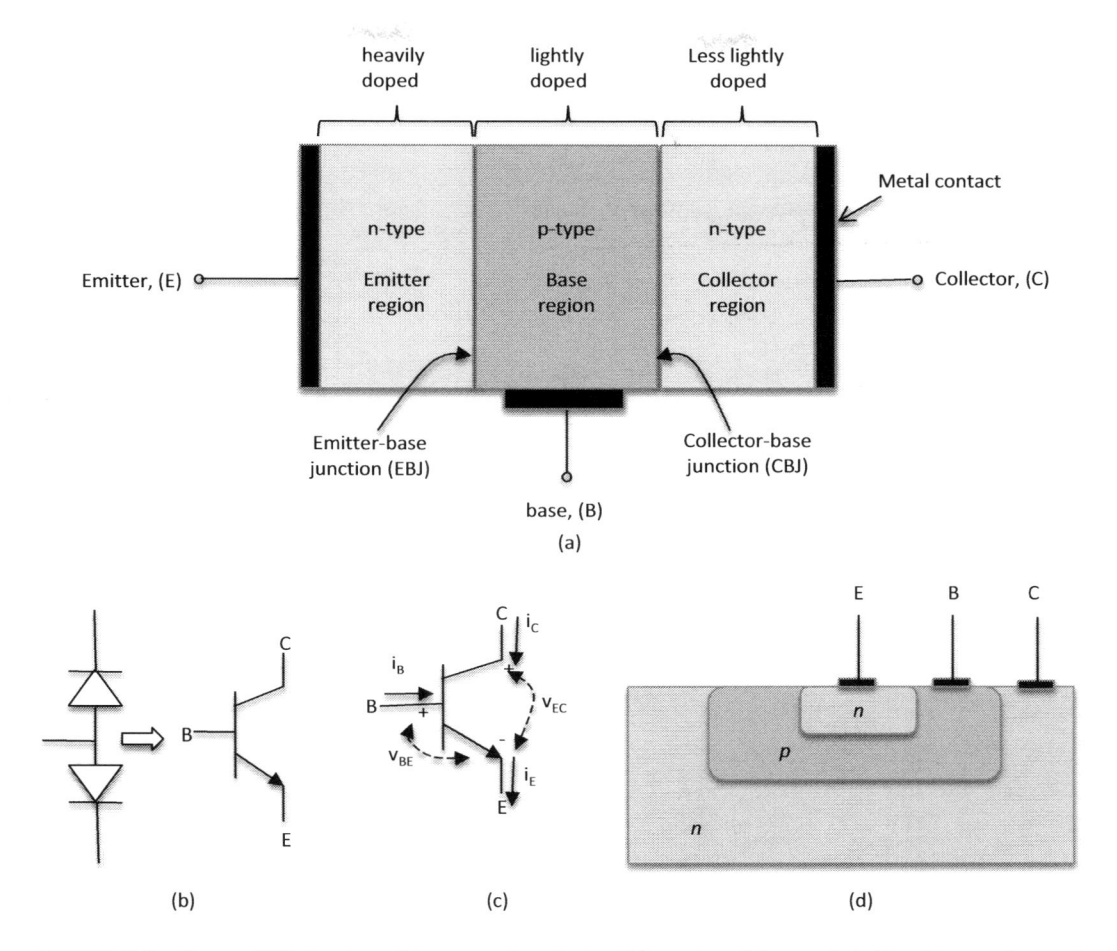

FIGURE 3.2 An *npn* BJT structure: (a) *npn* semiconductor, (b) *npn* transistor symbol, (c) voltage and current notations, and (d) fabrication geometry.

(the emitter terminal) is positive with respect to the second subscript (the base terminal). Likewise, for *npn* transistor of Figure 3.2c, the voltage notation v_{BE}, with the dual subscript, denotes the voltage between the B (base) and E (emitter) terminals. Implicit in the notation is that the first subscript (the base terminal) is positive with respect to the second subscript (the emitter terminal). As a consequence, the emitter current, i_E, is the total transistor current, which is the sum of the other terminal currents. Formally

$$i_E = i_B + i_C \tag{3.1}$$

Small letters are used to denote "variable" currents.

Figures 3.1d and 3.2d show the fabrication method where each device is not symmetrical electrically. This asymmetry occurs because the geometries of the emitter and collector regions are not the same, and the impurity doping concentrations in the three regions are substantially different. For example, the impurity doping concentrations in the emitter, base, and collector may be in the order of 10^{15}, 10^8, and $10^6 \mathrm{cm}^{-3}$, respectively. As a result, even though both ends are either p-type or n-type on a given transistor, switching the two ends makes the device act in drastically different ways.

The operation of transistors depends on the two pn junctions staying in close proximity; naturally the width of the base must be very narrow, typically in the range of few millimeters. One of the properties of the pn junction is that it rectifies allowing an electric current to pass only in one direction.

In essence, the *pn* junction is a versatile element, which can be used as a rectifier, as an isolation structure, and as a voltage-dependent capacitor. In addition, severally configured *pn junctions* can be used as solar cells, photodiodes, light-emitting diodes (LEDs) and even laser diodes. There are two general types of transistors: *bipolar* transistor and FET. These two device types are the basis of modern microelectronics of which *pn junctions* are an essential part. Each device type is equally important and each has particular advantages for specific applications. A brief description of bipolar type has been given above.

3.1.1 FIELD-EFFECT TRANSISTOR

A FET, also known as *junction FET* (JFET), is made from a continuous "channel" (or Body) of doped semiconductor material, either *n-type* or *p*-type. An oppositely doped semiconductor is then implanted or diffused on either side of the channel. The diffused regions are connected together by wire, which are called the "gate" of the transistor. Either of the channel's ends can be labeled as the source or the drain. The selection is based on the type of majority charge carrier present in the channel and the direction of those charge carriers' motion. The physical structures of the *n*- and *p*-type FETs are shown in Figures 3.3a and 3.3b, respectively, while the symbols of the *n*- and *p*-type FETs are shown in Figure 3.3c and d, respectively.

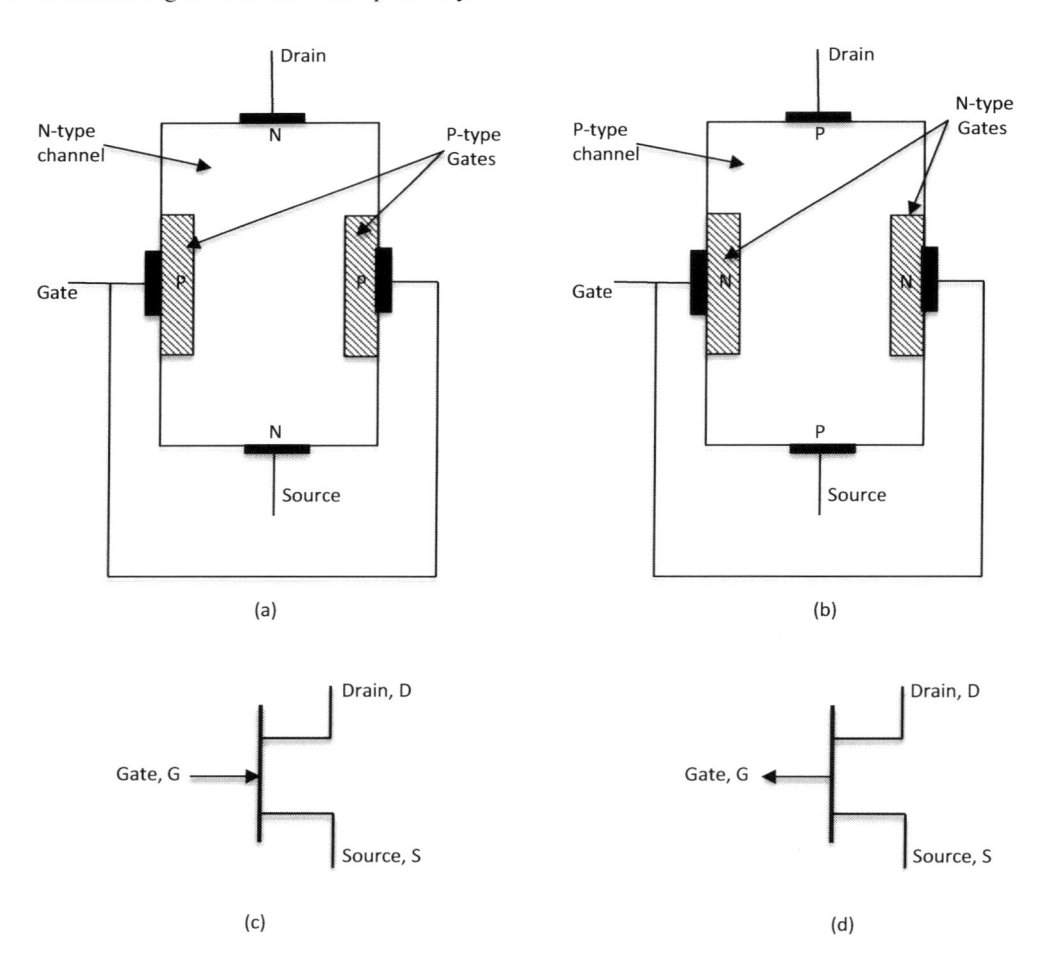

FIGURE 3.3 FETs: fabrication and symbols: (a) *n*-Channel FET, (b) *p*-Channel FET, (c) *n*-Channel FET symbol, and (d) *p*-Channel FET symbol.

All FETs have a *gate*, a *drain*, and a *source* terminal that correspond roughly to the *base*, *collector*, and *emitter* of bipolar transistors. In an FET, the width of a conducting channel is a semiconductor and, therefore, its current-carrying capability, is varied by the application of an electric field (thus, the name FET). As such, an FET is a "voltage-controlled" device. The difference between the bipolar and FETs is that for bipolar devices an input *current* controls the large current flow through the device, while for the FETs an input *voltage* provides the control. This control is the basic transistor action. The most widely used FETs are *metal oxide semiconductor* FETs (or MOSFET; or simply MOS).

In essence, the working principles of the device are based on sequential tunneling of single electrons between the basis material's atom and the source and drain leads of the transistor. Tunneling is a functioning concept that arises from quantum mechanics. Classically, an object hitting an impenetrable barrier will not pass through. In contrast, objects with very small mass with wave-like characteristics, for instance electrons, will penetrate such a barrier; such action is referred to as *tunneling*.

It is worth noting that the source and drain regions of the FET are interchangeable, in contrast to the bipolar transistors whose emitter and collector regions are not interchangeable, as their emitter is much more heavily doped than the collector. Almost all logic circuits, microprocessor and memory chips contain exclusively MOSFETs and their variants. More is said about MOSFET circuits in Chapter 5.

Advances in technology ensure that the variants improve on the previous meeting-specific requirements like low noise figures and high gain, particularly at millimeter-wave frequencies. Prominent variants are gallium arsenide (GaAs) substrate-based *metal-semiconductor FET* (MESFET), *high-electron mobility transistor* (HEMT), and *pseudomorphic* HEMT (or pHEMT). The main difference between HEMTs and MESFETs is the epitaxial layer structure. In the HEMT structure, compositionally different layers are grown in order to optimize and extend the performance of the FET. These different layers form heterojunctions since each layer has a different bandgap, see Figure 3.4. Structures grown with the same lattice constant but different bandgaps are referred to as lattice-matched HEMTs. Those structures grown with slightly different lattice constants are called pHEMTs. The MESFETs and HEMTs are grown on a semi-insulating GaAs substrate using molecular beam epitaxy (MBE), or metal–organic chemical vapor deposition (MOCVD), which is currently less common. Epitaxy is the natural or artificial growth of crystals on a crystalline substrate determining their orientation. Other commonly used names for HEMTs include MODFET (modulation-doped FET), TEGFET (two-dimensional electron gas FET), and SDHT (selectively doped heterojunction transistor).

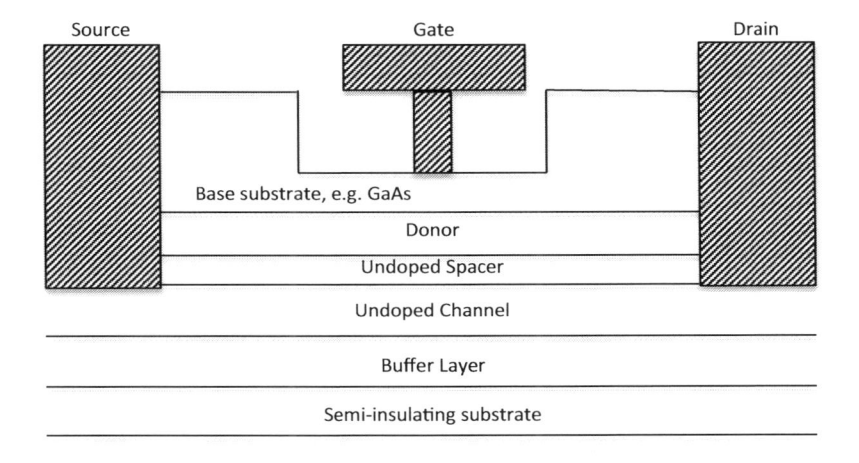

FIGURE 3.4 Structure of a basic HEMT.

3.1.2 TRANSISTOR FORWARD-ACTIVE MODE OPERATION

If the collector of a *pnp* transistor, as in Figure 3.5a, is open circuited, it would look like a diode of Figure 3.5b. Likewise, if the collector of an *npn* transistor, as in Figure 3.6a, is open circuited, it would look like a diode of Figure 3.6b. When forward biased, the current in the base–emitter junction would consist of electrons (PNP) or holes (NPN) injected into the emitter from the base and holes (PNP) or electrons (NPN) injected into the base from the emitter. But since there are many more holes in the emitter than electrons in the base, the vast majority of the current will be due to holes, as in the PNP case. Following similar logic, since there are many more electrons in the emitter than holes in the base, the vast majority of the current will be due to electrons, as in the NPN case. As a result, the base current is much smaller than the emitter and collector currents in *forward-active operating mode*, or simply the *active region*.

When the reverse-biased collector is added, as in Figure 3.7, it extracts the holes (for PNP) or electrons (for NPN) out of the base. As a result, the base–emitter current is due predominantly to electrons (PNP) or holes (NPN) current (as the smaller current component) while the collector–emitter current is due to holes (PNP) or electrons (NPN) (as the larger current component due to more holes or electrons from the $p-$ or $n+$ emitter doping respectively).

Given that diodes form the basis of transistor (as shown in Figures 3.1b and 3.2b for two *pn* junctions), we can develop transistor equations through knowledge of diode equation, which hopefully would enable us to understand the law of the junction.

3.2 DIODE EQUATION

A *pn*-junction structure constitutes a diode, as shown in Figure 3.8a. In this *pn-junction structure*, there exists two distinct regions, as seen in Figure 3.8b, namely: (i) the depletion region is the transition region where charge redistribution has taken place. Technically, this region is empty of

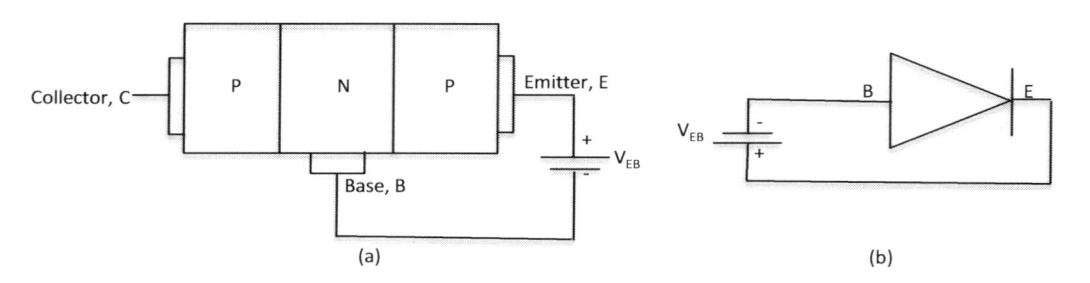

FIGURE 3.5 Understanding forward-active mode of a PNP transistor: (a) PNP transistor and (b) open-circuited collector, PNP, as a diode.

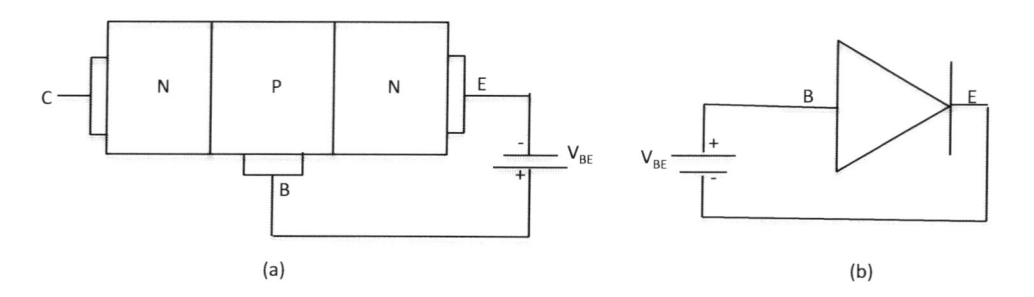

FIGURE 3.6 Understanding forward-active mode of an NPN transistor: (a) NPN transistor and (b) open-circuited collector, NPN, as a diode.

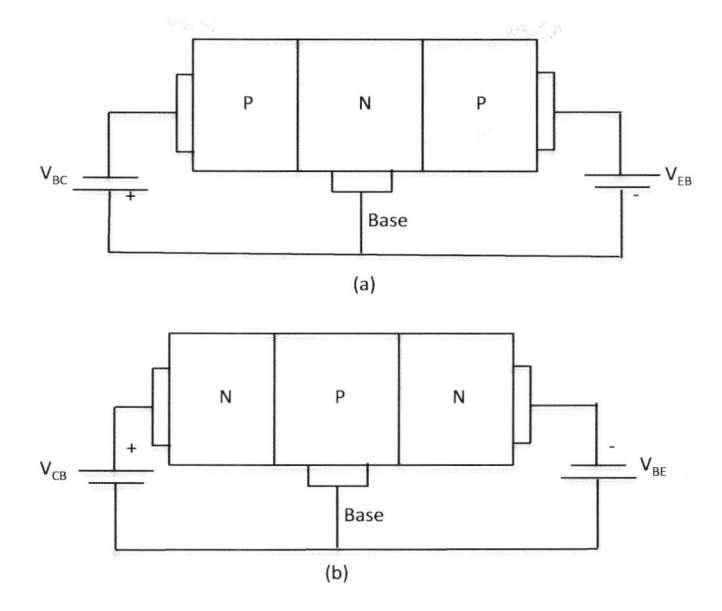

(a)

(b)

FIGURE 3.7 Understanding reverse biasing of transistors: (a) reverse-biased collector of a PNP transistor and (b) reverse-biased collector of an NPN transistor.

free careers, and (ii) the neutral regions are those which are essentially unaffected by the charge redistribution or carrier exchange.

Semiconductor diodes always include special marks on the packages to help identify the leads, as shown in Figure 3.8c. The circuit schematic symbol, as shown in Figure 3.8d, forms an arrow that points in the direction of current flow when the anode is biased positively with respect to the cathode, the "forward conduction region."

When calculating the current in a *pn* diode one needs to know the carrier density and the electric field throughout the *pn* diode which can then be used to obtain the drift and diffusion currents. Unfortunately, this requires the knowledge of the quasi-Fermi energies, which is only known if the currents are known. To avoid this problem, we will assume that the electron and hole quasi-Fermi energies in the depletion region equal those in the adjacent *n*-type and *p*-type quasi-neutral regions, and the Maxwell–Boltzmann statistics applied to diffusion of charge carriers can predict current

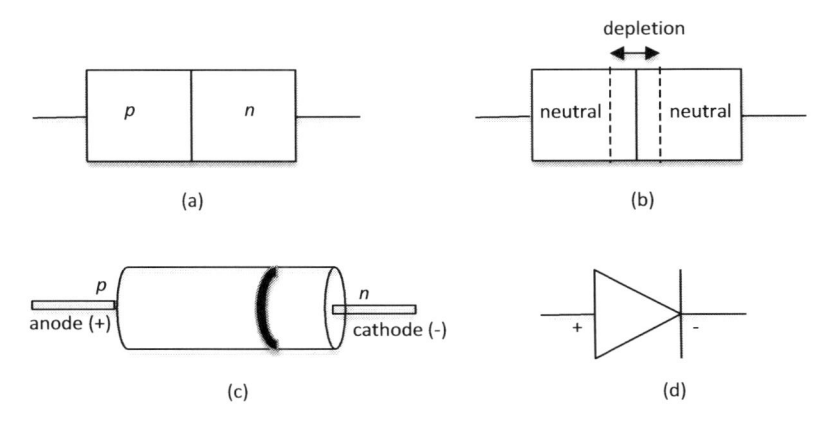

(a)

(b)

(c)

(d)

FIGURE 3.8 A semiconductor diode and distinct regions: (a) a *pn* junction, (b) regions, (c) diode package with identifying mark, and (d) schematic symbol.

density across junction in forward- and reverse-bias situations. Consequently, the current I through a diode can be expressed as

$$I = I_s \left[\exp\left(\frac{qV}{kT}\right) - 1 \right] \quad A \tag{3.2}$$

where
- V = the applied voltage across the junction in Volts (V);
- q = the electron charge or unit charge (= 1.60218×10^{-19} C);
- k = Boltzmann's constant [= 1.38065×10^{-23} JK^{-1}];
- T = the temperature (in kelvin, K). Note that 1 K = 273.15 + t (measured temperature in degree Celsius, °C); and
- I_s = the maximum current of the diode (also called *saturation current* or *leakage current*, in Amps, A).

The saturation current, in a *pn* semiconductor diode, is due to the diffusive flow of minority electrons from the *p*-side to the *n*-side and the minority holes from the *n*-side to the *p*-side. For this reason, the saturation current depends on the diffusion coefficient of electrons and holes. The minority carriers are thermally generated, so the reverse saturation current is almost unaffected by the reverse bias but is highly sensitive to temperature changes. The saturation current, I_s, in terms of the physical components, is expressed as

$$I_s = A_D \left(J_n + J_p \right) = q A_D \left(\frac{D_n n_{po}}{N_d W_n} + \frac{D_p n_{no}}{N_a W_p} \right) \quad A \tag{3.3}$$

where
- J_n is the electron's carrier (current) density;
- J_p is the hole's injection carrier (current) density;
- A_D is the cross-sectional area;
- W_n and W_p are the widths of the *n*-side and the *p*-side, respectively;
- D_n and D_p are diffusion coefficients of holes and electrons, respectively;
- N_d and N_a are the donor and acceptor concentrations at the *n*-side and the *p*-side, respectively;
- n_{po} and n_{no} are the intrinsic carrier concentrations in the *n* and *p* regions, respectively, of the semiconductor material. The intrinsic carrier concentration is temperature (T) dependent, i.e. $n_{po,no} = f(T)$.

If we represent the thermal voltage V_T by

$$V_T = \frac{kT}{q} \quad V \tag{3.4}$$

and substitute it in Equation (3.2) we write

$$I = I_s \left[\exp\left(\frac{V}{V_T}\right) - 1 \right] \quad A \tag{3.5}$$

The expression (3.5) says nothing about the possibility of reverse-bias breakdown.

When the *p*-junction of the diode is connected to the positive terminal of the voltage source, and the *n*-junction is connected to the negative terminal of the voltage, then the diode is forward biased, as seen in Figure 3.9a. The electrons and holes are pushed towards the junction, thereby reducing

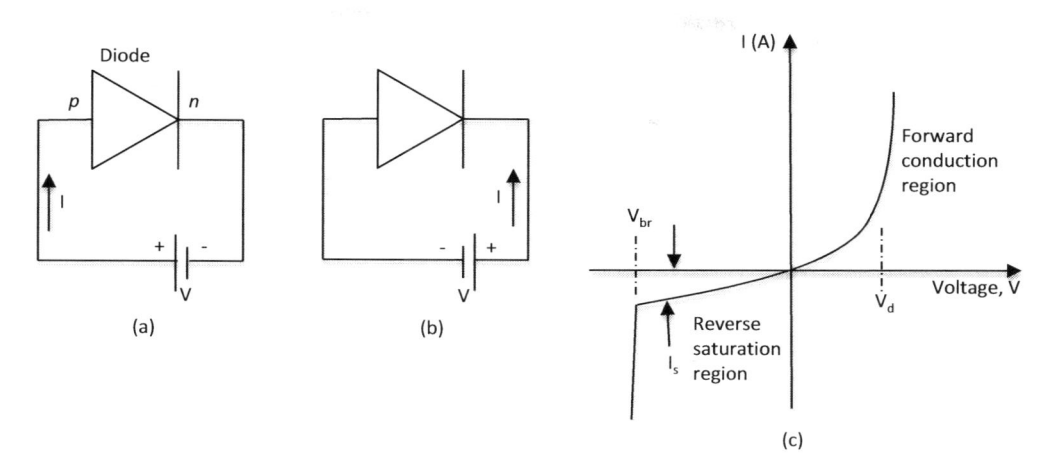

FIGURE 3.9 Diode bias types and its characteristic curve: (a) forward bias, (b) reverse bias, and (c) characteristic curve.

the width of the depletion zone. These effects make it easier for free electrons and holes with modest amounts of thermal (kinetic) energy to cross the junction.

When forward biased, the forward current $I_F (= I)$ becomes much larger than the leakage current I_s. So, the current approximates to

$$I \approx I_s e^{V/V_T} \quad A \tag{3.6}$$

which is represented by Figure 3.9c. Above the forward threshold voltage V_d the current increases exponentially. V_d, also called the *forward turn-on* voltage, represents the height of the diode barrier when no voltage is applied. This means that the energy required for an electron (or hole) to be able to cross the barrier is qV_d.

The turn-on voltage V_d is largely governed by the choice of semiconductor material and the device design. Silicon-based PN-junction diodes have threshold voltages V_d of ~0.6–0.8 V. Germanium diodes have $V_d \approx 0.2$ V. The current-handling capacity is often limited by thermal considerations (too much current and/or poor heatsinking and the device can be destroyed), but this can be manipulated by geometrical factors such as the device cross section, so devices of various sizes with a wide range of current-handling capacities are available.

On the other hand, if p-junction of the diode is connected to the negative terminal of the voltage source, and the n-junction is connected to the positive terminal of the voltage, then the diode is reverse biased, as seen in Figure 3.9b. In this situation, the electrons and holes are pulled apart, thereby increasing the height of the energy barrier between the two sides of the diode. As a result, it is almost impossible for any electrons or holes to cross the depletion zone and the diode current produced is virtually zero. A few lucky electrons and holes may happen to pick up a lot of thermal energy, enabling the few to cross the barrier, thereby making the reverse-biased current to be very small. As a result, the voltage V is negative and hence the exponential term in the current expression of Equation (3.5) is very small. Hence,

$$I \approx I_s \tag{3.7}$$

Under reverse bias only a tiny reverse saturation current I_s flows until breakdown is encountered, at which point the current increases rapidly, as shown in Figure 3.9c. Note that the reverse saturation current is exaggerated for clarity in Figure 3.9c, and the reverse breakdown voltage V_{br} is usually much larger than the forward turn-on V_d. The diode characteristic for both forward-biased and reverse-biased situations is shown in Figure 3.9c.

By substituting values of the electron charge (q) and the Boltzmann's constant (k) in Equation (3.4) and at room temperature (when $T = 23°C$), the thermal voltage V_T becomes

$$V_T = \frac{kT}{q} = \frac{1.3807 \times 10^{-23} \times (273.15 + 23)}{1.6022 \times 10^{-19}} \approx 0.025\,\text{V}$$

and the diode current when forward biased can be expressed as

$$I \approx I_s\, e^{V/0.25} \approx I_s\, e^{40V} \quad \text{A} \tag{3.8}$$

We know from Ohms law that, for a linear resistor (resistance $R = V/I$), but, as seen in Equation (3.8), the relationship between voltage V and current I is not constant or linear but exponential. Hence, the slope of the line at any point gives the "resistance" of the forward-bias junction. If we differentiate the current I in Equation (3.8) with respect to voltage V, we have

$$\frac{dI}{dV} = 40 I_s e^{40\,V}$$
$$= 40 I \tag{3.9}$$

which is the conductance (in siemens, S). Inverting Equation (3.9) gives the resistance. Hence,

$$\frac{dV}{dI} = \frac{1}{40 I} \quad \Omega$$
$$= r \tag{3.10}$$

where r is the "dynamic resistance" of forward-biased junction. Alternately,

$$r = \frac{0.025}{I} \quad \Omega \tag{3.11}$$

In most circuit analyses we can often get by with a much simpler phenomenological model that treats turn-on and breakdown as abrupt transitions and neglects the reverse saturation current. In this case there are basically three parameters that describe the operation, namely: V_d, V_{br}, and the maximum current that the device can sustain during operation.

3.2.1 LIGHT-EMITTING DIODES

LEDs are diodes made with direct bandgap semiconductors (such as GaAs, *gallium nitride;* GaN, etc.) and designed to generate light when enough current passes through the device. LEDs are functionally similar to any diode, but they typically have a much larger threshold voltage (about 2 V for red and green LEDs, and upwards of 4 V for blue and white LEDs).

Theoretically, by arranging two diodes, as shown in Figure 3.10, which are in parallel but with an opposite orientation of cathodes, we can create a polarity indicator that gives a visual display of the polarity of the supply voltage, V_s: as "red" if diode D_1 is painted red; and as "green" if diode D_2 is painted green. Such devices are packaged and sold as "bi-color" LEDs (denoted by dashed box in Figure 3.10).

You might think this topology would be useful for splitting a large current among devices with small current-handling capacity. It can be used for splitting unless the diodes are very well matched; otherwise one device ends up carrying most of the current. This is because, as seen in

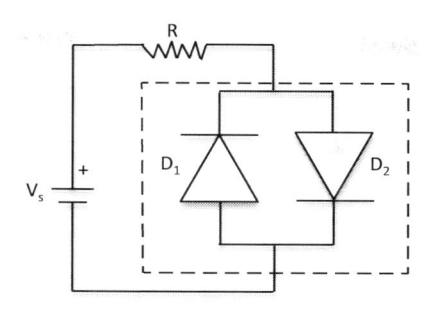

FIGURE 3.10 Anti-parallel diodes (LEDs).

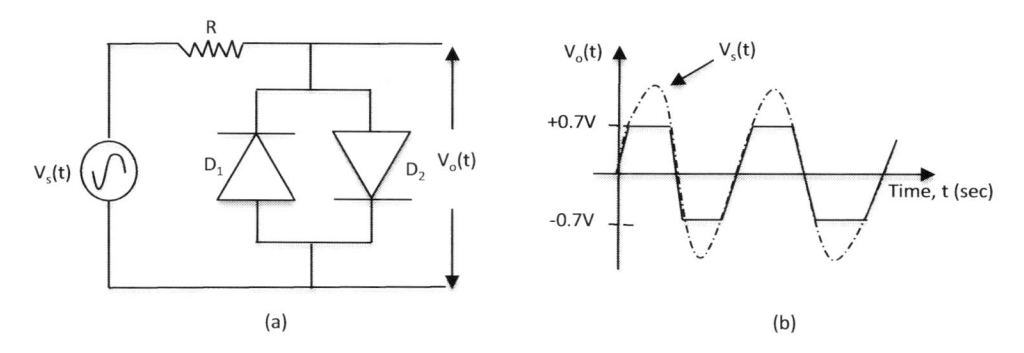

FIGURE 3.11 A simple bipolar limiter: (a) bipolar limiter circuit and (b) bipolar output signal.

Equation (3.2), the diode current depends exponentially on several factors including temperature; so, small changes in device or environmental parameters can lead to rather large changes in current from one device to the next. As a general rule, *diodes or LEDs in parallel each one should have its own bias resistor.* An exception is within an IC where we can often count on the devices being well matched, but, even in that case, it is probably better to simply use one large diode than a bunch of smaller ones in parallel.

Instead of a dc voltage as source voltage, we replace it with a sinusoidal input voltage as in Figure 3.11a. If the input-source signal exceeds the diode forward voltage drop, the diodes will conduct and clamp the output to ±0.7 V, as shown in Figure 3.11b. This is sometimes called a "clipper" circuit because the tops of the sinusoidal waveforms (signals exceeding 0.7 V in magnitude) appear to be clipped off. It is also called a "limiter" for obvious reasons. This is useful in some applications, for example at the input of a sensitive high-gain amplifier, where the input signal needs to be limited to prevent overdriving the amplifier. Other application of diodes in conjunction with transistors and ICs will be discussed later in this chapter and Chapter 4, respectively.

Having understood the behavior of a diode, we are now in a better position understanding the behavior and characteristics of a transistor. A transistor is formed from the fusion of two diodes in different configurations.

3.3 TRANSISTOR CHARACTERISTICS

In dc operation, a bipolar transistor has three regions of operation, namely: the active region, the cutoff region, and the saturation region, as shown in Figure 3.12b.

Figure 3.12b shows bipolar transistor current–voltage (*I–V*) characteristic curves depicting its (Figure 3.12a) behavior of an *npn transistor*. The characteristic curves of a *pnp* transistor will look

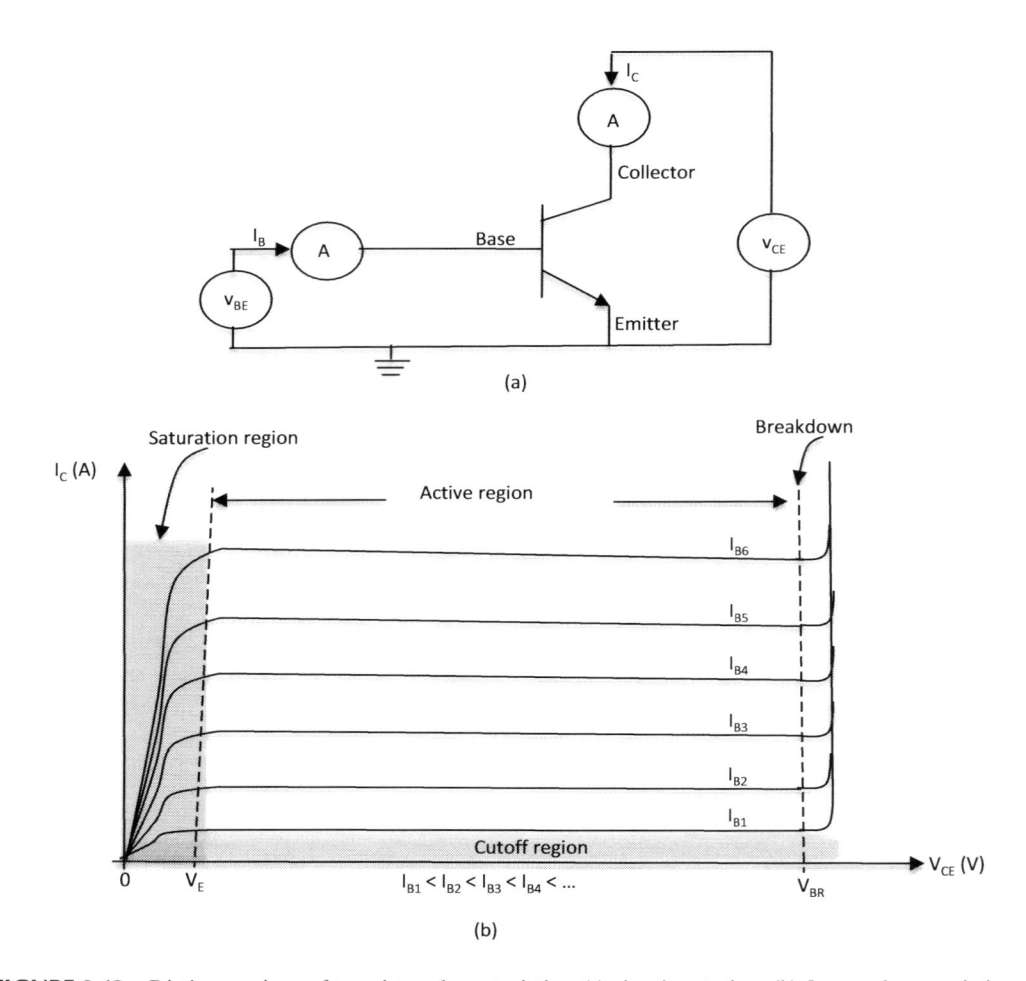

FIGURE 3.12 Distinct regions of transistor characteristics: (a) circuit notation, (b) I_C–v_{CE} characteristics.

the same but the polarity on the base–emitter voltage will be switched; that is, v_{BE} becomes v_{EB}. Each of the curves, in Figure 3.12b, illustrates the dependence of the collector current (I_C) on the collector–emitter voltage (v_{CE}) when the base current (I_B) has a constant value, i.e. when v_{BE} is held constant.

Ideally, Figure 3.13 shows the base current (I_B) versus voltage (V_{BE}) characteristics. The potential drop across the emitter junction is low, often less than one volt (e.g. for silicon-doped device, $V_{BE} \approx V_o \approx 0.7\,\text{V}$). The conditions under which these three operational regions exist are described as follows.

In the **active** region, as seen in Figure 3.12b, the base–emitter junction is forward biased and the base–collector junction is reverse biased. In this region of operation, the transistor has gain and acts like a current controlled current source (CCCS). A small base current I_B (typically a few microamps, µA) controls a large collector current. Ideally, the curves are horizontal straight lines, indicating that the collector behaves as a constant current source independent of the collector voltage. This becomes evident in Equation (3.8). The terminal equations for the transistor operating in the active region can be developed. As we know from diode study, the forward bias causes excess minority carriers on the base–emitter side, while the reverse bias at the base–collector end causes the minority concentration to approach zero. We assume (without loss of generality) that the short-base diode model is valid for all junctions.

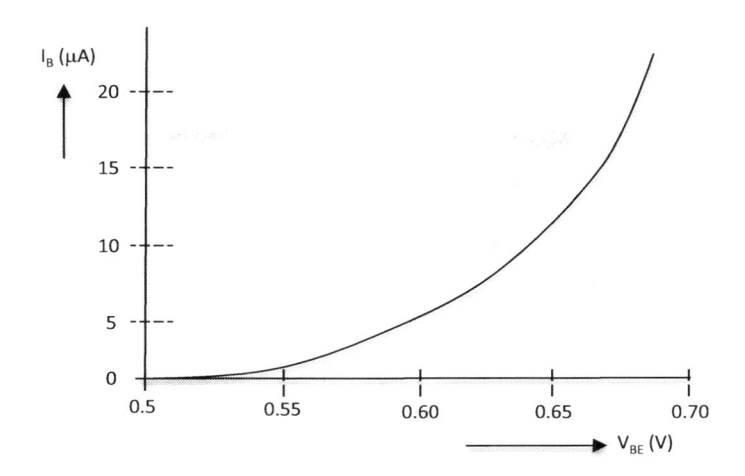

FIGURE 3.13   An ideal common-emitter I_B–v_{BE} (base) characteristic.

The *law of the junction* is still valid and can be used to evaluate the minority carrier concentrations at the boundaries of the base region of the transistor. Since the width of the base region, which is substantially below one micrometer ($<1.0\,\mu m$) in contemporary technologies, is sufficiently smaller than diffusion length, the minority carriers in the base region would display a linear gradient, similar to the case of the "short" diode. Since we do not know the current due to recombination in the depletion region we will simply assume that it can be ignored. As such, the value of this current is the same as that given by Equation (3.5), i.e.

$$I = I_S \left[\exp\left(\frac{V}{V_T} \right) - 1 \right] \tag{3.5}$$

where I_S is the saturation current (also called maximum current) of the transistor. This current varies with the semiconductor material used. For Silicon, Si, $10^{-12} < I_s < 10^{-10}$ (Ampere, A) and for Germanium, Ge, $I_s \cong 10^{-4}$ (A). The I_s expression for diode given by Equation (3.3) is also valid for transistor. In some textbooks, the V_T in the exponent of Equation (3.4) is written as ηV_T where η is the ideality factor, which varies between 1 and 2, depending on the fabrication process and semiconductor material. In many cases, η is assumed to be approximately equal to 1.

For a bipolar transistor, forward biased, Equation (3.5) applies by replacing V with the base–emitter voltage V_{BE}, and approximates to

$$I = I_S \exp\left(\frac{V_{BE}}{V_T} \right) \tag{3.12}$$

since $V_{BE} \gg V_T$ in most cases.

The terminal equations for the transistor operating in the active region are

$$I_C = I_S \exp\left(\frac{V_{BE}}{V_T} \right) \tag{3.13}$$

The base current I_B relates to the collector current I_C by a constant ratio called the *forward current gain, β*:

$$\beta = \frac{I_C}{I_B} \tag{3.14}$$

So,

$$I_B = \frac{I_s}{\beta} \exp\left(\frac{V_{BE}}{V_T}\right) \tag{3.15}$$

This demonstrates that the base current I_B is relatively small: the smaller the better, actually. For typical digital bipolar processes, β varies between 50 and 100. This means that a small hole current into the base sustains a large electron current at the collector; hence the current gain.

The relationship between I_E and I_B is obtained by enforcing the current conservation law; for instance, by treating the bipolar transistor as a single node, and thence applying Kirchhoff's current law (KCL), to have

$$I_E = I_C + I_B = I_B(\beta + 1) \tag{3.16}$$

By combining Equations (3.8) and (3.10), we can relate I_C and I_E.

$$\frac{I_C}{I_E} = \frac{\beta}{\beta + 1} = \alpha \tag{3.17}$$

where α is called the *forward common-base current gain*, and is always slightly less than 1. The emitter current can thus be expressed as

$$I_E = \left(\frac{\beta + 1}{\beta}\right) I_S \exp\left(\frac{V_{BE}}{V_T}\right) = \frac{I_S}{\alpha} \exp\left(\frac{V_{BE}}{V_T}\right) \tag{3.18}$$

From Equation (3.17), we can state the common-emitter current gain, β, in terms of the common-base current gain, α, thus

$$\beta = \frac{\alpha}{1 - \alpha} \tag{3.19}$$

The preceding analysis has demonstrated that when the transistor is forward biased, the forward-biased voltage V_{BE} causes an exponentially related flow of electrons from the emitter into the base where they diffuse across the base region and are collected in the collector region. The collector current, I_C, is independent of the base–collector voltage as long as the base–collector junction is reverse biased. The collector, therefore, behaves as an ideal current source. The collector current I_C is a fraction α of the emitter current I_E, and the base current I_B is a fraction $1/\beta$ of the collector current I_C. If $\beta \gg 1$, then $\alpha \approx 1$, and $I_C \approx I_E$. Equations (3.13) through to (3.18) apply to *pnp* transistors, except that the polarity on the base–emitter voltage will be switched (v_{BE} becomes v_{EB}). In essence, it is advantageous to reduce the transition time for the collector current to decrease the switching losses. While this can be achieved by decreasing the width of the *p*-base region, which also increases the current gain β, it can also lead to a reduction of the breakdown voltage due to the reach-through phenomenon.

In the **saturation** region both junctions (i.e. the emitter–base junction and the base–collector junction) are forward biased. Both the collector and emitter inject minority carriers into the base [1]. As in Figure 3.12b, as the magnitude of v_{CE} decreases, there comes a point when the collector voltage becomes less than the base voltage. When this happens, the transistor leaves the linear region of operation and enters a highly nonlinear and unstable region, which is the saturation region. When the transistor saturates, a substantial drop in current gain occurs. In digital circuits, operation in the saturation or reverse regions is, in general, avoided, as the circuit performance in those regions tends to deteriorate.

The **cutoff** region, as seen in Figure 3.12b, occurs when the emitter–base junction and the base–collector junctions are reverse biased. In the region, no excess base charge is present; the transistor is turned off and no collector current flows, and the base currents are near zero. The currents into the terminals are limited to the saturation currents of the reverse-biased diodes and are extremely small. The transistor is considered to be in the *off state*. In digital circuits, the transistor is operated by preference in the cutoff or forward-active mode.

Both the **cutoff** and **saturation regions** are unusable for amplification.

3.3.1 BREAKDOWN MECHANISM

Breakdown is characterized by the rapid increase of the current under reverse-bias condition, that is, when the base–collector junction (BJT) is reverse biased. The corresponding applied voltage is referred to as the *breakdown voltage*, V_{BR} (denoted on the V_{CE}-axis of Figure 3.12b). The breakdown voltage is a key parameter of power devices.

Breakdown is caused by two mechanisms, namely: avalanche multiplication and quantum mechanical tunneling of carriers through the bandgap. Whilst neither of the two breakdown mechanisms is destructive, large heat dissipation caused by the large breakdown current and high breakdown voltage can cause permanent damage to the semiconductor or the contacts. Typically, breakdown is dominated by avalanche multiplication, which means that the large electric field in the base–collector depletion region causes carrier multiplication due to impact ionization. The ionization causes generation of additional electrons and holes. As a result, when avalanche breakdown occurs at the base–collector junction, the holes generated by impact ionization are pulled back into the base region resulting in an additional base current. This additional base current causes an even larger additional flow of electrons through the base and into the collector due to the current gain of the BJT. This larger flow of electrons in the base–collector junction causes an even larger generation of electrons and holes.

Breakdown in silicon, at room temperature, can be predicted using the following empirical expression for the electric field at breakdown:

$$|V_{BR}| = \frac{4 \times 10^5}{1 - \frac{1}{3}\log\left(N \times 10^{-16}\right)} \text{ V/cm} \tag{3.20}$$

where N = doping concentration.

Consider a case of the transistor's emitter node being open and the base being grounded as shown in Figure 3.14. If we assume that between point x_1 and point x_2, within the depletion region, there is sufficient electric field, although assumed constant, but large enough to cause impact ionization. Outside this x_1 and x_2 range, the electric field is assumed to be too low to cause impact ionization.

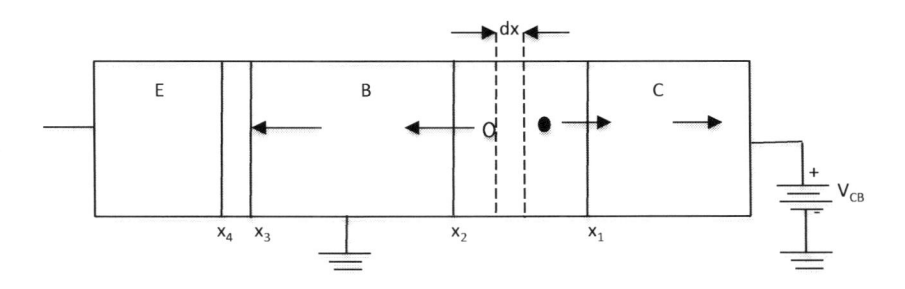

FIGURE 3.14   Open emitter, grounded base, and reverse-biased collector.

Assuming that the ionization coefficients of electrons and holes are the same (i.e. $\alpha_n = \alpha_p = \alpha$) and within a narrow region dx, the gain factor, G_{ion}, as a result of ionization that causes a generation of additional electrons and holes, is defined as

$$G_{\text{ion}} = \int_{x_1}^{x_2} \alpha \, dx \tag{3.21}$$

Thus, the multiplication factor M can be calculated from

$$M = \frac{1}{1 - G_{\text{ion}}} \tag{3.22}$$

The current due to onset of ionization $I_{C(\text{ion})}$ will equal to the product of M and the normally generated current, I_C, i.e.

$$I_{C(\text{ion})} = M \times I_C \tag{3.23}$$

When M tends to infinity (i.e. $M \to \infty$), an avalanche breakdown would occur. The condition for this is

$$G_{\text{ion}} = \int_{x_1}^{x_2} \alpha \, dx = 1 \tag{3.24}$$

In fact, the condition for breakdown of a diode in isolation is identical to that of breakdown in the transistor; in this configuration for example, it is simply the breakdown of the collector–base diode. The multiplication factor is commonly expressed as a function of the applied voltage, V, and the breakdown voltage, V_{BR}, using the following empirical relation:

$$M = \frac{1}{1 - \left| \dfrac{V}{V_{BR}} \right|^n} \quad \text{where} \quad 2 < n < 6 \tag{3.25}$$

where n varies with the structure of the PN junction, whether it is p and n, or n and p.

On the other hand, if the base is open, and the emitter is grounded, as in Figure 3.15, the holes generated in the collector depletion region have nowhere to flow into but to the emitter. This will result in an injection of electrons by the emitter into the base, thus ensuring the electron's flow into the collector. As a result, the collector current when the emitter is open, $I_{C(EO)}$, would have three components:

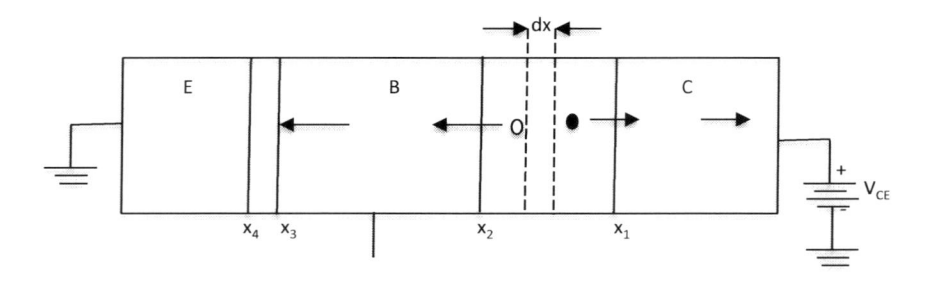

FIGURE 3.15 Open base, grounded emitter, and reverse-biased collector.

i. the normally generated thermal current when the base is open, $I_{C(BO)}$;
ii. the current due to ionization, $I_{C(ion)}$, which is Equation (3.23); and
iii. the current due to electrons injected by the collector. The current injected by the emitter would be transistor gain β times the hole current, which is simply the sum of the currents of (i) and (ii), i.e. $\left(I_{C(BO)} + I_{C(ion)}\right)\beta$.

By rearranging Equation (3.22); i.e. as $G_{ion} = \dfrac{M-1}{M}$, the net current can be written as

$$I_{C(EO)} = I_{C(BO)} + I_{C(ion)} + \left(I_{C(BO)} + I_{C(ion)}\right)\beta$$

$$= (\beta+1)\left[I_{C(BO)} + I_{C(EO)}\right] \tag{3.26}$$

$$= \frac{(\beta+1)I_{C(BO)}}{1-(\beta+1)G_{ion}} = \frac{(\beta+1)I_{C(BO)}}{1-(\beta+1)\dfrac{M-1}{M}}$$

When avalanche multiplication is small, the collector current is simply β times the current under emitter-open condition. In this case, breakdown would occur when the denominator of Equation (3.26) equates to zero, i.e.

$$1-(\beta+1)\frac{M-1}{M} = 0 \tag{3.27}$$

Rearranging to have

$$M = \frac{\beta+1}{\beta} \tag{3.28}$$

Using Equation (3.25) for the multiplication factor (or coefficient), we can write

$$V_{BR(CEO)} = \frac{V_{BR(CBO)}}{(\beta+1)^{1/n}} \quad \text{for} \quad 2 < n < 6 \tag{3.29}$$

again, where n varies with the structure of the PN junction, whether it is p and n, or n and p. This expression, Equation (3.29), shows that the common emitter breakdown voltage can be characterized by the open base breakdown voltage, $V_{BR(CEO)}$, to be significantly less than the open emitter breakdown voltage, $V_{BR(CBO)}$.

To further analyze the ionization effect quantitatively, we first write the total collector current, I_C, in response to an applied base current, I_B:

$$I_C = \beta\left[I_B + (M-1)I_C\right] \tag{3.30}$$

The term $[(M-1)I_C]$ was added to the base current to include the holes generated due to impact ionization. If $M = 1$, Equation (3.30) reverts to Equation (3.14). We can rearrange Equation (3.30) to yield

$$I_C = \frac{\beta I_B}{1+\beta(1-M)} \tag{3.31}$$

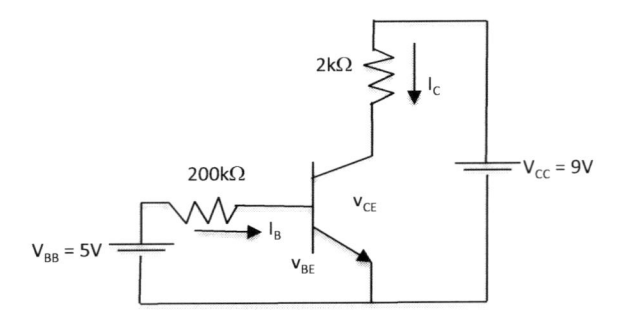

FIGURE 3.16 A common-emitter circuit.

Example 3.1

Calculate the base, collector, and emitter currents and the collector–emitter voltage for a common-emitter circuit of Figure 3.16. Also calculate the transistor power dissipation. Assume a silicon-doped device and a current gain of 200.

Solution:

Since the transistor is silicon doped, when ON, $V_{BE} = 0.7\,\text{V}$, and given gain $\beta = 200$. Calculating the currents:

i. The base current:

$$I_B = \frac{V_{BB} - V_{BE}}{200 \times 10^3} = \frac{5 - 0.7}{200 \times 10^3} = 21.5\,\mu\text{A}$$

ii. The collector current can be calculated using Equation (3.14):

$$I_C = \beta I_B = 21.5 \times 200 \times 10^{-6} = 4.3\,\text{mA}$$

iii. The collector current can be calculated using Equation (3.16):

$$I_E = (\beta + 1)I_B = (200 + 1) \times 21.5 \times 10^{-6} = 4.3215\,\text{mA}.$$

iv. The collector–emitter voltage v_{CE} is

$$V_{CE} = V_{CC} - I_C (2 \times 10^3) = 9 - (4.3 \times 10^{-3} \times 2 \times 10^3) = 0.4\,\text{V}$$

v. The power dissipated, P_T, in the transistor is found to be

$$P_T = I_B V_{BE} + I_C V_{CE} = (21.5 \times 10^{-6} \times 0.7) + (4.3 \times 10^{-3} \times 0.4) = 1.735\,\text{mW}$$

The power dissipated in the transistor can be approximated as the product of the collector current and voltage drop across the collector–emitter junction; i.e. $P_T \cong I_C V_{CE}$. When substituted with values, $P_T \cong I_C V_{CE} \cong 4.3 \times 10^{-3} \times 0.4 = 1.72\,\text{mW}$.

Example 3.2

Calculate the base, collector, and emitter currents and the voltages of Figure 2.15 when the transistor is driven into saturation assuming $V_{CE}(\text{sat}) = 0.5\,\text{V}$.

Solution:

i. The base current:

$$I_B = \frac{V_{BB} - V_{BE}}{200 \times 10^3} = \frac{5 - 0.7}{200 \times 10^3} = 21.5\,\mu A$$

ii. The collector current is, in this case, calculated as

$$I_C = I_C\,(\text{sat}) = \frac{V_{CC} - V_{CE}\,(\text{sat})}{2 \times 10^3} = 4.85\,\text{mA}$$

With the saturation condition, the gain can be calculated using equation (3.14) and compared with the given gain β, i.e.

$$\beta_{\text{sat}} = \frac{I_C}{I_B} = 197.67 < \beta$$

The emitter current is calculated using KCL:

$$I_E = I_C + I_B = (4.85 + 0.0215)\,\text{mA} = 4.8715\,\text{mA}.$$

$$P_T = I_B V_{BE} + I_C V_{CE} = \left(21.5 \times 10^{-6} \times 0.7\right) + \left(4.85 \times 10^{-3} \times 0.5\right) = 2.44\,\text{mW}$$

When a transistor is driven into saturation, we use V_{CE}(sat) as another piecewise linear parameter. Furthermore, when a transistor is biased in the saturation mode, the collector current is less than the linear expression of gain multiply by base current; i.e. $I_C < \beta I_{B(\text{sat})}$. This condition is very often used to prove that a transistor is indeed biased in the saturation mode.

In the instance when the transistor is operating in the *saturation region,* and is said to be *saturated* (ON), no matter how much current I_B we put into the base, V_{CE} cannot drop below $V_{CE(\text{sat})}$, and the collector current I_C is determined mainly by the load resistor R_C:

$$I_C = \frac{V_{cc} - V_{CE(\text{sat})}}{R_C + R_{CE(\text{sat})}} \tag{3.32}$$

where $R_{CE(\text{sat})}$ is the saturation resistance of the transistor. Typically, $R_{CE(\text{sat})}$ is 50 Ω or less and is insignificant compared with R_C.

Example 3.3

Explain what happens if the power supply in the base circuit is 0.15 V, (i.e. $V_{BB} = 0.15$ V) in Figure 3.16, assuming all parameters are the same.

Solution:

Since the power supply voltage in the base circuit is smaller than the turn-on base–emitter voltage; that is, then $V_{BB} < V_{BE}$; the base and emitter currents are equal to zero; i.e. $I_B = I_C = 0$. In this situation, the transistor is in the **cutoff** mode. In this cutoff mode, all transistor currents are zero, neglecting leakage currents, $V_{CC} = V_{CE} = 10$ V.

3.4 BASIC TRANSISTOR APPLICATIONS

Transistors can be used for a variety of applications. For example, it can be used to switch currents, voltages, and power; to perform digital logic functions; and to amplify time-varying signals. Some of these applications are briefly described in this section.

3.4.1 Switch

A transistor acts like a switch in a circuit when the transistor is switched between cutoff and saturation. To illustrate these observations, we first examine a simple bipolar gate, called the resistor-transistor logic (RTL) gate, schematically shown in Figure 3.17a, and Figure 3.17b is its symbolic representation. Figure 3.17a is the same as Figure 3.16, except that load is used instead of collector resistor, i.e. R_L instead of R_C to let the readers know that physically, the load R_L could be any electrical device, for example, such as a motor, an LED, or simply a resistor.

It can be readily seen from the circuit that when the input voltage V_{in} is less than the turn-on voltage $V_{BE(on)}$ of the transistor, (i.e. $V_{in} < V_{BE(on)}$), the transistor is in cutoff mode; the collector current I_C essentially equals zero, and the output voltage V_o equals to the supply voltage V_{CC}. Typically, $V_{BE(on)} \approx 0.8\,V$. However, when the input voltage is increased above $V_{BE(on)}$, the transistor turns on and enters the forward-active mode. If we increase the input voltage further the output voltage would drop due to the increasing collector current I_C. This drop is swift, as the bipolar transistor is characterized by a large gain in the forward-active mode. With sufficient input voltage, when the output voltage has fallen sufficiently, the transistor will enter the saturation region. The output voltage V_o remains fixed at the saturation collector–emitter voltage, $V_{CE(sat)}$. Typically, $V_{CE(sat)} \approx 0.1\,V$. To enter this saturation mode of operation, we require (as a first approximation) that $V_{BE} \geq V_{BE(sat)}$ ($\approx 0.8\,V$). This preceding discussion shows that the RTL gate of Figure 3.17 acts as an *inverter*. A sketch of the approximate voltage-transfer characteristic of the inverter is shown in Figure 3.18, where the two major breakpoints of the graph are joined together by linear interpolations.

The quiescent point (Q-point) is the switching threshold when setting the $V_o = V_{in}$. Q-point represents the dc bias, or starting point, for the transistor circuit and that the variations about this point carry the information (ac signals) in the circuit. Despite the tremendous advancement in semiconductor technology, there are still changes in the transistor parameters among different units of the same type, same number, same construction, etc. These changes induce changes in the values of *forward current gain*, β, in practice, which may shift the operating point (Q-point). Therefore, to avoid thermal instability, the biasing circuit must be designed to provide a degree of temperature stability, by ensuring that the operating point operates in the middle of the active region even though there are temperature changes, which may cause Q-point to shift. Designing the biasing circuit to stabilize the Q-point is known as *bias stability*. Thermal stability of a circuit is assessed by deriving a stability factor. More is said about the stability factor in Section 3.6.

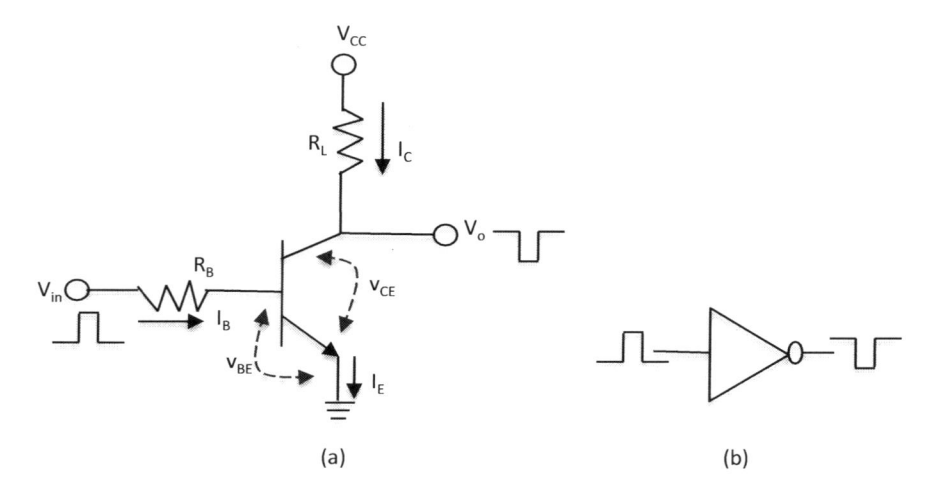

(a)   (b)

FIGURE 3.17 Representation of an RTL inverter: (a) resistor-transistor logic of inverter and (b) inverter symbol.

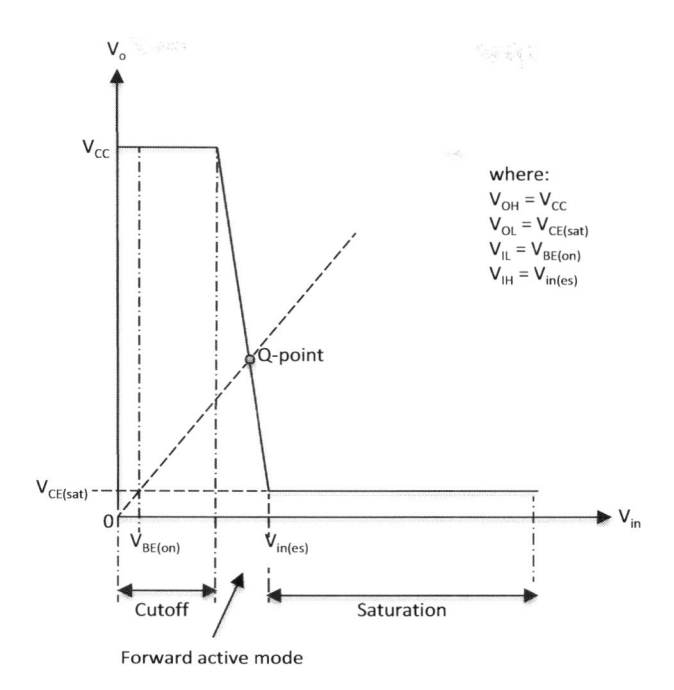

where:
$V_{OH} = V_{CC}$
$V_{OL} = V_{CE(sat)}$
$V_{IL} = V_{BE(on)}$
$V_{IH} = V_{in(es)}$

FIGURE 3.18 Approximate voltage characteristic of gate inverter.

The dc voltage-transfer characteristic of an inverter is derived as follows:

i. First, assume that the *fan-out* of the inverter is set to 0. Fan-out of a gate is the number of gates driven by that gate, i.e. the maximum number of gates (or load) that can exist without impairing the normal operation of the gate. For example, when there is no load, fan-out is zero, as in Figure 3.19a. When there is load connected to the inverter, as in Figures 3.19b and 3.19d, the fan-out is 1, whereas in Figure 3.19c the fan-out is 3. Physically, a large fan-out means a large number of connections, which is a large load capacitance.

ii. For $V_{in} < V_{BE(on)}$, Q is in cutoff, $I_C = 0$, and $V_o = V_{OH} = V_{CC}$.

iii. In the low output mode, Q is in saturation, if the condition $V_{BE} \geq V_{BE(sat)}$ is fulfilled, or when $V_o = V_{CE(sat)} \approx 0.1$ V. The boundary condition on V_{in} for this to be valid is derivable by analyzing the logic gate with its switching threshold, Q, at the edge of saturation. This translates to the following condition on V_{in}:

$$I_B = \frac{I_C}{\beta_F} = \frac{V_{CC} - V_{CE(sat)}}{R_L \beta_F} \tag{3.33}$$

$$V_{in(es)} = V_{BE(on)} + I_B R_B = V_{BE(on)} + \frac{R_B}{R_L}\left(\frac{V_{CC} - V_{CE(sat)}}{\beta_F}\right) \tag{3.34}$$

From Figure 3.19, we can map the RTL between analog and digital signals, as in Figure 3.20. The "ON" condition relates to digital "1" in the band V_{IH} and V_{OH}; and the "OFF" condition relates to digital "0" in the band V_{IL} and V_{OL}, as shown in Figure 3.20a. Within the *bands*, two types of *noise margins* occur, namely: *noise margin high* (NM_H) and *noise margin low* (NM_L), as shown in Figure 3.20b). More is said about noise margins in Chapter 6, Section 6.11.

These noise margins (NM_H and NM_L) should not be confused with switching noise. *Switching noise* is the noise present on the supply rails resulting from the switching of large currents in the

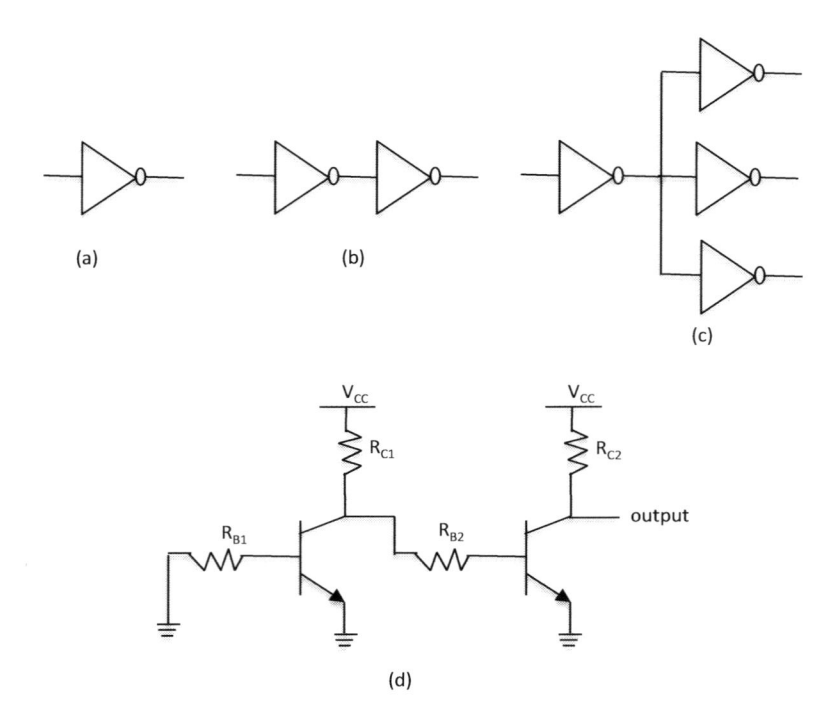

FIGURE 3.19 Number of inverter fan-outs: (a) fan-out = 0, (b) fan-out = 1, (c) fan-out = 3, and (d) fan-out = 1.

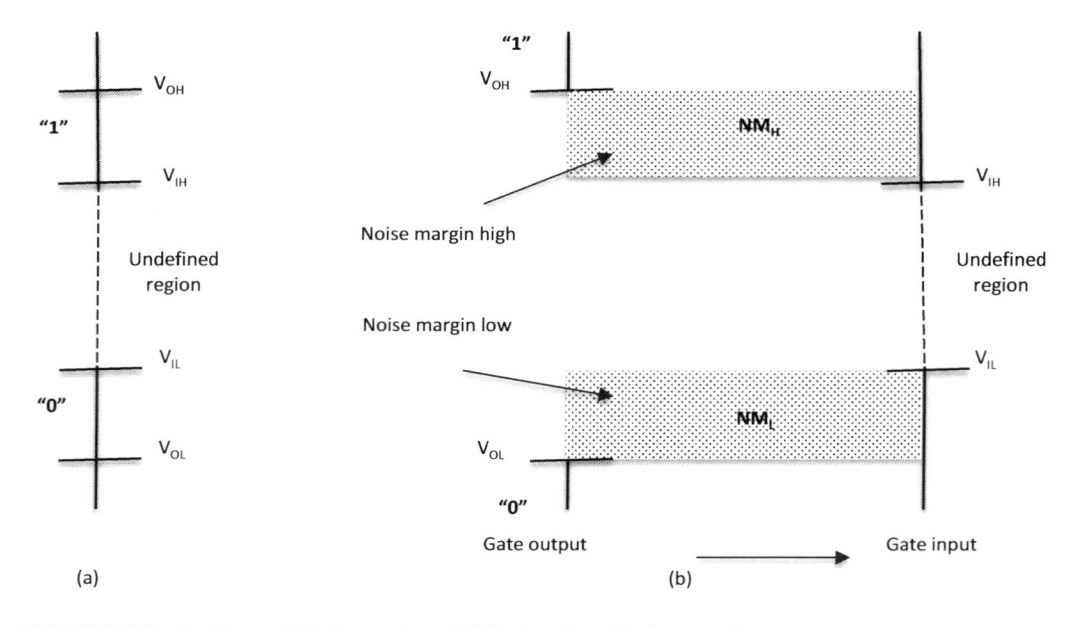

FIGURE 3.20 Analog to digital mapping of RTL signals and noise margins.

presence of parasitic resistances and inductors. The noise margin of a logic family is a very important consideration in system design. With regard to their magnitude, it is desirable to keep both noise margins (NM_H and NM_L) as large as possible and the undefined range between them as narrow as possible. If the noise margin is not large enough in some applications, internally or externally sourced interference can modify (i.e. falsify) a signal to fall within the undefined range. Internal noise is caused by inductive or ohmic drops or by inductive and capacitive couplings with other signaling lines. The coupling between signal lines is the more critical aspect in most cases.

In conclusion, BJTs can operate as amplifiers in the active mode, or switches in the saturation mode as "ON," and in the cutoff mode as "OFF".

3.4.2 BIASING

As seen in Figure 3.12b, transistor has highly nonlinear characteristics. In order to bias the transistor, we must use external dc sources for the transistor to operate in the region of the characteristic curves where behaviors are approximately linear. This is because proper biasing of the transistor in the linear region of the operation will allow us to apply a time-varying input signal to the input and get an amplified, undistorted signal at the output.

Figure 3.21 shows a practical common-emitter bias circuit, where the biasing is provided by three resistors: R_1, R_2, and R_E. The resistors R_1 and R_2 act as a potential divider giving a fixed voltage to the base.

The emitter resistor R_E works with the base resistors to stabilize the operating point (Q-point) with respect to variations in gain β due to component variation and temperature by providing negative feedback. If the collector current I_C increases due to change in β or temperature, the emitter current I_E also changes, as well as the voltage drop across R_E. In this process the voltage difference between base and emitter (V_{BE}) reduces. As a consequence of reduction in V_{BE}, the base current I_B, as well as the collector current I_C reduces.

Capacitors C_1 and C_2 are coupling or dc blocking capacitors. The reactance (X_C) of these capacitors is negligible at the operating signal frequency, f. Note that reactance is

$$X_C = |Z_C| = \frac{1}{2\pi f C} \tag{3.35}$$

Capacitor C_E is a bypass capacitor, which provides a low impedance path for ac current from emitter to ground. It effectively removes R_E (required for good Q-point stability) from the circuit when ac signals are considered. These capacitors—C_1, C_2, and C_E—are usually large in practice, effectively look like open circuits (infinite impedance) for dc conditions, while their impedance is close to zero (short circuits) at desired frequencies in AC operation.

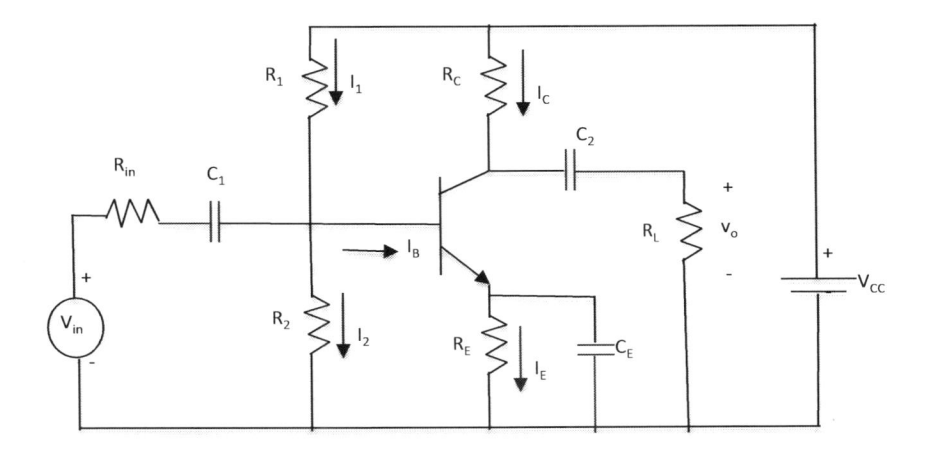

FIGURE 3.21 A bias circuit.

For **dc** situation, capacitor C_E is an open circuit and R_E can participate in the definition of **dc load line** and maintain a good Q-point stability. For ac situation, capacitor C_E has essentially zero impedance. This shorts out (or bypasses) R_E for ac signals and ties the emitter directly to the ground. Capacitors C_1 and C_2 are ac *coupling* capacitors. Their function in the circuit is dc blocking for the input and output signals—that is, they allow only ac signals to pass for the frequency range of interest and they do not affect the dc biasing.

3.4.2.1   DC Biasing Analysis

The first step is to find the ac equivalent circuit by replacing all capacitors by open circuits and inductors (if any) by short circuits. The second step is to find the dc Q-point from the equivalent circuit by using the appropriate large-signal transistor model (to be developed). In this instance, all the capacitors in Figure 3.21 are open-circuited as shown in Figure 3.22a.

Figure 3.22b analyses the dc voltage-divider bias circuit using Thevenin's approach. Thevenin rule states that *a linear two-terminal circuit or network is equivalent to a voltage source in series with a resistor*. So, we have

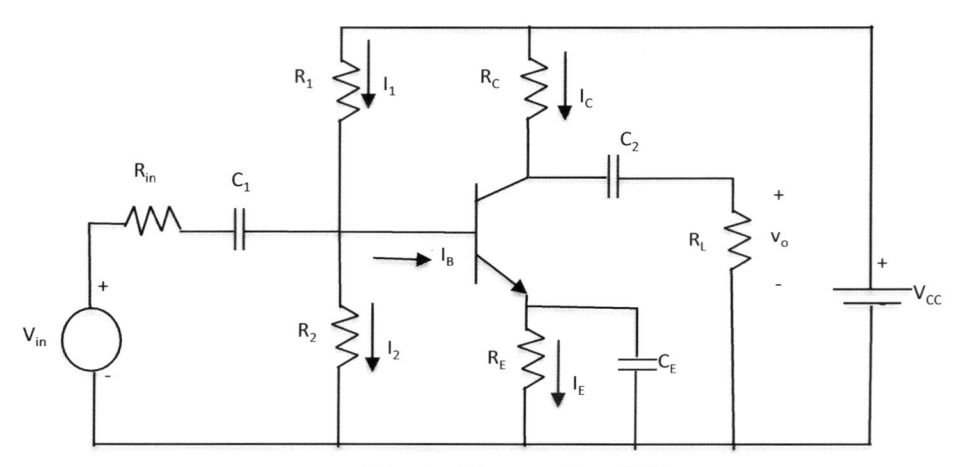

A bias circuit (same as Figure 3.21).

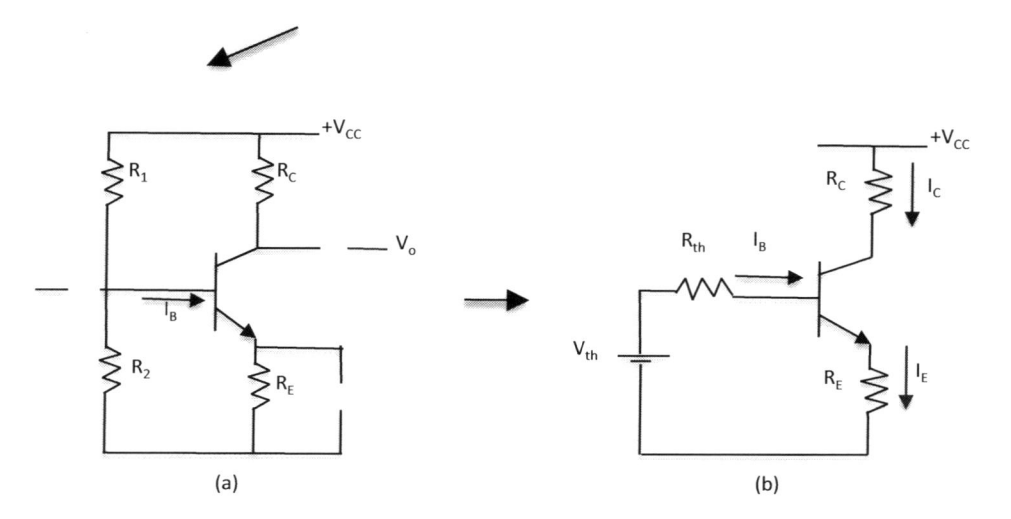

(a)   (b)

FIGURE 3.22   DC biasing of Figure 3.21: (a) capacitors open circuited and (b) Thevenin's equivalent circuit.

i. Thevenin resistance, $R_{\mathrm{th}} = R_1 \parallel R_2 = \dfrac{R_1 R_2}{R_1 + R_2}$ (3.36)

ii. Thevenin voltage, $V_{\mathrm{th}} = \dfrac{R_2}{R_1 + R_2} V_{CC}$ (3.37)

iii. From the base loop, solve for base current, I_B:

$$V_{\mathrm{th}} - V_{BE} - I_B R_{\mathrm{th}} - I_E R_E = 0 \tag{3.38}$$

We know from Equation (3.16) that $I_E = (\beta + 1) I_B$. So, substitute this expression in Equation (3.38) to have

$$V_{\mathrm{th}} - V_{BE} - I_B R_{\mathrm{th}} - I_B (\beta + 1) R_E = 0$$

and rearrange in terms of I_B

$$I_B = \frac{V_{\mathrm{th}} - V_{BE}}{R_{\mathrm{th}} + (\beta + 1) R_E} \tag{3.39}$$

Consider the outer loop comprising of R_C and R_E.

$$V_{CC} - V_{CE} - I_C R_C - I_E R_E = 0 \tag{3.40}$$

We also know from Equation (3.17) the relationship between I_E and I_C; that is,

$$I_E = \frac{\beta + 1}{\beta} I_C \tag{3.41}$$

and substitute this expression in Equation (3.36) as

$$V_{CC} - V_{CE} - I_C R_C - I_C \left(\frac{\beta + 1}{\beta} \right) R_E = 0 \tag{3.42}$$

and rearrange in terms of I_C, we have

$$I_C = \frac{V_{CC} - V_{CE}}{R_C + \left(\dfrac{\beta + 1}{\beta} \right) R_E} \tag{3.43}$$

If $I_B > 0$ and $V_{CE} > 0.2$, the transistor will be considered to be in active region of operation. As such, the Q-points (V_{CEQ} and I_{CQ}) will lie at V_{CE} and I_C, which can be calculated from Equations (3.39) and (3.43), respectively.

To draw the **dc load line,** we need to consider two other points, namely: the ideal cutoff voltage, $V_{CE(\mathrm{off})}$; and ideal saturation current, $I_{C(\mathrm{sat})}$; as follows.

For an ideal cutoff:

$$V_{CE(\mathrm{off})} = V_{CC} \tag{3.44}$$

For an ideal saturation:

$$I_{C(\mathrm{sat})} = \frac{V_{CC}}{R_C + R_E} \tag{3.45}$$

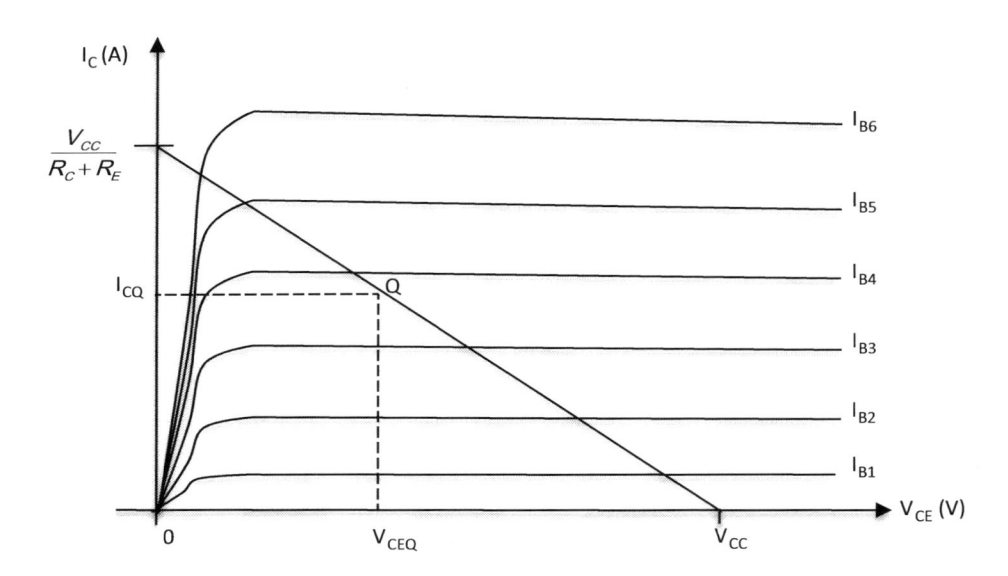

FIGURE 3.23 DC load line.

The plot of **dc** load line is shown in Figure 3.23.

In essence, dc bias is provided to stabilize the operating point in the desired operation region.

3.4.2.2 AC Analysis

In the case of ac analysis, several steps are taken, as follows:

   i. Find the ac equivalent circuit by replacing all capacitors by short circuits, inductors (if any) by open circuits, dc voltage sources by ground connections, and dc current sources by open circuits.
  ii. Replace the transistor by its "small-signal model" (to be developed in Section 3.5.1).
 iii. Use this equivalent circuit to analyze the ac characteristics of the amplifier.
  iv. Combine the results of dc and ac analysis (superposition) to yield the total voltages and currents in the circuit.

By replacing all capacitors by short circuits, dc voltage sources by ground connections, and dc current sources by open circuits, the ac equivalent circuit is shown in Figure 3.24, where

$$R_{12} = R_1 \parallel R_2 = \frac{R_1 R_2}{R_1 + R_2} \tag{3.46}$$

Notice the "ac ground" in the circuit of Figure 3.24 because it is an **extremely important concept**. Since the voltage at this terminal is held constant at V_{CC}, there is no time variation of the voltage. Consequently, the terminal is set to be an "ac ground" in the small-signal circuit. Generally, for "ac grounds", we "kill" the dc sources at that terminal: short circuit voltage sources and open circuit current sources.

By combining parallel resistors into equivalent R_{12} and R_{CL}—($R_C \parallel R_L$)—the equivalent AC circuit above is constructed. These resistors would play another role when the transistors are replaced by their "equivalent small-signal AC models," (to be developed).

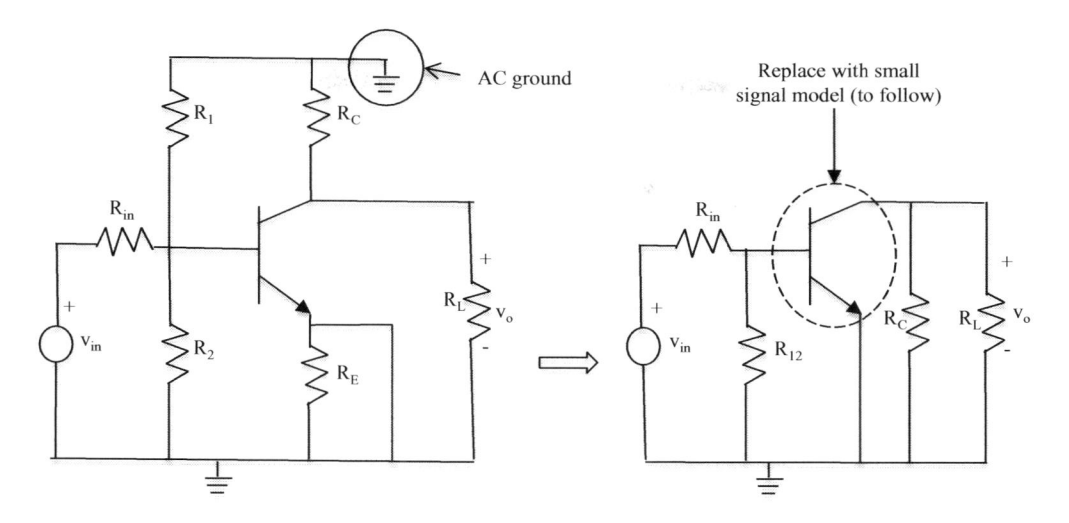

FIGURE 3.24 AC equivalent of BJT amplifier of Figure 3.21.

3.5 BJT SMALL- AND LARGE-SIGNAL MODELS

A large variety of BJT models have been developed in the literature leading to an understanding of the transistor's behavior and making it easy to describe its frequency dependence. However, one distinguishes between small-signal and large-signal models. The small-signal model lends itself well to small-signal design and analysis. Small-signal models are a linear representation of the nonlinear transistor electrical behavior. They are also a linear approximation to the large-signal behavior. Whereas, the charge control model is particularly well suited to analyze the large-signal transient behavior of a bipolar transistor. These two types are discussed under appropriate subheadings in this section.

3.5.1 SMALL-SIGNAL MODEL

In order to develop the BJT small-signal models, there are small-signal resistances that we must first determine. These resistances, denoted in Figure 3.25, are

- r_e: the active mode output resistance between the base and emitter; that is, the resistance seen looking into the emitter.
- r_π: the active mode input resistance between the base and emitter; that is, the resistance seen looking into the base.
- r_o: the output resistance is included for completeness, as well as it models the fact that collector current varies as v_{ce} varies.

The resistances r_π and r_e are not the same because the transistor is not a reciprocal device. For instance, like a diode shown in Figure 3.26, the behavior of the BJT in the circuit changes if the terminals are interchanged. In the forward-bias situation, as in Figure 3.26a, the pn junction drives holes to the junction from the p-type material and electrons to the junction from the n-type material; thus, combining electrons and holes to maintain a continuous *drain* current, I_D. The diode will have a low resistance, that is, $r_{df} \Rightarrow$ low. However, in the reverse-biased situation as in Figure 3.26b, the pn junction will cause a transient current to flow as both electrons and holes are pulled away from the junction. In this case, the diode will have a high resistance, that is, $r_{dr} \Rightarrow$ high.

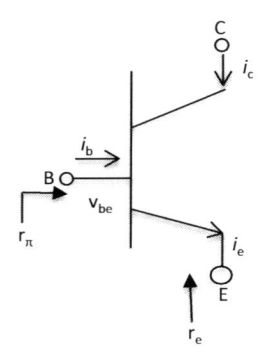

FIGURE 3.25   Small-signal resistances for an *npn* variant.

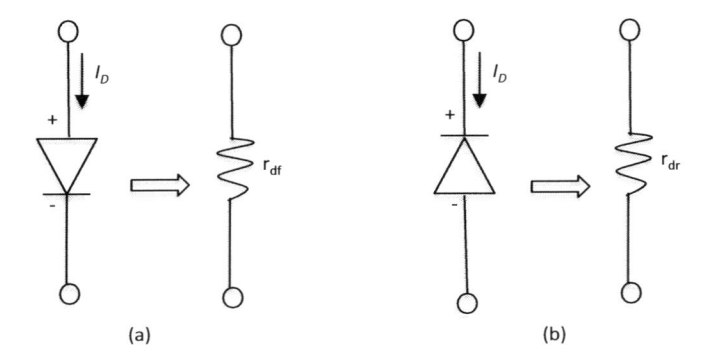

(a)   (b)

FIGURE 3.26   Equivalent diode resistances for forward and reverse bias: (a) forward bias and (b) reverse bias.

3.5.2   DETERMINING RESISTANCES: R_Π, R_E, AND R_O

From the small-signal equivalent circuit of Figure 3.25, we can define

$$r_\pi = \frac{v_{be}}{i_b} \tag{3.47}$$

where the base current, when operating in the active mode, is

$$i_B = \frac{i_C}{\beta} = \frac{1}{\beta}\left(\underbrace{I_C}_{\text{dc}} + \underbrace{i_b}_{\text{ac}}\right)$$

$$= \frac{1}{\beta}\left(\underbrace{I_C}_{\text{dc}} + \underbrace{\frac{I_C}{V_T}v_{be}}_{\text{ac}}\right) \tag{3.48}$$

So, the AC component of the base current is

$$i_b = \frac{1}{\beta}\frac{I_C}{V_T}v_{be} = \frac{g_m}{\beta}v_{be} \tag{3.49}$$

The ratio (I_C/V_T) is called *transconductance* (denoted by g_m) whose unit is in Siemens, S. Consequently,

$$i_c = \beta i_b = g_m v_{be} \tag{3.50}$$

Rearranging Equation (3.49) in view of this ratio and substituting in Equation (3.47), we obtain

$$r_\pi = \frac{\beta}{g_m} \quad [\Omega] \tag{3.51}$$

implying that we are fictitiously separating the source from the base of the BJT and observing the input resistance, where β is most often taken as *forward current gain*, and β_F as a constant.

We can determine r_e following similar procedure as for r_π, thus

$$r_e = \frac{v_e}{-i_e} \quad [\Omega] \tag{3.52}$$

Assuming an ideal signal voltage source, then $v_e = -v_{be}$ and

$$r_e = \frac{v_{be}}{i_e} \quad [\Omega] \tag{3.53}$$

And

$$i_E = \frac{i_C}{\alpha} = \frac{1}{\alpha}\left(\underbrace{I_C}_{dc} + \underbrace{i_c}_{ac} \right) \quad [A]$$

$$= \frac{1}{\alpha}\left(\underbrace{I_C}_{dc} + \underbrace{\frac{I_C}{V_T} v_{be}}_{ac} \right) \tag{3.54}$$

So, the AC component of emitter current is

$$i_e = \frac{1}{\alpha}\frac{I_C}{V_T} v_{be} \quad [A] \tag{3.55}$$

We noted in Equation (3.17) that $I_E = I_C/\alpha$, and when substituted in Equations (3.55) and (3.53), we obtain the resistance between the base and emitter as

$$r_e = \frac{V_T}{I_E} = \frac{\alpha}{g_m} = \frac{r_\pi}{\beta+1} \quad [\Omega] \tag{3.56}$$

Alternately

$$r_\pi = (\beta+1)r_e \quad [\Omega] \tag{3.57}$$

Comparing Equation (3.56) with (3.51), the result shows that r_e is not the same r_π (i.e. $r_e \neq r_\pi$). This is expected because the active mode BJT is a non-reciprocal device.

Finally, the output resistance, r_o, is defined as

$$r_o = \frac{v_{cb}}{i_c} \quad [\Omega] \tag{3.58}$$

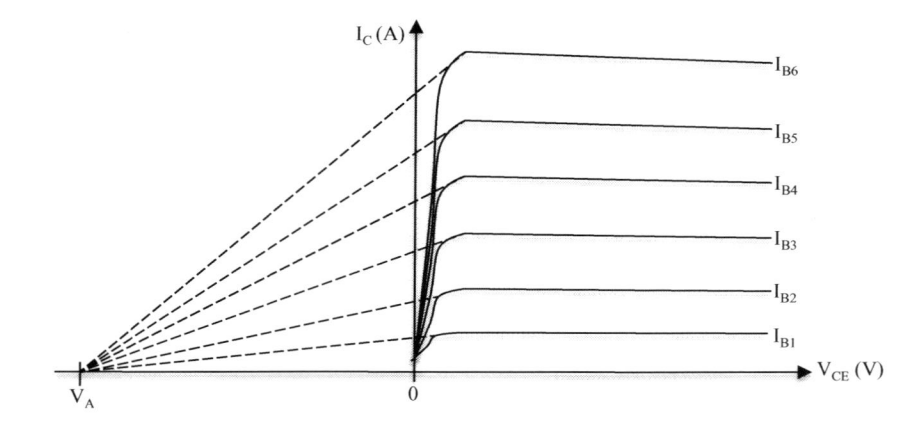

FIGURE 3.27   Non-ideal collector characteristic curve with trace lines to V_{CE} axis.

which, at a fixed i_B point, represents the slope of the line tangent to that point on the curve, and can be further expanded as

$$r_o = \frac{V_{CE} + |V_A|}{I_C} \quad [\Omega] \tag{3.59}$$

where V_A is an *early voltage*. An early voltage is the variable base current I_B traced back to the collector–emitter voltage line, as shown in Figure 3.27. This is the *early effect*, which is due to *base-width modulation*. This effect can be minimized by placing a heavily doped layer buried underneath the collector. This is necessary, since the collector must be lightly doped for the transistor to have high gain. The finite slope in the active region due to decreasing base width can be approximated by

$$i_c = I_s \exp\left(\frac{v_{be}}{V_T}\right)\left[1 + \frac{v_{ce}}{V_A}\right] \tag{3.60}$$

This means that the output resistance between the collector and emitter is not infinite—very important for analog design.

Even though the point of intersection is on the negative half of the V_{CE} axis, V_A is universally quoted as a positive number, and the magnitude bars (||) dropped. Typically, V_A ranges from about 50– to 250 V. Since $V_A \gg V_{CE}$, for practical circuits,

$$r_o \approx \frac{V_A}{I_C} \tag{3.61}$$

Generally, r_o is not considered for dc bias point calculations, it can have a significant impact on transistor amplifier gain.

Having determined these resistances, we can now discuss two small-signal models—called the T-model and hybrid-π model—used to model BJTs.

3.5.3   HYBRID-π MODEL

The hybrid-π model has two variants, namely: Voltage-Controlled Current source (VCCS) version and CCCS version. Let us look at how these models are constructed, as shown in Figure 3.28. For example, the VCCS type of Figure 3.28a:

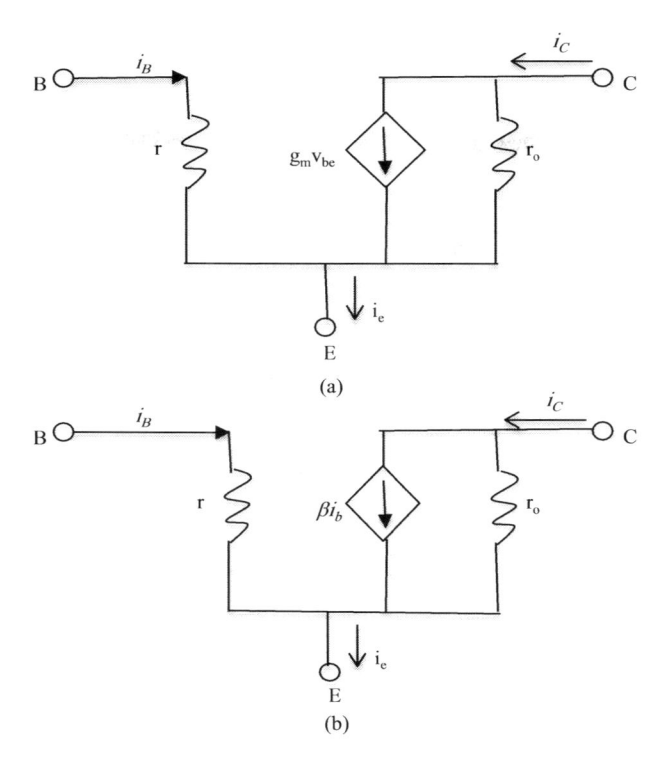

FIGURE 3.28 Alternative representations of hybrid-π models: (a) VCCS hybrid-π model and (b) CCCS hybrid-π model.

- the base current i_b satisfies the expression given by Equation (3.14);
- the collector current i_c satisfies the expression $i_c = g_m v_{be}$ given by Equation (3.50); and
- the emitter current i_e, by KCL, $i_e = i_{cb} + i_c = v_{be}/r_e$, satisfying Equation (3.53).

The VCCS $g_m v_{be}$ can be transformed into a CCCS, as drawn in Figure 3.28b, by drawing from Equation (3.50), i.e. $g_m v_{be} = g_m (i_b r_\pi) = \beta i_b$. The basic relationship $i_c = \beta i_b$ is useful in both dc and ac analysis when the BJT is biased in the forward-active region.

3.5.3.1 Amplification Factor

If changes in operating currents and voltages were small enough, then I_C and V_{CE} waveforms would be undistorted replicas of the input signal. As such, a small voltage change at the base would cause a large voltage change at the collector. The amplification factor μ_F is expressed as

$$\mu_F = \frac{v_{ce}}{v_{be}} \tag{3.62}$$

Gain factor is responsible for amplifying the signal. We know that

$$v_{ce} = i_e r_o \tag{3.63}$$

Substituting Equations (3.55) and (3.59) in Equation (3.63), we obtain the expression for the collector–emitter voltage as

$$v_{ce} = \frac{1}{\alpha} \frac{I_C}{V_T} \frac{|V_A|}{I_C} v_{be} \tag{3.64}$$

So, by rearranging Equation (3.64), the full expression for the amplification factor is

$$\mu_F = \frac{v_{ce}}{v_{be}} = \frac{1}{\alpha}\frac{I_C}{I_C}\frac{|V_A|}{V_T} = \frac{1}{\alpha}\frac{|V_A|}{V_T} \tag{3.65}$$

Nominally $\alpha \approx 1$, therefore

$$\mu_F \approx \frac{|V_A|}{V_T} \tag{3.66}$$

This expression represents the maximum voltage gain an individual BJT can provide, independent of the operating point.

3.5.4 T-MODEL

Like the hybrid-π model, the T-model has two versions—called Version A and Version B, as shown as Figure 3.29a and b, respectively. Their construction uses the parameters earlier derived.

The small-signal models for *pnp* BJTs are identically the same as those shown here for the *npn* transistors. It is important to note that there is no change in any polarities (voltage or current) for the *pnp* models relative to the *npn* models. Again, these small-signal models are identically the same.

The hybrid-π model is a more popular small-signal model for the BJT, although the T-model is useful in certain situations.

We are now in a position to solve the AC analysis part of a typical bias circuit of Figure 3.22. The reduced component is shown in Figure 3.30, and the complete circuit for evaluation is shown in Figure 3.31.

From Figure 3.31, we solve the following components to obtain the gain.

$$v_{thB} = v_{in} \times \left[\frac{R_{12}}{R_{in} + R_{12}} \right] \tag{3.67a}$$

$$r_{thB} = R_{12} \parallel R_{in} = \frac{R_{12} R_{in}}{R_{12} + R_{in}} = \frac{R_1 R_2 R_{in}}{R_1 + R_2 + R_{in}} \tag{3.67b}$$

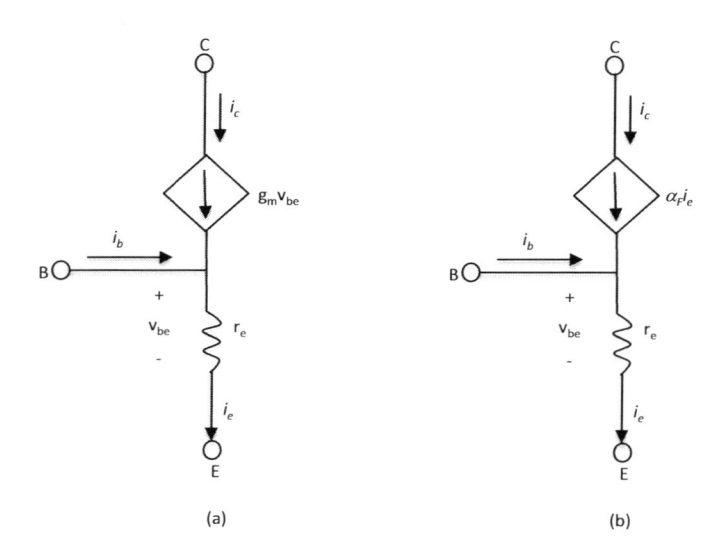

(a) (b)

FIGURE 3.29 Alternative representations of T-models: (a) Version A and (b) Version B.

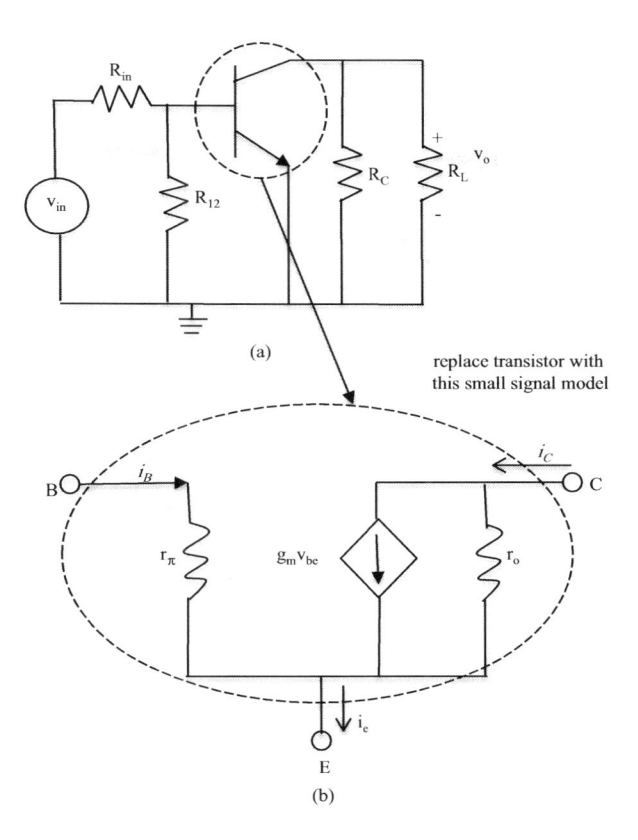

FIGURE 3.30 Solving ac analysis of circuit of Figure 3.21: (a) from Figure 3.24 and (b) small-signal model (hybrid-π) of transistor.

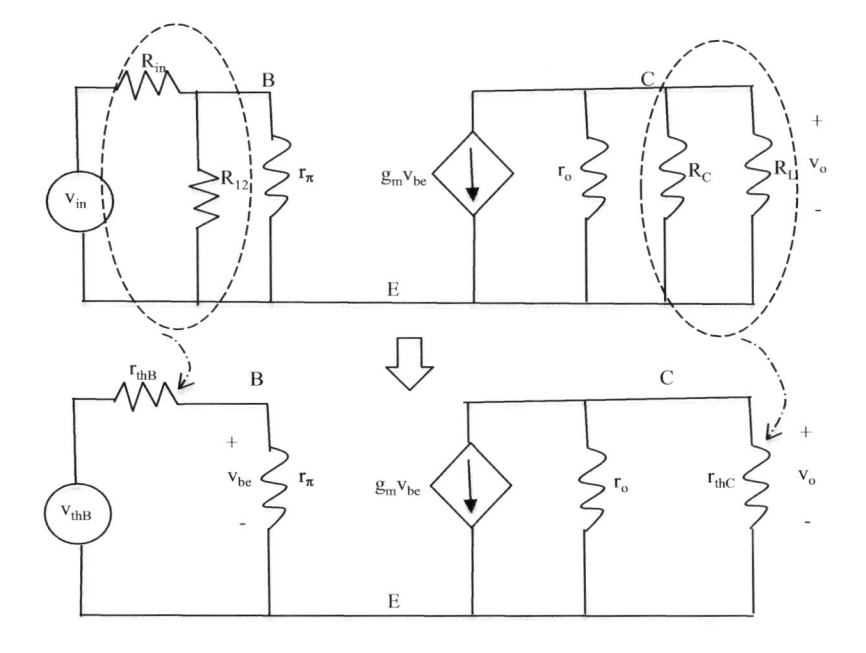

FIGURE 3.31 Solving the ac analysis of circuit of Figure 3.21.

$$r_{\mathrm{thC}} = R_c \parallel R_L = \frac{R_C R_L}{R_C + R_L} \qquad (3.67c)$$

$$v_{be} = \frac{r_\pi}{r_\pi + r_{\mathrm{thB}}} v_{\mathrm{thB}} \qquad (3.67d)$$

To find the circuit output v_o, apply Ohm's law, that is:

$$v_o = -IR = \left(-g_m v_{be}\right)\left(r_o \parallel r_{\mathrm{thC}}\right)$$

$$= \left(-g_m v_{be}\right)\left(\frac{r_o r_{\mathrm{thC}}}{r_o + r_{\mathrm{thC}}}\right) \qquad (3.68)$$

From Equation (3.68), the *gain factor*, A_{vf} of the transistor is the ratio of transistor output (v_o) to input (v_{be}):

$$A_{vf} = \frac{v_o}{v_{be}} = -g_m \left(\frac{r_o r_{\mathrm{thC}}}{r_o + r_{\mathrm{thC}}}\right) \qquad (3.69)$$

The overall gain of the circuit is the ratio of transistor output (v_o) to circuit input (v_{in}), which is calculated thus as:

$$\mathrm{Gain} = \frac{v_o}{v_{\mathrm{in}}} = \left(\frac{v_{\mathrm{thB}}}{v_{\mathrm{in}}}\right)\left(\frac{v_{be}}{v_{\mathrm{thB}}}\right)\left(\frac{v_o}{v_{be}}\right) \qquad (3.70)$$

The terms $\left(\dfrac{v_{\mathrm{thB}}}{v_{\mathrm{in}}}\right)$ and $\left(\dfrac{v_{be}}{v_{\mathrm{thB}}}\right)$ are called *loss factors*, because they can never be greater than unity in magnitude and thus cause the gain to decrease.

Rearranging Equations (3.67a), (3.67d), and (3.68) and substituting in Equation (3.70), the overall (voltage) gain is obtained:

$$\mathrm{Gain} = A_v = \frac{v_o}{v_{\mathrm{in}}} = -\left(\frac{R_{12}}{R_{12} + R_{\mathrm{in}}}\right)\left(\frac{r_\pi}{r_\pi + r_{\mathrm{thB}}}\right)\left(g_m \frac{r_o r_{\mathrm{thC}}}{r_o + r_{\mathrm{thC}}}\right) \qquad (3.71)$$

The *negative* sign indicates that the transistor is an *inverting* amplifier; meaning that there has been a phase shift of 180°. Note that $R_{12} = R_1 \parallel R_2 = \dfrac{R_1 R_2}{R_1 + R_2}$.

3.5.4.1 Impact of Coupling Capacitors

Up to now, we have neglected the impact of the coupling capacitors in the circuit; the coupling capacitors have been assumed short-circuited. This is not a valid assumption at low frequencies. The coupling capacitor, particularly C_1, results in a lower cutoff frequency for the transistor amplifiers. The impact of coupling capacitor C_2 is factored in the impedance analysis of the output resistance and capacitive load. In order to find the cutoff frequency, we need to repeat the above analysis and include the coupling capacitor impedance in the calculation. In most cases, however, the impact of the coupling capacitor and the lower cutoff frequency can be deduced by reconfiguring Figure 3.31 and accounting for the impedance of the capacitors as in Figure 3.32 and examining the amplifier circuit model. For instance, when we account for impedance of the coupling capacitor C_1, as in Figure 3.32, we set up a high-pass filter in the input part of the circuit (combination of the coupling capacitor and the input resistance of the amplifier). This combination introduces a lower cutoff frequency, f_l, for the amplifier, which ironically is the same as the cutoff frequency of the high-pass filter:

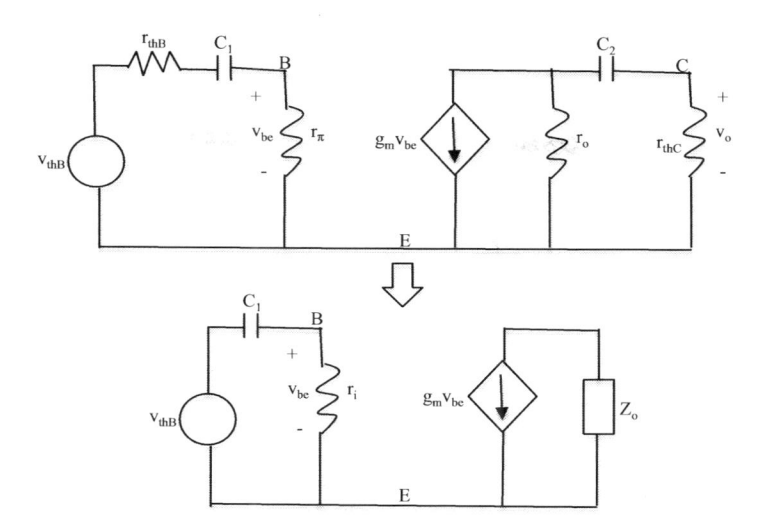

FIGURE 3.32 Complete low frequency, ac circuit analysis of Figure 3.21.

$$\omega_l \left(\text{coupling}\right) = 2\pi f_l = \frac{1}{r_i C_1} \tag{3.72}$$

where

$$r_i = r_{\text{thB}} \parallel r_\pi = \frac{r_{\text{thB}} r_\pi}{r_{\text{thB}} + r_\pi} = \frac{R_1 R_2 R_{\text{in}} r_\pi}{R_1 + R_2 + R_{\text{in}} + r_\pi} \tag{3.73}$$

In reality, both the coupling and bypass capacitors contribute to setting the lower cutoff frequency for the amplifier, both act as a high-pass filter with

$$\omega_l \left(\text{coupling}\right) = 2\pi f_l = \frac{1}{r_i C_1} \tag{3.74}$$

$$\omega_l \left(\text{bypass}\right) = 2\pi f_l = \frac{1}{R_e C_E} \tag{3.75}$$

where $R_e = R_E \parallel r_e = \dfrac{R_E r_e}{R_E + r_e}$. Usually, $R_E \gg r_e$, and, therefore, $R_e \approx r_e$,

Selecting the cutoff frequency depends on the magnitudes of $\omega_l \left(\text{coupling}\right)$ and $\omega_l \left(\text{bypass}\right)$. For example, if these two frequencies are far apart, choosing the larger of the two cutoff frequencies sets the low cutoff frequency of the amplifier. Specifically, if

$$\omega_l \left(\text{coupling}\right) \gg \omega_l \left(\text{bypass}\right) \quad \text{cut-off frequency,} \quad f_l = \frac{1}{2\pi r_i C_1} \tag{3.76}$$

$$\omega_l \left(\text{bypass}\right) \gg \omega_l \left(\text{coupling}\right) \quad \text{cut-off frequency,} \quad f_l = \frac{1}{2\pi R_e C_E} \tag{3.77}$$

However, when the two frequencies are close to each other, there is no exact analytical formula to choosing the low cutoff frequency. An approximate formula for the cutoff frequency (accurate within a factor of two and exact at the limits) is

$$f_l = \frac{1}{2\pi}\left(\frac{1}{r_i C_1} + \frac{1}{R_e C_E}\right) \tag{3.78}$$

So far, the small-signal model discussed is a *low-frequency* model.

3.5.4.2 Small-Signal Model for Large Frequency

At *high frequencies*, the transistor's parasitic (diffusion and depletion) capacitances, which are responsible for charging and discharging across the junctions, become important. In the bipolar devices, these parasitic capacitive effects originate from the following sources: the base–emitter and base–collector depletion regions, the collector–substrate diode, and the excess minority carrier charge in the base. These parasitic capacitances are briefly discussed in this section.

In active mode when the emitter–base is forward biased, the capacitance of the emitter–base junction is dominated by the diffusion capacitance, C_B (not depletion capacitance). Thus, the diffusion capacitance is

$$C_B = \left.\frac{\partial q_B}{\partial v_{BE}}\right|_{Q-\text{point}} = \left(\frac{W_B^2}{2D_B}\right)\left.\frac{\partial i_C}{\partial v_{BE}}\right|_{Q-\text{point}} \tag{3.79}$$

The "forward" base transport time, τ_F, being the average time the carriers spend in the base, has been shown that [2]

$$\tau_F = \frac{W_B^2}{2D_B} \tag{3.80}$$

where
$\quad W_B$ = Base width in the quasi-neutral region.
$\quad D_B$ = Base minority carrier diffusion coefficient.

And, as preciously noted, the transconductance, g_m, of a bipolar transistor is defined as the change in the collector current divided by the change of the base–emitter voltage, that is,

$$g_m = \frac{\partial i_C}{\partial v_{BE}} = \frac{I_C}{V_T} \tag{3.81}$$

In view of Equations (3.80) and (3.81) in Equation (3.79), the diffusion capacitance is written as

$$C_B = \tau_F g_m \tag{3.82}$$

The upper operational frequency, f, of the transistor is limited by the forward base transport time, i.e.

$$f \le \frac{1}{2\pi\tau_F} \tag{3.83}$$

In the active mode for small forward biases, the depletion capacitance C_{jE} of the base–emitter junction can contribute to the total base–emitter capacitance C_π:

$$C_{jE} = \frac{C_{jEo}}{\sqrt{1 - \dfrac{V_{BE}}{V_{i,BE}}}} \tag{3.84}$$

where

C_{jEo} = zero bias depletion capacitance
$V_{i,BE}$ = built-in voltage for the base–emitter junction.

So, the total emitter–base capacitance is

$$C_\pi = C_B + C_{jE} \tag{3.85}$$

In active mode when the collector–base is reverse biased, the capacitance of the collector–base junction is dominated by the depletion capacitance (not diffusion capacitance).

$$C_{jBC} = \frac{C_{jBCo}}{\sqrt{1 - \dfrac{V_{BC}}{V_{i,BC}}}} \tag{3.86}$$

where

C_{jBCo} = zero bias depletion capacitance
$V_{i,BC}$ = built-in voltage for the base–collector junction.

The base–collector junction capacitance, C_{jBC}, is also referred to as the Miller capacitance [3]. By including these capacitances, the complete VCCS small-signal (hybrid-π) model can be drawn as shown in Figure 3.33. Although the value of the collector zero bias depletion capacitance C_{jBCo} is typically a trifle larger than its counterpart at the emitter side, C_{jEo}, both values are comparable [4].

In some IC BJTs (in particular, lateral BJTs) the device has a capacitance to the substrate wafer it is fabricated in resulting from a "buried" reverse-biased junction. The collector–substrate junction is reverse biased and the capacitance of the collector–substrate junction is dominated by the depletion capacitance. This collector–substrate junction capacitance, C_{jCS}, is defined thus

$$C_{jCS} = \frac{C_{jCSo}}{\sqrt{1 + \dfrac{V_{CS}}{V_{i,CS}}}} \tag{3.87}$$

where

C_{jCSo} = zero bias depletion capacitance of the collector substrate.
$V_{i,CS}$ = built-in voltage for the collector–substrate junction.

If this capacitance, C_{jCS}, is appreciative enough, the small-signal (hybrid-π) model of Figure 3.33 is modified accordingly as shown in Figure 3.34.

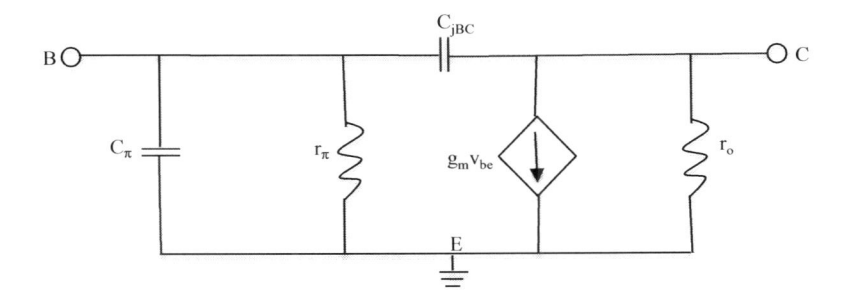

FIGURE 3.33 Small-signal model (hybrid-π model) with junction capacitances.

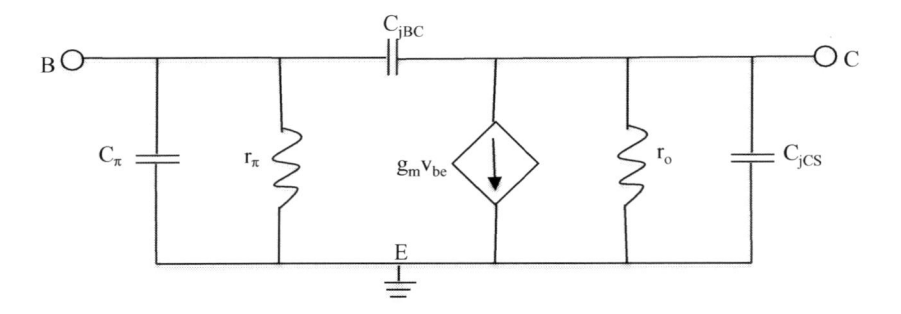

FIGURE 3.34   Complete large frequency, small-signal (hybrid-π) model.

3.5.4.3   Selection of Coupling and Bypass Capacitors

The key objective in design is to make the capacitive reactance much smaller at the operating frequency, f, than the associated resistance that must be coupled or bypassed, ensuring that the gain error A_δ is very small, preferably less than one percentage (i.e. $A_\delta \leq 0.01$). As a consequence,

 i. For capacitor C_1, make its capacitive reactance (X_{C_1}) less than or equal to $0.01(r_i)$; that is,

$$X_{C_1} \leq \frac{1}{2\pi f C_1} \leq 0.01 r_i \tag{3.88}$$

Consequently,

$$C_1 \geq \frac{50}{\pi f}\left(\frac{R_1 + R_2 + R_{in} + r_\pi}{R_1 R_2 R_{in} r_\pi}\right) \ \ [\text{F}] \tag{3.89}$$

Note that resistance r_i is defined by Equation (3.73).
 ii. For capacitor C_2, make its reactance (X_{C_2}) less than or equal to 1 Ω; that is,

$$X_{C_2} \leq \frac{1}{2\pi f C_2} \leq 1 \tag{3.90}$$

So,

$$C_2 \geq \frac{1}{2\pi f} \ \ [\text{F}] \tag{3.91}$$

 iii. For emitter capacitor C_E, make its capacitive reactance (X_{CE}) less than or equal to $0.01 R_E$; i.e.

$$X_{CE} \leq \frac{1}{2\pi f C_E} \leq 0.01 R_E \tag{3.92}$$

So,

$$C_E \geq \frac{50}{\pi f R_E} \ \ [\text{F}] \tag{3.93}$$

Example 3.4

The common-emitter BJT amplifier circuit, shown in Figure 3.35, is designed to operate up to 32°C at 15 V supply voltage for a forward current gain of 150. The circuit resistive components have the following values: $R_{in} = 0.1$ kΩ, $R_1 = 45$ kΩ, $R_2 = 10$ kΩ, $R_C = 4.5$ kΩ, $R_E = 1$ kΩ, and $R_L = 45$ kΩ. The amplifier is nominally doped, having base–emitter voltage of 0.7 V and 0.2 V collector–emitter voltage at saturation. Suppose the source voltage has a relationship defined by $v_s(t) = 0.01 \sin(1000\pi t)$ and the transistor's transconductance is rated as 25 μS at Q-point, determine the following:

a. The Q-point.
b. Plotting the dc load line determines what the maximum (peak-to-peak) output voltage swing available to this amplifier would be.
c. Determine the ac model parameters.
d. Estimate the gain factor of the transistor as well as the overall gain of the circuit.
e. Determine whether clipping occurs if $v_s(t) = 0.025 \sin(1000\pi t)$.
f. Design suitable values for the circuit's capacitors.

Solution:

Analyzing the dc voltage-divider bias circuit, we have

i. Thevenin resistance, $R_{12} = R_1 \parallel R_2 = \dfrac{R_1 R_2}{R_1 + R_2} = \dfrac{45 \times 10 \times 10^6}{(45+10) \times 10^3} = 8.182\,\text{k}\Omega$

ii. Thevenin voltage, $V_{th} = \dfrac{R_2}{R_1 + R_2} V_{CC} = \dfrac{10 \times 10^3}{(45+10) \times 10^3} 15 = 2.727\,\text{V}$

iii. Base current, $I_B = \dfrac{V_{th} - V_{BE}}{R_{12} + (\beta + 1) R_E} = \dfrac{2.727 - 0.7}{[8.182 + (151 \times 1)] \times 10^3} = 12.73\,\mu\text{A}$

iv. Emitter current, $I_E = (\beta + 1) I_B = (150 + 1) \times 62.58 \times 10^{-6} = 1.923\,\text{mA}$

v. Collector current, $I_C = \beta I_B = 150 \times 62.58 \times 10^{-6} = 1.91\,\text{mA}$

vi. Emitter voltage, $V_E = I_E R_E = 1.923 \times 10^{-3} \times 1.0 \times 10^3 = 1.923\,\text{V}$

vii. Collector voltage, $V_C = I_C R_C = 1.91 \times 10^{-3} \times 4.5 \times 10^3 = 8.595\,\text{V}$

viii. Common-emitter voltage, $V_{CE} = V_{CC} - I_C R_C - I_E R_E = 4.482\,\text{V}$

To be able to plot the dc load-line, we must know the ideal cutoff voltage $V_{CE(off)}$; ideal saturation current $I_{C(sat)}$; and Q-points (V_{CEQ} and I_{CQ}).

For an ideal cutoff voltage, from Equation (3.21): $V_{CE(off)} = V_{CC} = 15\,\text{V}$

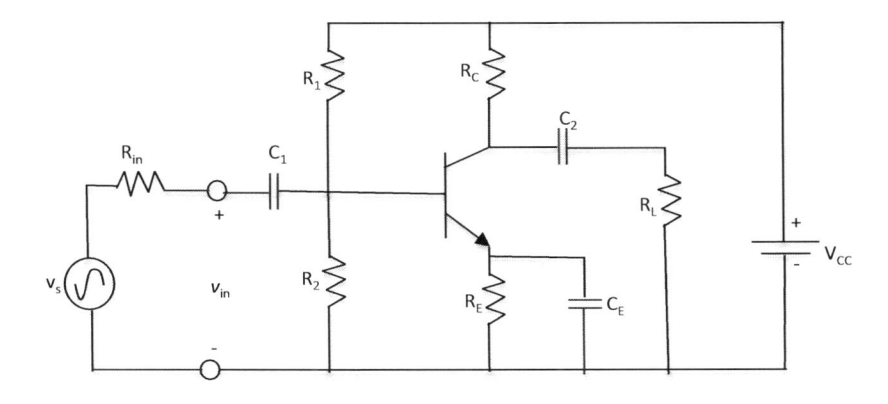

FIGURE 3.35 Common-emitter BJT amplifier circuit.

For an ideal saturation current, from Equation (3.22): $I_{C(sat)} = 2.73\,\text{mA}$
As $I_B > 0$ and $V_{CE} > 0.2$, the transistor is considered to be in the active region of operation.

a. So, the Q-points lie at

$$V_{CEQ} = V_{CE} = 4.482\,\text{V};$$

$$I_{CQ} = I_C = 1.91\,\text{mA}.$$

b. Determining the dc load line
    i.   We are now in a position to plot the dc load line as shown in Figure 3.36a.
    ii. To obtain the maximum (peak-to-peak) output voltage swing, observe where the Q-point leans closer to the saturation or cutoff. From Figure 3.36, it is observed that the Q-point lies closer to saturation ($V_{CE(sat)} = 0.2\,\text{V}$) than cutoff ($V_{CE(off)} = 15\,\text{V}$). Given that $V_{CE(sat)} = 0.2\,\text{V}$, the maximum available peak to peak output voltage swing $V_{o(swing)}$ is estimated thus:

$$V_{o(swing)} = 2\left(V_{CEQ} - V_{CE(sat)}\right) = 8.564\,\text{V}$$

Figure 3.36b gives the overall load line: $V_{cc} = v_{CE} + i_C R_C$. The AC voltage v_{CE} and current i_C will be calculated under the AC analysis.

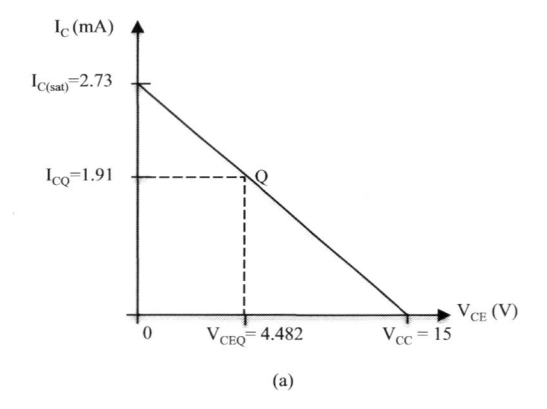

(a)

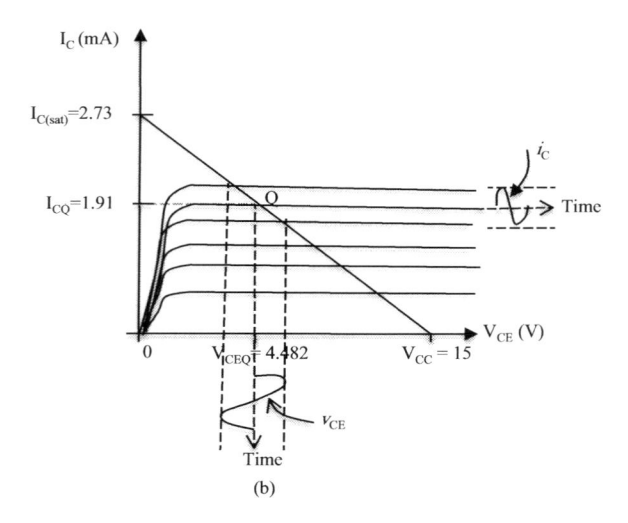

(b)

FIGURE 3.36 Load line for Example 3.4: (a) DC load line and (b) complete load line.

AC ANALYSIS

By replacing the capacitors by short circuits and the supply voltage V_{CC} by virtual AC ground, the small-signal AC equivalent circuit is shown in Figure 3.37.

c. Determining small-signal AC parameters:
   i. Using Equation (3.4), V_T at 32°C $= 8.6171 \times 10^{-5} \times (273+32) = 26.28\,\text{mV}$
   ii. Using Equation (3.49), $g_m = I_C/V_T = 0.00191/0.02628 = 0.07267\ \Omega^{-1}$.
   iii. Since $\beta = 150$, $r_\pi = \beta/g_m = 150/0.07267 = 2.06\ \text{k}\Omega$
   iv. $r_{thB} = R_{in} \parallel R_{12} = \dfrac{R_{in} R_{12}}{R_{in} + R_{12}} = \dfrac{0.1 \times 8.182}{0.1 + 8.182} = 98.8\,\Omega$
   v. $r_{thC} = R_C \parallel R_L = \dfrac{R_C R_L}{R_C + R_L} = \dfrac{4.5 \times 45}{4.5 + 45} = 4.1\,\text{k}\Omega$

d. Given the transistor's transconductance at Q-point, its output resistance is computed as

$$r_o = \left.\frac{1}{g_m}\right|_{Q-\text{point}} = \frac{1}{25 \times 10^{-6}} = 40\,\text{k}\Omega$$

Therefore, transistor gain factor is

$$A_{vf} = \frac{v_o}{v_{BE}} = g_m \left(\frac{r_o r_{thC}}{r_o + r_{thC}} \right) = 0.07267 \left(\frac{40 \times 4.1}{40 + 4.1} \right) \times 10^3 = 240.8 \approx 241.$$

So, the overall circuit gain A_v can be defined as

$$A_v = \frac{v_o}{v_{in}} = -A_{vf} \left(\frac{R_{12}}{R_{12} + R_{in}} \right) \left(\frac{r_\pi}{r_\pi + r_{thB}} \right)$$

$$A_v = -240.8 \left(\frac{8.182}{8.182 + 0.1} \right) \left(\frac{2.06}{2.06 + 0.0988} \right) = -227$$

Note that the negative sign indicates that the output voltage is 180° out of phase with input voltage (inverting amplifier).

e. Determining the effect of input signal on output swing:
   Given the supply signal $v_s(t) = 0.025 \sin(1000\pi t)$ where $\omega = 2\pi f = 1000\pi$, we can express the equation for output voltage with load, as follows:
   i. $v_o = A_v v_{in}$
     where $v_{in} = \dfrac{R_B}{R_B + R_{in}} v_s$

$$R_B = R_{12} \parallel r_\pi = \frac{R_{12} r_\pi}{R_{12} + r_\pi}$$

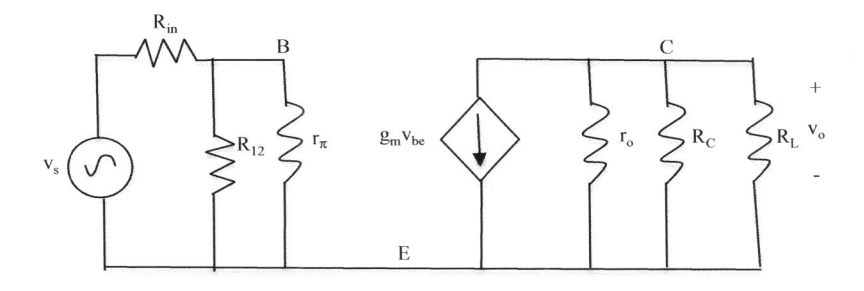

FIGURE 3.37 Small-signal ac circuit of Example 3.4: (a) T model and (b) hybrid-π model.

Substituting the values: $r_\pi = 2.06$ kΩ, $R_{12} = 8.182$ kΩ, and $R_{in} = 0.1$ kΩ, we have

$$R_B = \frac{8.182 \times 2.06}{8.182 + 2.06} = 1.646 \text{ k}\Omega$$

$$v_{in} = \left(\frac{1.626}{1.626 + 0.1}\right)v_s$$

$$v_{in} = 0.9427 v_s$$

Consequently,

$$v_o = A_v v_{in} = A_v\left(0.9427 v_s\right)$$

$$= (-227) \times \left[0.9427 \times 0.025 \sin(1000\pi t)\right] = 5.35 \sin(1000\pi t)$$

Note that the negative sign in the v_o expression indicates that output voltage is 180° out of phase with input voltage (inverting amplifier). The required peak-to-peak output voltage swing, $v_{o(swing)} = 2v_o = 2 \times 5.35 = 11.7$ V. The calculated maximum available peak-to-peak output voltage swing, from (b), i.e. $V_{o(swing)} = 8.564$ V which is less than $v_{o(swing)}$ (= 11.7 V). Hence, clipping will take place. Had $v_{o(swing)}$ been less than $V_{o(swing)}$ (i.e. $v_{o(swing)} < V_{o(swing)}$) clipping will not take place.

f. Designing suitable values for the circuit's capacitors.

From Equations (3.89), (3.91), and (3.93) and noting that $\omega = 2\pi f = 1000\pi$,

i. Calculating capacitor C_1: from Equation (3.89);

$$C_1 \geq \frac{100}{1000\pi}\left(\frac{r_{thB} + r_\pi}{r_{thB} r_\pi}\right) = \frac{1}{10\pi}\left(\frac{r_{thB} + r_\pi}{r_{thB} r_\pi}\right)$$

$$\geq \frac{1}{10\pi}\left(\frac{0.0988 + 2.06}{0.0988 \times 2.06}\right)\frac{10^3}{10^6} \geq 337.6 \text{ }\mu\text{F}$$

ii. Calculating capacitor C_2: from Equation (3.91);

$$C_2 \geq \frac{1}{2\pi f} \geq \frac{1}{1000\pi} \geq 318.3 \text{ }\mu\text{F}$$

iii. Calculating capacitor C_E: from Equation (3.93) and noting that $R_E = 1$ kΩ;

$$C_E \geq \frac{50}{\pi f R_E} \geq \frac{100}{2\pi f R_E}$$

$$\geq \frac{100}{1000\pi R_E} \geq \frac{1}{10\pi R_E} \geq \frac{1}{10\pi R_E} \geq \frac{1}{10\pi * 10^3} \geq 31.83 \text{ }\mu\text{F}$$

In essence, small-signal models are a linear representation of the transistor electrical behavior. The ac analysis is done differently from that of dc biasing. In the ac analysis case, the transistor is replaced by the small-signal ac equivalent circuit, using the following procedure:

i. Kill all dc sources, i.e. short out dc sources and open dc current sources.
ii. Assume coupling capacitors are short circuit. The effect of these capacitors is to set a lower cutoff frequency for the circuit. This is analyzed in the last step.
iii. Inspect the circuit. If you identify the circuit as a prototype circuit, you can directly use the formulas for that circuit. Otherwise, go to step (iv).
iv. Replace the BJT with its small-signal model.
v. Analyze the small-signal circuit for the desired quantities such as voltage and small-signal voltage gain.

3.5.5 LARGE-SIGNAL MODEL

There are subtle differences between the small- and large-signal models. Two equivalent large-signal circuit models—called the T-model and hybrid-π model—are used to model BJT. The forward-active mode NPN variant is considered in Figure 3.38. However, for a PNP transistor, the difference is the reversal of the diode D_E direction. In this case, the base current has an opposite polarity. In the T model, Figure 3.38a, the diode current I_s is represented in terms of the emitter current. Whereas, in the hybrid-π model, as in Figure 3.38b, the diode current is represented in terms of the base current, i_B, looking as if a diode is between base and emitter.

The model's currents can be expressed as follows:

Collector current:

$$i_C = I_S \left[\exp\left(\frac{V_{BE}}{V_T} \right) - 1 \right] \tag{3.94}$$

Base current:

$$i_B = \frac{i_C}{\beta} = \frac{I_S}{\beta} \left[\exp\left(\frac{V_{BE}}{V_T} \right) - 1 \right] \approx \frac{I_S}{\beta} \exp\left(\frac{V_{BE}}{V_T} \right) \tag{3.95}$$

Emitter current:

$$i_E = i_B + i_C = \left(\frac{\beta+1}{\beta} \right) I_S \exp\left(\frac{V_{BE}}{V_T} \right) = \frac{I_S}{\alpha} \exp\left(\frac{V_{BE}}{V_T} \right) \tag{3.96}$$

3.5.6 LARGE-SIGNAL MODEL IN SATURATION

In saturation, both junctions are forward biased, and the impedance level looking into the emitter or collector is very low. The simplified model is as shown in Figure 3.39 and the junction voltages become $V_{BE(\text{on})} \approx 0.6$–$0.7\,\text{V}$; and $V_{CE(\text{sat})} \approx 0.2\,\text{V}$.

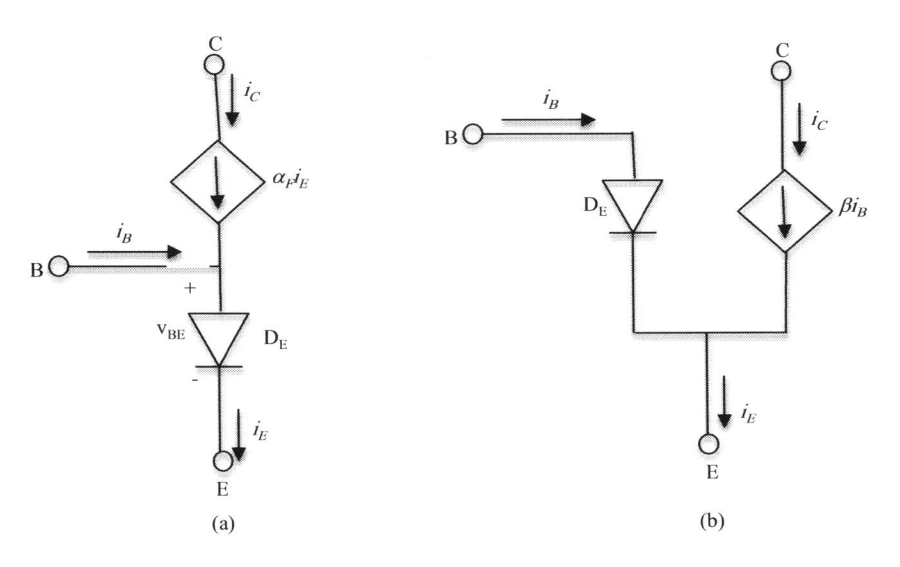

(a) (b)

FIGURE 3.38 NPN BJT large-signal models: (a) T model and (b) hybrid-π model.

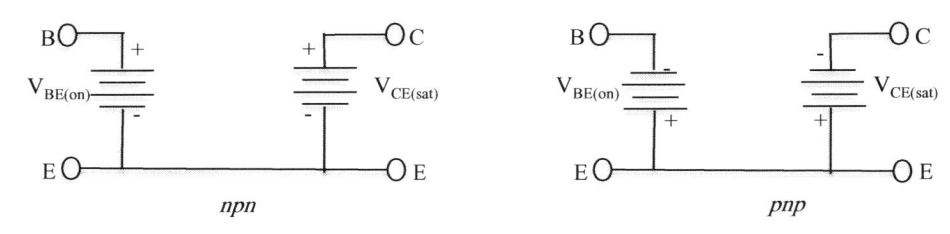

FIGURE 3.39 Large-signal model in saturation.

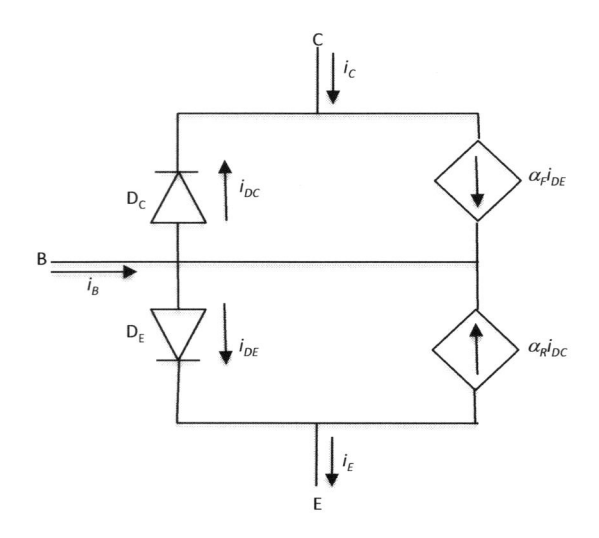

FIGURE 3.40 Unified Ebers–Moll model of bipolar transistor.

3.5.7 EBERS–MOLL MODEL

Another transistor model, known as the *Ebers–Moll model*, unifies into a single set of equations the different operation modes of the bipolar transistors. This model is represented by Figure 3.40, which combines both forward and reverse currents.

In its most general form, these currents are expressed by the following set of equations:

$$I_C = I_F - \frac{I_R}{\alpha_R}, \quad I_E = \frac{I_F}{\alpha_F} - I_R, \quad I_B = I_E - I_C \tag{3.97}$$

where the subscripts "*F*" and "*R*" denote "forward-mode" and "reverse-mode," and double subscripts "*BE*" and "*BC*" denote "base–emitter" and "base–collector." And

$$I_F = I_S \left[\exp\left(\frac{V_{BE}}{V_T} \right) - 1 \right] \tag{3.98}$$

$$I_R = I_S \left[\exp\left(\frac{V_{BC}}{V_T} \right) - 1 \right] \tag{3.99}$$

$$\alpha_F = \frac{\beta_F}{\beta_F + 1} \quad \alpha_R = \frac{\beta_R}{\beta_R + 1} \tag{3.100}$$

A transistor operates in reverse-active mode if the emitter–base junction is reverse biased and collector–base junction is forward biased. In this instance, the collector injects electrons into the base and the emitter collects most of them. The gain of the transistor is very low, usually less than unity. The *Ebers–Moll model* is often used for the computer representation (e.g. SPICE computer simulation program) of bipolar transistor with large-signal behavior. Improvement to systems and components characteristics modeling is continuously being made enriching our body of knowledge. As always, the actual transistor is somewhat more complex, and some second-order parameters, which represent secondary effects, have to be considered. Of recent, a scalable model, known as Mextram, has been introduced, which describes the various intrinsic and extrinsic regions of bipolar transistors [5].

Conservative voltage bias for best operating point stability and signal swing works. But the vexing question is: How does one obtain operating point stability, and simultaneously achieve a respectable voltage gain?

3.6 STABILITY FACTOR

The operating point (Q-point) usually shifts with change in temperature as the transistor parameters, namely: reverse saturation current, I_{CO}; base–emitter voltage, V_{BE}; and the forward gain, β. These parameters are all functions of temperature. The performance of discrete type biasing circuits is usually measured in terms of stability factors. A stability factor gives the change in the collector current, I_C, due to variation in parameters I_{CO}, V_{BE}, and β. Stabilization techniques give different biasing circuits. In practice, a bias circuit that is less sensitive to temperature variations is often preferred. To compare the stability provided by different biasing circuits, a term called *stability factor, S*, which indicates degree of change in operating point due to variation in temperature, is used to compare their performances. Stability factor is defined as follows for each of the parameters affecting bias.

a. Current stability factor $S(I_{CO})$ is the rate of change of I_C with reverse current I_{CO} keeping V_{BE}, and β constant, that is,

$$S\left(I_{CO}\right)=\left.\frac{\partial I_C}{\partial I_{CO}}\right|_{V_{BE},\beta}=\frac{\Delta I_C}{\Delta I_{CO}} \tag{3.101}$$

As earlier indicated in the text, the behavior of the *npn*-transistor is determined by its two *pn*-junctions. For instance, the forward-biased base–emitter (*BE*) junction allows the free electrons to flow from the emitter through the PN junction to form the emitter current I_E. Whereas, when reverse biased, the collector–base (*CB*) junction blocks the majority carriers (i.e. holes in the *p*-type base and electrons in *n*-type collector), but allows the minority carriers to go through, including the free electrons in the base coming from the emitter resulting in *leakage current, I_{CEO}*, and the reverse saturation current of the collector–base junction I_{CBO}. The reverse saturation current, I_{CBO}, produces heat at the junctions. This heat increases the temperature at the collector–emitter (*CE*) and *CB* junctions. The increase in the minority carriers increases the leakage current I_{CEO}:

$$I_{CEO} =\left(1+\beta\right)I_{CBO} \tag{3.102}$$

In practice, it has been observed that I_{CBO} doubles for every 10°C rise in temperature in silicon transistor [6]; see Table 3.1.

Increase in I_{CEO}, in turn, increases the collector current, I_C:

$$I_C = \beta I_B + I_{CEO} \tag{3.103}$$

TABLE 3.1

Variation of Silicon Transistor Parameters with Temperature [6]

Temperature T (°C)	Reverse Saturation Current I_{CO} (nA)	Base–Emitter Voltage V_{BE} (V)	Forward Gain β
−65	0.2×10^{-3}	0.85	20
25	0.1	0.65	50
100	20	0.48	80
175	3.3×10^{3}	0.30	120

For a common emitter configuration, the collector current is given by

$$I_C = \beta I_B + I_{CEO} \tag{3.104}$$

or

$$I_C = \beta I_B + (1 + \beta) I_{CBO} \tag{3.105}$$

In conformity with notations in Equation (3.101) and Table 3.1 (Column 2), the subscript "CBO" is simply written as "CO". It is important to recognize that β is the same as forward gain β_F when transistor is in the forward-active region. Taking partial derivative of Equation (3.105) with respect to I_C, assuming β constant, we have

$$1 = \beta \frac{dI_B}{dI_C} + (\beta + 1) \frac{dI_{CO}}{dI_C} = \beta \frac{dI_B}{dI_C} + (\beta + 1) \frac{1}{S(I_{CO})} \tag{3.106}$$

Rearranging Equation (3.82) to have

$$S(I_{CO}) = \frac{(\beta + 1)}{1 - \beta \dfrac{dI_B}{dI_C}} \tag{3.107}$$

The value of dI_B / dI_C depends on the biasing arrangement.

    b. Voltage stability factor, $S(V_{BE})$, is the rate of change of I_C with base–emitter voltage V_{BE} keeping I_{CO}, and β constant; that is,

$$S(V_{BE}) = \frac{\partial I_C}{\partial V_{BE}} \bigg|_{I_{CO}, \beta} = \frac{\Delta I_C}{\Delta V_{BE}} \tag{3.108}$$

Again, the value of the stability factor depends on the biasing arrangement. For any given circuit configuration, obtain the standard equation for IC and replace where found IB in terms of V_{BE} to get $S(V_{BE})$.

    c. Gain Stability factor, $S(\beta)$, is the rate of change of I_C with gain β keeping V_{BE} and I_{CO} constant, that is,

$$S(\beta) = \frac{\partial I_C}{\partial \beta} \bigg|_{I_{CO}, V_{BE}} = \frac{\Delta I_C}{\Delta \beta} \tag{3.109}$$

As above, the value of this stability factor depends on the biasing arrangement. For any given circuit configuration, obtain the standard equation for I_C in section (b) and differentiate with respect to β to get $S(\beta)$.

In essence, Equation (3.107) can be considered as a standard equation for derivation of stability factors of other biasing circuits.

We can summarize the general procedure to obtaining stability factors for various biasing circuits as follows [7]:

Step 1: Obtain the expression for I_B.

Step 2: Obtain $\partial I_B/\partial I_C$ and use it in Equation (3.101) to get $S(I_{CO})$.

Step 3: In standard equation of I_C, replace I_B in terms of V_{BE} to get $S(V_{BE})$.

Step 4: Differentiate the equation obtained in step 3 with respect to β to get $S(\beta)$.

Many circuits used for automatic, military, and aerospace applications operate at different temperatures, ranging from $-60°C$ to $160°C$. Device manufacturers attempt to design their devices to conveniently accommodate this operational range. Often the manufacturers provide data justifying their components operationally performance range. An example is Table 3.1, which shows drift with temperature of a typical silicon transistor's parameters.

In essence, when embarking on any amplifier design it is very important to spend time checking on the stability of the device chosen; otherwise, the amplifier may well turn into unintended operation such as an oscillator. Of course, oscillators have widespread applications in electronics. For instance, oscillators are used for erasing and biasing in magnetic recording and time clock pulses in computers. Oscillators are also used in transmitters to provide the carrier source. Many electronic measuring instruments use oscillators forming part of a phase-locked-loop detector.

Example 3.5

Consider the transistor circuit shown in Figure 3.41 as having a forward gain of 50 at 25°C. It is desired that the collector–emitter voltage be 4.5 V. Determine (a) an appropriate value for resistor R_1. Also, (b) determine the stability factors and the value of collector current if the circuit is designed to perform effectively up to 100°C. Use the values for the reverse saturation current, I_{CO}, and base–emitter voltage, V_{BE}, at appropriate temperatures specified in Table 3.1.

Solution:

Let us relabel the transistor circuit of Figure 3.41a, as shown in Figure 3.41b: the currents (I_1, I_C, I_E, I) flowing through the circuit, supply voltage ($V_{CC} = 15$ V), and the collector–emitter voltage (V_{CE}).

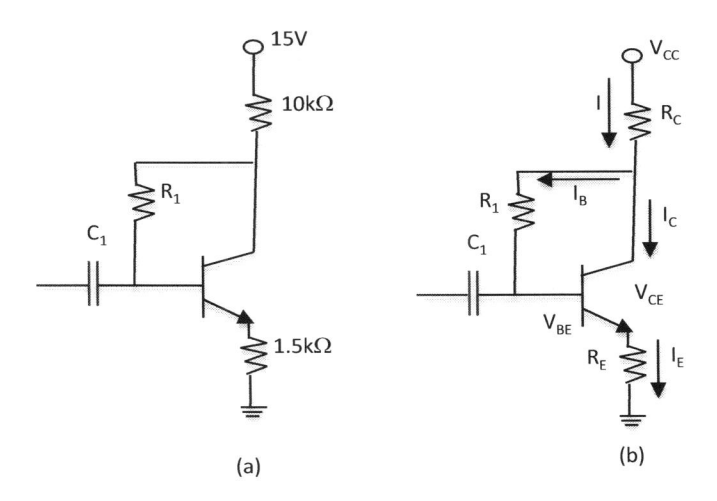

(a)   (b)

FIGURE 3.41   A transistor circuit.

a. Estimating resistor, R_1:

By applying the Kirchhoff Voltage Law (KVL) to the circuit, we can write

$$V_{CC} = IR_C + V_{CE} + R_E I_E = R_C (I_B + I_C) + V_{CE} + R_E (I_B + I_C) \tag{3.110a}$$

$$V_{CE} = I_B R_1 + V_{BE} \tag{3.110b}$$

We know that

$$I_E = I_B + I_C = (\beta + 1) I_B \tag{3.110c}$$

In view of Equation (3.110c) in Equation (3.110a), we write the base current I_B:

$$I_B = \frac{V_{CC} - V_{CE}}{(\beta + 1)(R_C + R_E)} \tag{3.110d}$$

Substitute actual values in Equation (3.110d) to get

$$I_B = \frac{15 - 4.5}{(50 + 1)(10 + 1.5)} = 0.913 \, \text{mA} \tag{3.110e}$$

Collector current at the Q-point, implying $I_{CQ} = I_C$; that is,

$$I_{CQ} = I_C = b I_B = 50 \times 0.913 = 45.65 \, \text{mA} \tag{3.110f}$$

Given that, from Table 3.1, $V_{BE} = 0.65 \, \text{V}$ at 25°C, then from (ii), we obtain the value of R_1 as

$$R_1 = \frac{V_{CE} - V_{BE}}{I_B} = \frac{4.5 - 0.65}{0.913} = 4.22 \, \text{k}\Omega \tag{3.110g}$$

b. Solving stability factors

We know from Equation (3.105) that

$$I_C = \beta I_B + (1 + \beta) I_{CBO} \tag{3.110h}$$

For collector–base circuit, we can express

$$V_{CC} = (R_C + R_E)(I_B + I_C) + V_{BE} + I_B R_1 \tag{3.110i}$$

Rewriting (3.110i) in terms of I_B:

$$I_B = \frac{V_{CC} - V_{BE}}{R_1 + R_C + R_E} - \frac{I_C (R_C + R_E)}{R_1 + R_C + R_E} \tag{3.110j}$$

By substituting I_B in Equation (3.110h), we write

$$I_C = \beta \left[\frac{V_{CC} - V_{BE}}{R_1 + R_C + R_E} - \frac{I_C (R_C + R_E)}{R_1 + R_C + R_E} \right] + (\beta + 1) I_{CO} \tag{3.110k}$$

Collecting terms:

$$I_C \left[1 + \frac{\beta (R_C + R_E)}{R_1 + R_C + R_E} \right] = \frac{\beta (V_{CC} - V_{BE})}{R_1 + R_C + R_E} + (\beta + 1) I_{CO} \tag{3.110l}$$

Differentiating I_C in Equation (3.110l) with respect to I_{CO}, we get

$$S(I_{CO}) = \frac{dI_C}{dI_{CO}} = \frac{(\beta + 1)}{1 + \dfrac{\beta(R_C + R_E)}{R_1 + R_C + R_E}} \qquad (3.110m)$$

Substituting values in Equation (3.110m), we have

$$S(I_{CO}) = 1.36 \qquad (3.110n)$$

Incremental change in I_C for a change in I_{CO} from 25°C to 100°C, collecting I_{CO} values at these temperatures from Table 3.1, is

$$\Delta I_C = S(I_{CO}) \times \Delta I_{CO}$$

$$= S(I_{CO}) \times \left[I_{CO}|_{100°C} - I_{CO}|_{25°C} \right] = 1.36 \times [20 - 0.1] = 27.06\,\text{nA} \qquad (3.110p)$$

Consequently, the value of I_C at 100°C is found by adding Equation (3.110p) to Equation (3.110f): that is,

$$I_C|_{100°C} = I_{CQ} + \Delta I_C = (45.65 + 0.00002706) = 45.652\,\text{mA} \qquad (3.110q)$$

Differentiating I_C in Equation (3.110l) with respect to V_{BE}, we get

$$S(V_{BE}) = \frac{dI_C}{dV_{BE}} = \frac{-\beta\big/R_1 + R_C + R_E}{1 + \dfrac{\beta(R_C + R_E)}{R_1 + R_C + R_E}} \quad [\text{S}] \qquad (3.110r)$$

Substituting values in Equation (3.110r), we have

$$S(V_{BE}) = -11.5\,\text{mS} \qquad (3.110s)$$

Incremental change in I_C for a change in V_{BE} from 25°C to 100°C, collecting V_{BE} values at these temperatures from Table 3.1, is

$$\Delta I_C = S(V_{BE}) \times \Delta V_{BE}$$

$$= S(V_{BE}) \times \left[V_{BE}|_{100°C} - V_{BE}|_{25°C} \right] = -0.0115 \times [0.48 - 0.65] = 1.95\,\text{mA} \qquad (3.110t)$$

Consequently, following the definition in Equation (3.110q), the value of I_C at 100°C with the base–emitter shifted is found by adding Equation (3.110t) to Equation (3.110f):

$$I_C = (45.65 + 1.95) = 47.6\,\text{mA} \qquad (3.110u)$$

Differentiating I_C in Equation (3.110l) with respect to β, we get

$$S(\beta) = \frac{dI_C}{d\beta} = \frac{I_{CO} + \left[\dfrac{V_{CC} - V_{BE}}{R_1 + R_C + R_E}\right]}{1 + \left[\dfrac{R_C + R_E}{R_1 + R_C + R_E}\right]} \qquad (3.110v)$$

Substituting values in Equation (3.110v), we have

$$S(\beta) = 0.5272\,\text{mA} \tag{3.110w}$$

Incremental change in I_C for a change in β from 25°C to 100°C, collecting β values at these temperatures from Table 3.1, is

$$\Delta I_C = S(\beta) \times \Delta\beta$$

$$= S(\beta) \times \left[\beta\big|_{100°C} - \beta\big|_{25°C} \right] = 0.5272 \times [80 - 25] \times 10^{-3} = 29\,mA \tag{3.110x}$$

Consequently, following definition in Equation (3.110q), the value of I_C at 100°C with the forward gain shifted is found by adding Equation (3.110x) to Equation (3.110f):

$$I_C = (45.65 + 0.029) = 45.679\,\text{mA} \tag{3.110y}$$

In conclusion, this analysis has demonstrated the effect of temperature on the transistor and indicates the nominal change on the amount of current that flows through the collector.

3.7   POWER DISSIPATION AND EFFICIENCY

The input signal applied to amplifiers is alternating in nature. The basic features of any alternating signal are amplitude, frequency, and phase. Basic electrical theory emphasizes the relationship between the *peak* values and root-mean-square (rms) values of an AC circuit. The peak current I_{peak} and peak voltage V_{peak} are defined thus

$$V_{\text{peak}} = \sqrt{2}V_{\text{rms}} \tag{3.111}$$

$$I_{\text{peak}} = \sqrt{2}I_{\text{rms}} \tag{3.112}$$

For an amplifier output load R_L, the AC power output, P_{ac}, can be expressed as

$$P_{\text{ac}} = V_{\text{rms}}I_{\text{rms}} = I_{\text{rms}}^2 R_L = \frac{V_{\text{rms}}^2}{R_L} \tag{3.113}$$

Alternately, in terms of peak values

$$P_{\text{ac}} = \frac{1}{2}V_{\text{peak}}I_{\text{peak}} = \frac{1}{2}I_{\text{peak}}^2 R_L = \frac{V_{\text{peak}}^2}{2R_L} \tag{3.114}$$

Power dissipated by the transistor is the difference between AC power output P_{ac} and DC power input P_d. Specifically,

$$P_{\text{dc}} = V_{CC}I_{CQ} \tag{3.115}$$

Subtract Equation (3.114) from Equation (3.115) to have power dissipation:

$$P_d = P_{\text{dc}} - P_{\text{ac}} = V_{CC}I_{CQ} - \frac{V_{\text{peak}}I_{\text{peak}}}{2} \tag{3.116}$$

The AC components of Equation (3.116) can be expressed in terms of *rms* if the reader wishes to solve a given problem in *rms* values.

Efficiency, η, is defined as the ratio of AC power to DC power, i.e.

$$\eta = \frac{P_{ac}}{P_{dc}} \qquad (3.117)$$

One can express efficiency as a percentage:

$$\eta = \frac{P_{ac}}{P_{dc}} \times 100\% \qquad (3.118)$$

Example 3.6

An 8-ohm loudspeaker is connected to the secondary of the output transformer of an amplifier shown in Figure 3.42. The transformer is assumed ideal with a turns ratio of 4:1. The transistor is forward biased with a gain of 120. The base resistor is adjusted for maximum input swings. Calculate (i) the collector current at the Q-point, (ii) the base current, (iii) maximum AC power, (iv) the maximum power dissipated, and (v) maximum efficiency.

Solution:

Given: $R_L = 8\ \Omega$, $\beta = 120$, $n_1:n_2 = 4:1$, $V_{CC} = 25\ V$

Turns ratio, $n = n_2/n_1$, being the number of turns on the secondary (n_2) of the transformer to the number of turns on primary (n_1). The effect of the turns ratio is to transform the loudspeaker's load to the primary side of the amplifier, i.e. the transformed load becomes R_{L*}, defined as

$$R_{L*} = \frac{R_L}{n^2} = \frac{8}{\left(\frac{1}{4}\right)^2} = 128\ \Omega \qquad (3.119a)$$

Since R_B is adjusted for maximum input swings, the AC power P_{ac} output is also maximum. Hence,

$$P_{ac}\big|_{max} = \frac{V_{CC}^2}{2R_{L*}} = \frac{25 \times 25}{2 \times 128} = 2.441\ W \qquad (3.119b)$$

whilst the collector current at the Q-point I_{CQ} is

$$I_{CQ} = \frac{V_{CC}}{R_{L*}} = \frac{25}{128} = 0.1953\ A \qquad (3.119c)$$

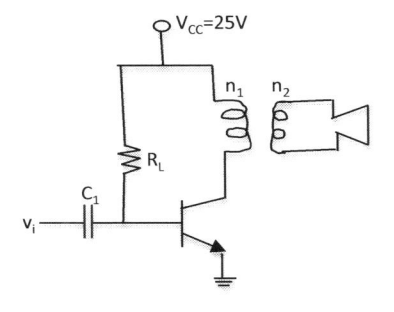

FIGURE 3.42 Class A amplifier circuit.

Base current is therefore calculated as

$$I_B = \frac{I_{CQ}}{\beta} = \frac{0.1953}{120} = 1.63\,\text{mA} \tag{3.119d}$$

From Equation (3.115) the DC power is expressed as

$$P_{dc} = V_{CC} I_{CQ} = 25 \times 0.1953 = 4.883\,\text{W} \tag{3.119e}$$

$$\text{Power dissipated } P_d = P_{dc} - P_{ac} = 4.883 - 2.441 = 2.442\,\text{W} \tag{3.119f}$$

Finally, from Equation (3.118) the amplifier maximum efficiency is

$$\eta = \frac{P_{ac}}{P_{dc}} = \frac{2.441}{4.883} \times 100 = 50\% \tag{3.119g}$$

Example 3.7

Figure 3.43 shows a circuit of Class B amplifier. The circuit's two transistors Q1 and Q2 are assumed identical and ideal.

    i. Calculate the maximum AC power that can be delivered to the load R_L.
    ii. If the circuit is designed to increase the power delivered to the load by 38%,
        a. by what percent must the supply voltage be increased?
        b. what should be the transistor's maximum breakdown voltage when the supply voltage is increased?

Solution:

    i. Maximum AC power delivered at the load

$$P_{ac(max)} = \frac{V_{cc}^2}{2R_L} = \frac{25 \times 25}{2 \times 50} = 6.25\,\text{W}$$

    ii. Let us denote the new power to be delivered by $P_{ac(new)}$ and the corresponding voltage $V_{cc(new)}$, then

$$P_{ac(new)} = \left(1 + \frac{38}{100}\right) P_{ac(max)} = \frac{V_{cc(new)}^2}{2R_L}$$

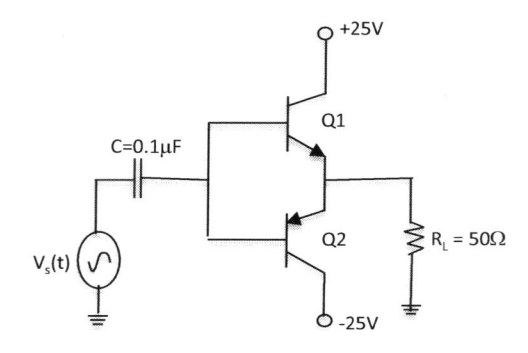

FIGURE 3.43 Class B amplifier circuit.

So, the voltage

$$V_{cc(\text{new})} = \sqrt{2R_L P_{\text{ac(new)}}} = \sqrt{2R_L(1+0.38)P_{\text{ac(max)}}} = \sqrt{2*50*1.38*6.25} = 29.37\,\text{V}$$

a. Hence % increase is

$$\Delta V(\%) = 100 \times \left(\frac{V_{cc(\text{new})} - V_{cc(\text{old})}}{V_{cc(\text{old})}}\right) = 100 \times \left(\frac{29.37-25}{25}\right) = 17.47\%$$

b. Maximum breakdown voltage per transistor under this condition is
$2V_{cc(\text{new})} = 2 \times 29.3 = 58.74\,\text{V}$

3.8 TIME-BASE GENERATOR

Time-base generators are circuits that provide an output waveform, a part of which is characterized by a linear variation of current or voltage with respect to time. Time-base generators have applications in radar sets, oscilloscopes, and computer circuits. Most common types of time-base generators use some type of switching action with either the charge or discharge of an RC or RL circuit.

Figure 3.44a shows a simple transistor current time-base generator circuit. The transistor Q1 is used as a switch and is driven by a trigger signal v_B. The trigger signal v_B operates between two levels, i.e. ±1 (as shown in Figure 3.44b(i)): where the negative going period keeps the transistor in the cutoff state, while the positive going period switches on the transistor and drives it into saturation.

When time is less than zero, i.e. $t < t_o$, the trigger signal is negative. So the transistor is cutoff. As a result, no current flows in the transistor and the current flowing through the inductor is zero, i.e. $i_L = 0$ and $v_{CE} = V_{cc}$.

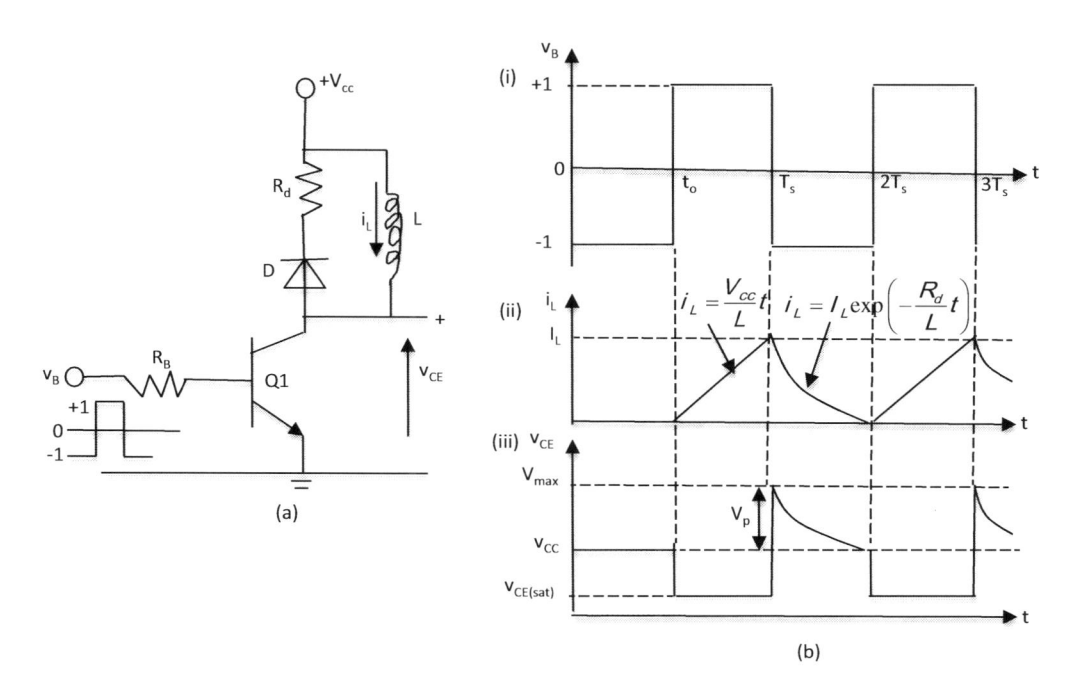

(a)

(b)

FIGURE 3.44 A simple time-base generator: (a) sweep generator circuit and (b) current and voltage responses.

At $t = t_o$, the trigger or gate signal goes positive. So the transistor conducts and goes into saturation, and the collector voltage falls to $V_{CE(SAT)}$ ensuring that the entire supply voltage V_{cc} is applied across the inductor. So, the current through the inductor is

$$i_L = \frac{1}{L} \int V_{cc} \, dt = \frac{V_{cc}}{L} t \qquad (3.120)$$

which increases linearly with time and continues till it reaches a maximum value I_L at $t = T_s$. During the sweep interval T_s (i.e. between $t = t_o$ and $t = T_s$) the diode is reverse biased and hence does not conduct. At $t = T_s$, the transistor is cutoff and there is no current flowing through the transistor. Since the current flowing through the inductor cannot change instantaneously at this cutoff mode, it flows through the diode and the diode conducts. Hence, there will be a voltage drop of V_p across R_d, which amounts to

$$V_p = I_L R_d \qquad (3.121)$$

So, at $t = T_s$, the potential at the collector terminal rises abruptly to V_{max}, causing a voltage spike of amplitude V_p. The duration of the spike depends on the inductance of L but the amplitude of the spike does not. For $t > T_s$, the inductor decays exponentially to zero with a time constant

$$\tau_L = \frac{L}{R_d} \qquad (3.122)$$

and inductor current

$$i_L = I_L \exp\left(-\frac{t}{\tau_L}\right) = I_L \exp\left(-\frac{R_d}{L}t\right) \qquad (3.123)$$

So, the voltage at the collector v_{CE} also decays exponentially until it reaches the saturation voltage V_{cc}. The waveforms of this sweep circuit are shown in Figure 3.44b: (i) for trigger signal, (ii) for sweep current, and (iii) for sweep voltage.

The voltage v_{CE} across the transistor is of special practical significance. The value of the spike voltage V_p has to be limited so that the transistor is below its collector-to-base *breakdown* voltage. Since I_L is determined by the deflection requirements, there is clearly an upper limit to the size of R_d. Examples below will shed some light on this concept.

Often in transistor circuit design, to eliminate sudden voltage spikes in the event the supply voltage is suddenly reduced or interrupted, a diode is connected across the inductor, as seen in Figure 3.44a. This diode is called *flyback* diode.

Example 3.8

For the circuit shown in Figure 3.44a at $t = 0$ having the driving waveform applied to the base of the transistor. Given that $R_B = 10$ kΩ, $R_d = 10$ Ω, $L = 10$ mH, $V_{CE(sat)} = 0.3$ V, $V_{BE} = 0.7$ V, and $V_{cc} = $ s10 V.

   a. Calculate the time required for the inductor to reach the maximum value, if the inductor is assumed to be ideal, the saturation resistance of the transistor to be zero and when the transistor is forward biased with a gain of 20.
   b. Calculate the time required for the inductor to decay to 10 mA.
   c. Calculate the collector–emitter voltage at the time when the transistor is reverse biased.

Solution:

a. At time $t = 0$, the base current is

$$i_B = I_B = \frac{v_B - V_{BE}}{R_B} = \frac{10 - 0.7}{10 \times 10^3} = 0.93\,\text{mA}$$

We know that the transistor gain $\beta = 20$, which is connected to the relation of Equation (3.14) thus

$$i_L = I_{L(\text{max})} = \beta I_B = 0.93 \times 20 = 18.06\,\text{mA}$$

Since the transistor is in saturation, the voltage available across the inductor is

$$v_L = V_{cc} - V_{CE(\text{sat})} = 10 - 0.3 = 9.7\,\text{V}$$

We know that the maximum inductor current from the relation $I_{L(\text{max})} = \dfrac{v_L}{L} T_s$, so the sweep time can be estimated as

$$T_s = \frac{I_{L(\text{max})} L}{v_L} = \frac{10 \times 10^{-3} \times 18.6 \times 10^{-3}}{9.7} = 19.1\,\mu\text{s}$$

b. At $t = T_s$, the transistor is reverse biased, as such, the current flowing through the inductor decays with a time constant τ:

$$\tau = \frac{L}{R_d} = \frac{10 \times 10^{-3}}{10} = 1\,\text{ms}$$

The time duration (i.e. $T_s \leq t \leq 2T_s$) in which the current in the inductor decays to $10\,\text{mA}$ can be found from using Equation (3.123), i.e. $i_L = I_L \exp\left(-\dfrac{t}{\tau}\right)$; expressing t in terms of other variables to have

$$t = -\tau \log_e\left(\frac{i_L}{I_{L(\text{max})}}\right) = 0.62\,\text{ms}$$

where $i_L = 10\,\text{mA}$, $I_{L(\text{max})} = 18.06\,\text{mA}$, and $\tau = 1\,\text{ms}$.

c. At the time when the transistor is reverse biased,

$$v_{CE} = V_{cc} + V_p = V_{cc} + I_L R_d = 10 + \left(10 \times 18.6 \times 10^{-3}\right) = 10.186\,\text{V}$$

meaning that the spike voltage is $0.186\,\text{V}$.

3.9 SUMMARY

This chapter has presented the structure of the bipolar transistor and showed how a three-layer structure with alternating n-type and p-type regions can provide current and voltage amplification. Semiconductor diodes form the basis of a transistor. The transistor equations were developed through knowledge of diode equations, thus enabling an understanding of the law of the junction.

Transistors are used in a great variety of circuits, which can be divided into two fairly simple classes: amplifiers and switches. Examples of these classes were given. For simpler analysis of the transistor circuits, two models were introduced called T- and hybrid-π enabling quantification of the effect of small and large signals. A method of quantifying the operating point stability was also discussed. It is imperative that when embarking on any amplifier design it is very important to spend time checking on the stability of the device chosen, otherwise the amplifier may well turn into an oscillator.

In most practical applications, it is better to use an operational amplifier (abbreviated op-amp) as a source of gain rather than building an amplifier from discrete transistors. A good understanding of transistor fundamentals is nevertheless essential, which this chapter has provided. The next chapter deals with operational amplifiers.

PROBLEMS

3.1 (i) Describe under what condition the emitter–base and collector–base junctions are forward- or reverse biased.

(ii) One of the applications of transistors is as a switch to close relay contacts by driving the coil. Why is a flyback diode needed when BJT is used as a switch?

3.2 Calculate the currents, voltages, and power dissipated in the transistor of Figure 3Q.1 circuit when the transistor is driven into saturation when the transistor parameters are $\beta = 100$, $V_{BE(on)} = 0.7\,V$, and $V_{CE(sat)} = 0.2\,V$.

3.3 Given $\beta = 200$, $V_{BE(on)} = 0.7\,V$, and $V_{CE(sat)} = 0.2\,V$, calculate the operating point conditions and the small-signal parameters (transconductance and the active mode input resistance) of Figure 3Q.2 when operated at room temperature.

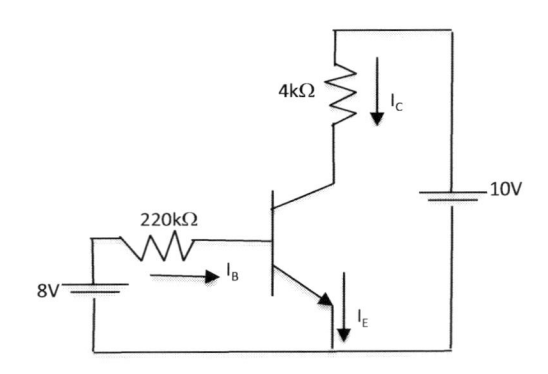

FIGURE 3Q.1

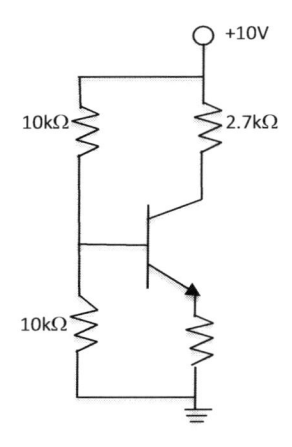

FIGURE 3Q.2

3.4 The transistor in the inverter circuit in Figure 3Q.3 is used to turn the LED on and off. Normally, LED is fabricated with compound semiconductor materials that have a larger cutin voltage $V\gamma$ compared to silicon diodes. Assume $V\gamma = 1.5\,\text{V}$. As a result, the required LED current is given as $12.5\,\text{mA}$ (i.e. $I_{LED} = 12.5\,\text{mA}$) to produce the required output light. Taking the transistor parameters of $\beta = 80$, $V_{BE(on)} = 0.7\,\text{V}$, and $V_{CE(sat)} = 0.2\,\text{V}$, design the appropriate resistance values and transistor power dissipation for the inverter switching configuration.

3.5 For the circuit shown in Figure 3Q.4, the common-emitter bipolar transistor has a forward current gain of 100, and base–emitter voltage of 0.7 V. Calculate the (a) Q-point current and voltage, (b) the base and emitter currents, (c) current flowing through 50-kΩ resistor, and (d) the voltage drop across the emitter load.

3.6 Calculate the Q-point of the two-stage amplifier of Figure 3Q.5, as well as all currents and voltages for a 15-V supply voltage with transistors Q1 and Q2 each having forward current gain of 100.

3.7 Write the stability factor $S(I_{CO})$ expression for a transistor circuit shown in Figure 3Q.6.

3.8 For circuit shown in Figure 3Q.7, determine V_o and $S(I_{CO})$ if $\beta = 120$.

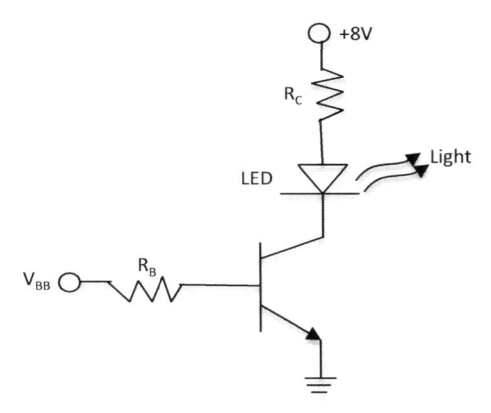

FIGURE 3Q.3

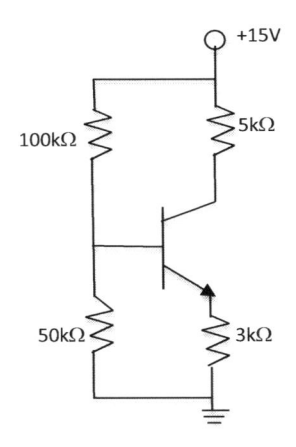

FIGURE 3Q.4

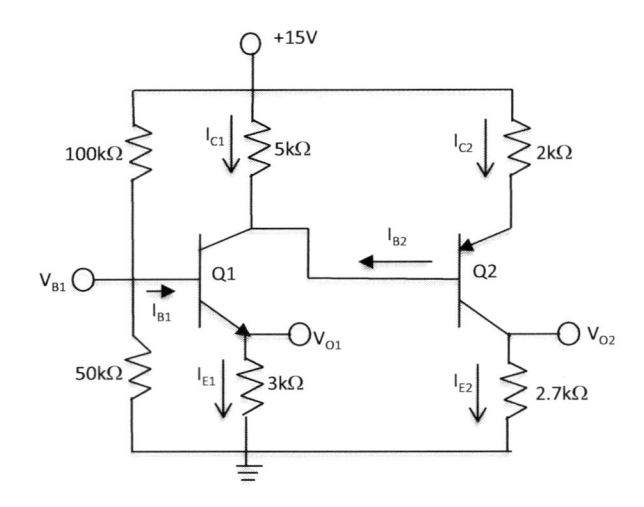

FIGURE 3Q.5

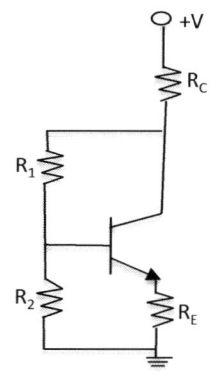

FIGURE 3Q.6

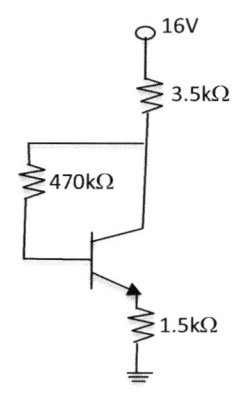

FIGURE 3Q.7

REFERENCES

1. Colinge, J.P. and Colinge, C.A. (2006). *Physics of Semiconductor Devices*. New York: Springer.
2. Pierret, R.F. and Neudeck, G. (1982). *Modular Series on Solid State Devices*. Reading, MA: Addison-Wesley.
3. Miller, J.M. (1920). Dependence of the input impedance of a three-electrode vacuum tube upon the load in the plate circuit. *Scientific Papers of the Bureau of Standards*, 15 (351), 367–385.
4. Yamaguchi, T., Yu, Y.C.S., Lane, E.E., Lee, J.S., Patton, E.E., Herman, R.D., Ahrendt, D.R., Drobny, V.F., Yuzuriha, T.H. and Garuts, V.E. (1988). Process and device performance of a high-speed double poly-si bipolar technology using borsenic-poly process with coupling-base implant. *IEEE Trans. Electron. Devices*, 35 (8), 1247–1255.
5. Paasschens, J.C.J., Kloosterman, W.J. and van der Toorn, R. (2002). Model derivation of Mextram 504. The physics behind the model. Unclassified Report NL-UR 2002/806, Philips Nat. Lab. [For the most recent model descriptions, source code, and documentation, see www.nxp.com/models.]
6. Kumar, B. and Jain, S.B. (2007). *Electronic Devices and Circuits*. New Delhi: Prentic-Hall.
7. Godse, A.P. and Bakshi, U.A. (2009). *Electronics Engineering*. Pune: Technical Publications.

4 Operational Amplifiers

As systems architectures evolve toward more complex hierarchical networks, requiring higher-performance body electronics modules and integrated circuits (ICs) to support gateway functions, as well as higher functional integration and centralized services, there is a need to explain the component parts from basic principles to empower future engineers and designers knowing and modifying the systems components as they see fit. This approach will enable the designers and engineers to migrate to higher levels of integration where essentially each electronic module connects well with the preferred system architecture. As a result, this chapter deals with the basic principles of core electronic devices starting from basic electronic operational amplifier (op-amp) characteristics and applications, including simple comparators, multivibrators, and oscillators.

4.1 BASIC AMPLIFIER DEFINITIONS

An electronic amplifier is a device that responds to a small input -signal (voltage, current, or power) and delivers a larger output signal that contains the essential waveform features of the input signal. This process is technically amplification. A signal is any time-varying, or spatial-varying, quantity. The amplification of weak signals into stronger signals is of fundamental importance in almost any electronic system.

The principal use of amplifiers is in the measurement process. In computer systems, for example, amplifier circuits are packaged as ICs (or *chips*), performing complex tasks including mathematical operations (addition, subtraction, multiplication, integration—calculating the areas under signals, and differentiation—calculating the slopes of signals). In communication systems, for example, it may be advantageous to replace a repeater unit in an optical fiber link by a direct amplification unit, or an amplifier unit could be used in a predetection capacity at the receiver. In power and control systems, amplifiers are used as a basic building block for phase shifting, filtering and signal conditioning, multiplexing, feedback operation, to name a few applications. Technically, amplifiers that perform this kind of operations are called *operational amplifiers* (*op-amps* for short).

The triangular block symbol of Figure 4.1 is used to represent an ideal op-amp. An amplifier has an *input port* and *an output port*. The input port consists of two terminals or pins called the *inverting* input (or *summing point*, V_-) and the *non-inverting* input (V_+). By "inverting" means a positive voltage produces a negative voltage at the output. Similarly, "non-inverting" means a positive voltage produces a positive voltage at the output. Note that these input symbols have nothing to do with

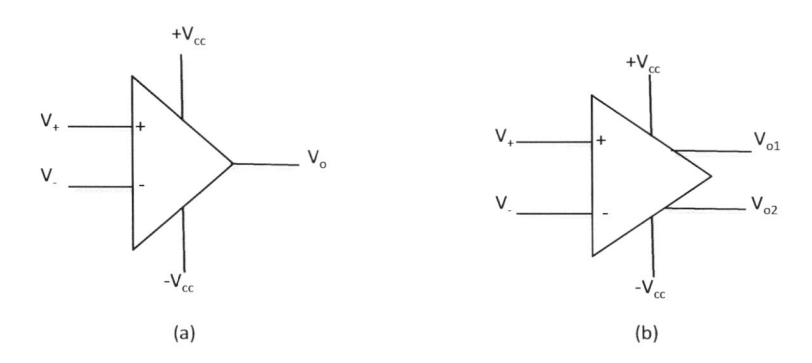

(a) (b)

FIGURE 4.1 A schematic of op-amps: (a) single output op-amp and (b) differential output op-amp.

the polarity of the applied input signals. As will be seen later in this book, we can use these pins together to create negative feedback loops, which make amplifier circuits behave very nicely.

The output port consists of two terminals. The way the output port is connected determines whether it is single-output (V_o as in Figure 4.1a, where one of which is usually connected to the ground node) or two-output terminals, where both terminals are used: this type of *two-input ports* and *two-output ports* amplifier arrangement is called *differential amplifier*, as seen in Figure 4.1b.

Generally, op-amps require two power supplies to energize or operate it: one supplying a positive voltage ($+V_c$) and another supplying a negative voltage ($-V_c$) with respect to circuit common. In some books, these supplying voltages are denoted $\pm V_{cc}$ or simply $+V, -V$. This bipolar power supply (i.e. $\pm V$ or $\pm V_c$) allows op-amps or chips to generate output results of either polarity. The power supply pins are sometimes called the *rails*.

The amplifier output voltage is limited to the range $-V_c$ to $+V_c$. This range is often called the *linear region* of the amplifier. However, when the output swings to $-V_c$ or $+V_c$, the op-amp is said to be *saturated*. As a result, we bound saturation mathematically thus

- If $V_+ > V_-$: it implies that $V_o = +V_c$, and the op-amp is said to be positively saturated.
- If $V_- > V_+$: it implies that $V_o = -V_c$, and the op-amp is said to be negatively saturated.

Circuit designers seldom forget these rails; nevertheless, they need to be acknowledged when actually building the circuit. In this text, we still retain these power supplies in the circuit.

4.2 AMPLIFIER GAIN

Amplifiers can be specified according to their input and output properties [1]. The relationship that exists between the signals measured at the output to that measured at the input is called *multiplication factor, amplification factor or open loop gain* (or simply *gain*), or *transfer function*. There is a difference though in the definitional aspect regarding *transfer function*. (More is said about this in Section 4.2.6.)

Generally speaking, the *gain* of an amplifier may be specified as "output voltage/input voltage," "output power/input power" or any other combination of current, voltage, and power. Depending on the nature of the input and output signals, we can have four types of *gain*:

- Voltage gain (output voltage/input voltage), i.e.

$$A_v = V_o / V_{in} \tag{4.1}$$

- Current gain (output current out/input current), i.e.

$$A_I = I_o / I_{in} \tag{4.2}$$

- Transresistance (output voltage/input current), i.e.

$$A_{tr} = V_o / I_{in} \tag{4.3}$$

- Transconductance (output current/input voltage).

$$A_{tc} = I_o / V_{in} \tag{4.4}$$

How these gains are obtained would be demonstrated with examples later in the text under appropriate subheadings. In electronics, signals, and communication engineering, the gains are expressed

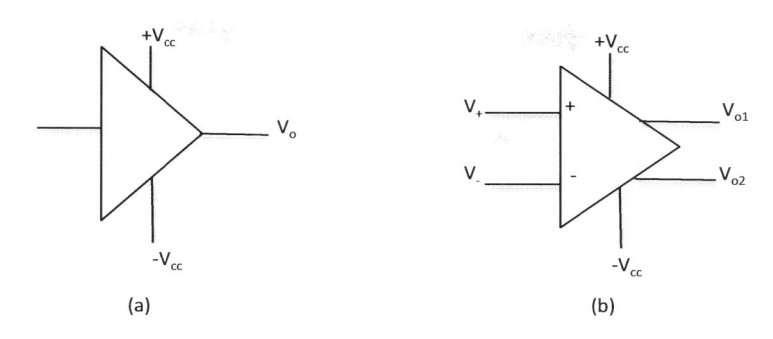

FIGURE 4.2 Standard symbols of op-amps: (a) buffer op-amp and (b) differential input, differential output op-amp.

in *decibel*, dB: a logarithmic unit is used to describe a ratio. As an example, the current gain (Equation (4.2)) in dB is $\{A_I = 10\log_{10}(I_o/I_{in})\}$.

Amplifier gain, in most cases, is a constant for any combination of input and output signals. When this occurs, the amplifier is considered *linear*. If the gain were not linear, the output signal would be distorted, e.g. by clipping the output signal at the limits of its capabilities.

Amplifiers are represented by three standard symbols:

1. Buffer op-amp; as shown in Figure 4.2a.
2. Differential input, single-ended output op-amp; the same as shown in Figure 4.1a. This type represents the most common types of op-amps, including voltage and current feedback. This differential-to-single-ended amplifier amplifies the voltage difference $(V_{in} = V_+ - V_-)$ at the input port and produces a voltage V_o at the output port that is referenced to the ground node of the circuit in which the op-amp is used.
3. Differential input, differential output op-amp, same as in Figure 4.1b but reproduced here as shown in Figure 4.2b. The outputs can be thought of as "inverting" and "non-inverting" and are shown across from the opposite polarity input for easy completion of feedback loops on schematics. The second output voltage is approximately equal and opposite in polarity to the other output voltage, each measured with respect to ground. When the two outputs are used as the output terminals without ground reference, they are known as "differential outputs" [2].

As noted earlier, a common practice is leaving out drawing the power supplies, $+V_c$ and $-V_c$ in the schematic diagram; in this text, however, we retain these power supplies in the circuit.

4.2.1 IDEAL OP-AMP CHARACTERISTICS

The equivalent electric circuit model of an ideal, single-ended op-amp is shown in Figure 4.3.

An ideal op-amp will have the following characteristic:

i. An infinite gain (i.e. $A = \infty$).
   This means that any output signal developed will be the result of an infinitesimally small input signal and the differential input voltage is zero. A consequence of this assumption is that, if the output voltage is within the finite linear region, we must have $V_- = V_+$. For most applications, we can get away by assuming $V_- \approx V_+$, because a real op-amp has an appreciable gain range (between 10^3 and 10^7; depending on the type), and will maintain a very small difference in input terminal voltages when operating in its linear region.

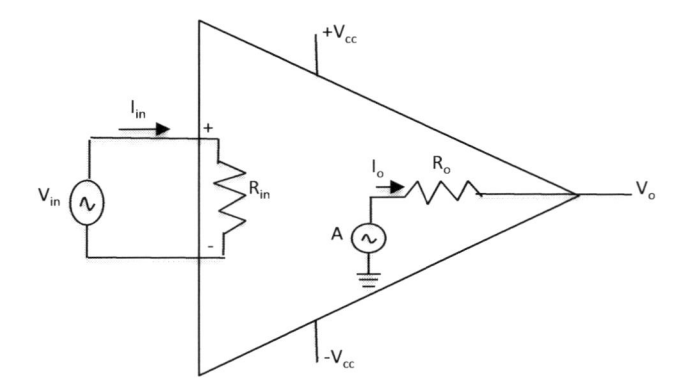

FIGURE 4.3   An electric circuit model of an op-amp with gain A, input resistance R_{in}, and output resistance R_o.

   ii. An infinite input resistance ($R_{in} = \infty$).

       This means that there is no current flow into either input terminal.

  iii. Zero output resistance ($R_o = 0$)

       As such, it can supply as much current as necessary to the load being driven.

  iv. An offset: the output will be zero when a zero signal appears between the inverting and non-inverting inputs.

   v. Zero response time.

       The output must occur at the same time as the inverting input so the response time is assumed to be zero. Phase shift will be 180°.

  vi. An infinite bandwidth ($BW = \infty$).

       This implies that the frequency response will be flat because an alternating current (ac) will be simply a rapidly varying direct current (dc) level to the amplifier.

The properties (i) and (ii) are the basics for op-amp circuit analysis and design.

4.2.1.1   Non-Inverting Amplifier

Figure 4.4a shows a basic, non-inverting amplifier. As noted earlier above, in the triangular symbol, the input terminal marked with a "+" (corresponding to V_+) is called the non-inverting input, while the input terminal marked with a "−" (corresponding to V_-) is called the inverting input. To ensure a non-inverted output signal, the input signal V_s is connected to the (+) input terminal of the amplifier. To understand how the non-inverting amplifier circuit works, we need to derive a relationship between the input voltage V_s and the output voltage V_o.

Using Figure 4.4b as a guide, for an ideal op-amp, there is no loading effect at the input terminals, so the voltages at (−) pin and (+) pin are the same; i.e.

$$V_+ = V_- \tag{4.5}$$

and the currents that flow through the (−) pin and (+) pin are zero; i.e.

$$I_+ = I_- = 0 \tag{4.6}$$

The input voltage source V_s is connected to the (+) pin, so

$$V_s = V_+ = V_- \tag{4.7}$$

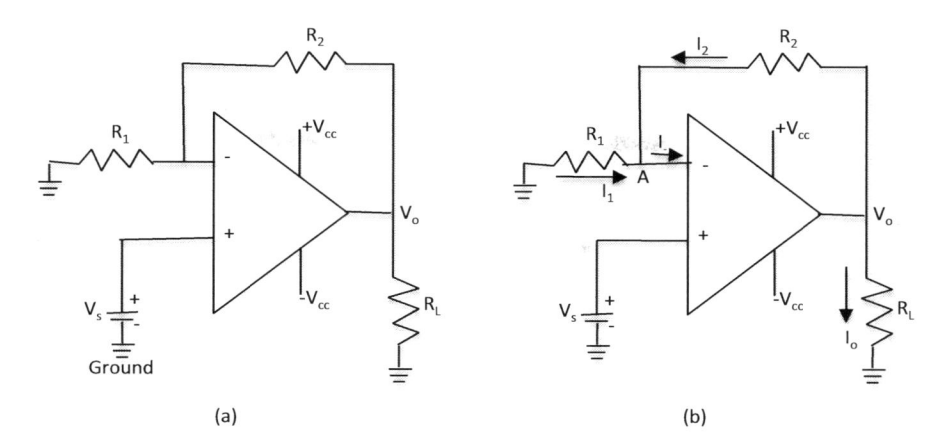

FIGURE 4.4   Analyzing a non-inverting op-amp: (a) non-inverting op-amp and (b) current and node notation.

Consider *node A* (of Figure 4.4b), and apply the Kirchhoff's current law (KCL), which states that *the sum of currents entering and leaving a node is zero*. We can write

$$I_1 + I_2 + I_- = 0 \tag{4.8}$$

which, in view of Equation (4.6), Equation (4.8) becomes

$$I_1 = -I_2 \tag{4.9}$$

From Figure 4.4b, we can write expressions for the following currents:

$$I_1 = \frac{0 - V_A}{R_1} \tag{4.10}$$

$$I_2 = \frac{V_o - V_A}{R_2} \tag{4.11}$$

Substitute Equations (4.10) and (4.11) in Equation (4.9), and rearrange terms:

$$V_A \left(\frac{1}{R_1} + \frac{1}{R_2} \right) = \frac{V_o}{R_2} \tag{4.12}$$

We note in Equation (4.7) that $V_+ = V_- = V_s = V_A$, so we rewrite Equation (4.12)

$$V_s \left(\frac{1}{R_1} + \frac{1}{R_2} \right) = \frac{V_o}{R_2} \tag{4.13}$$

Rearrange in terms of V_o,

$$V_o = R_2 \left(\frac{1}{R_1} + \frac{1}{R_2} \right) V_s \tag{4.14}$$

Rearranging in terms of *voltage gain*:

$$A_v = \frac{V_o}{V_s} = \frac{R_1 + R_2}{R_1} = 1 + \frac{R_2}{R_1} \tag{4.15}$$

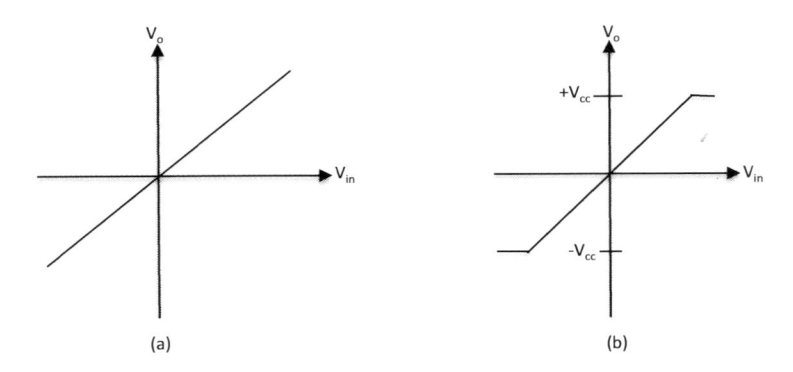

FIGURE 4.5 Voltage transfer function of non-inverting op-amp with gain, $A > 1$: (a) ideal situation and (b) realistic situation.

Equation (4.15) shows that the voltage gain is governed by external resistors.

Equivalently, if reactive elements of impedance Z are used, the op-amp gain can be expressed as

$$A = 1 + \frac{Z_2}{Z_1} \tag{4.16}$$

The voltage transfer curve of a non-inverting amplifier can be drawn as in Figure 4.5a for an ideal case. Figure 4.5b reflects a realistic case because the supply voltage $\pm V_c$ of the amplifier will clip the output voltage V_o.

When designing an op-amp, you need to know the expected amplifier gain. Let us say a gain of 10. From (4.16), we have

$$A = 10 = 1 + \frac{R_2}{R_1} \tag{4.17}$$

resulting $R_2 = 9R_1$.

It is now left to you to choose the value of input resistor to have. Rule of thumb is $10^3\,\Omega \leq R_{in} \leq 10^6\,\Omega$ to minimize resistor noise that creeps in.

4.2.1.1.1 Obtaining Other Amplifier Gains

To obtain other amplifier gains, namely *current gain*, *transresistance,* and *transconductance* expressed by Equations (4.2)–(4.4), respectively, we must first find the output current I_o. From Figure 4.4b,

$$I_o = \frac{V_o - 0}{R_L} \tag{4.18}$$

We already have expressions for the output voltage V_o {in Equation (4.14)} and input current I_{in} {which is the same thing as I_1 in Equation (4.10)}. As a result, we can write the following expressions for the other gains.

  a. Current gain

   Using Equations (4.10) and (4.18) we can write

$$A_I = \frac{I_o}{I_{in}} = \frac{I_o}{I_1} = \left(\frac{V_o}{R_L}\right)\left(\frac{R_1}{V_s}\right) = \frac{V_s R_2}{R_L}\left[\frac{1}{R_1} + \frac{1}{R_2}\right]\left(\frac{R_1}{V_s}\right)$$

$$= \frac{R_1 + R_2}{R_L} \tag{4.19}$$

b. Transconductance gain

Using Equations (4.7) and (4.11), we can write

$$A_{tc} = \frac{I_o}{V_s} = \frac{R_2}{R_L}\left(\frac{1}{R_1} + \frac{1}{R_2}\right) = \frac{R_1 + R_2}{R_1 R_L} \tag{4.20}$$

c. Transresistance gain

Using Equations (4.10) and (4.14), we can write

$$A_{tr} = \frac{V_o}{I_{in}} = \frac{V_o}{I_1} = R_1 R_2\left(\frac{1}{R_1} + \frac{1}{R_2}\right)\frac{V_s}{V_s}$$

$$= R_1 R_2\left(\frac{R_1 + R_2}{R_1 R_2}\right) = R_1 + R_2 \tag{4.21}$$

4.2.1.2 Buffer Amplifier

A buffer is a single-input device that has unity gain. Buffers are used as load isolators. We can make the differential op-amp of Figure 4.4a a *buffer* amplifier (as in Figure 4.6) by making the input resistance R_1 (or reactive elements of impedance Z_1) close to infinity and the output resistance R_2 (or reactive elements of impedance Z_2) very low: just a few ohms. As a result, the buffer gain becomes

$$A = 1 + \frac{R_2}{R_1} \cong 1 \tag{4.22}$$

Equivalently,

$$A = 1 + \frac{Z_2}{Z_1} \cong 1 \tag{4.23}$$

The buffer has an output that exactly mirrors the input. This arrangement is also called an *op-amp follower.*

In essence, buffers can be used to chain together sub-circuits in stages without worrying about impedance problems.

4.2.1.3 Inverting Amplifier

The easiest op-amp application to understand is the inverting amplifier, which is depicted in Figure 4.7a where the op-amp is powered with similar amplitude on the $+V_c$ rail and on the $-V_c$ rail,

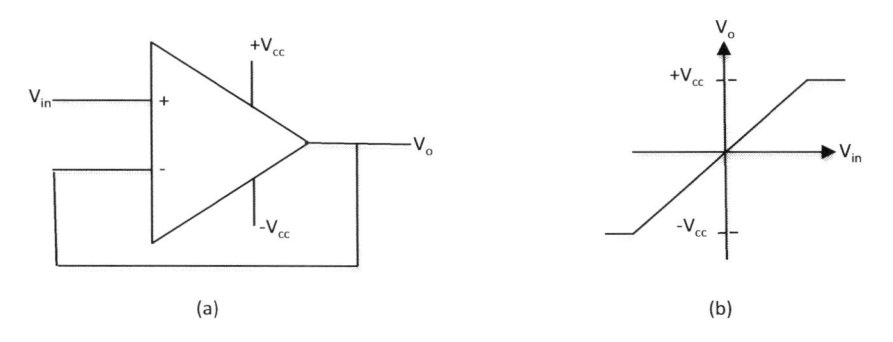

(a) (b)

FIGURE 4.6 Buffer amplifier and its voltage transfer function: (a) forming buffer op-amp from Figure 4.4a and (b) voltage transfer function, gain $A = 1$.

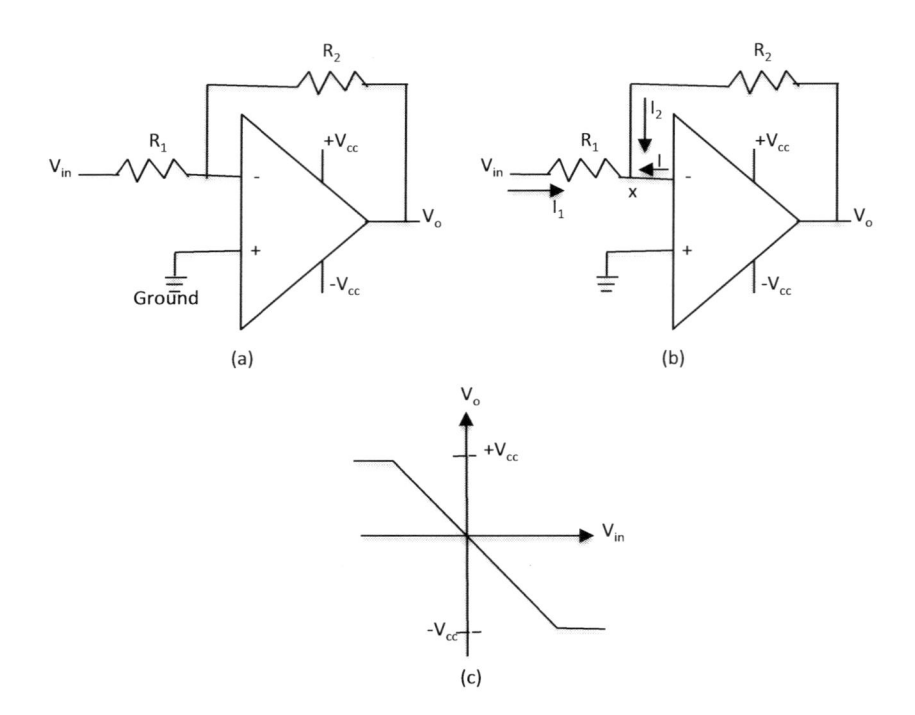

FIGURE 4.7 Inverting amplifier and its voltage transfer function: (a) inverting amplifier, (b) same as (a) with notations, and (c) voltage transfer function, gain $A < 1$.

and the input signal is connected to the (−) terminal of the triangular symbol, i.e. the inverting input. A portion of the output is fed back into the inverting input forming the basis of what we refer to as negative feedback loop, which trades off a reduced gain for clean amplification.

Like in the non-inverting amplifier discussion, we know that for an ideal op-amp there is no loading effect at the input terminals, so the voltages at (−) pin and (+) pin are the same and equals to zero since (+) pin is grounded; i.e.

$$V_+ = V_- = 0 \tag{4.24}$$

and the currents that flow through the (−) pin and (+) pin are zero; i.e.

$$I_+ = I_- = 0 \tag{4.25}$$

Consider *node x* (of Figure 4.7b), and apply the KCL, which states that *the sum of currents entering and leaving a node is zero*. We can write

$$I_1 + I_2 + I_- = 0 \tag{4.26}$$

which, in view of Equation (4.25), Equation (4.26) becomes

$$I_1 = -I_2 \tag{4.27}$$

From Figure 4.7b, we can write expressions for the following currents:

$$I_1 = \frac{V_{in} - V_D}{R_1} \tag{4.28a}$$

$$I_2 = \frac{V_D - V_o}{R_2} \tag{4.28b}$$

where $V_D = V_- = V_+ = 0$. Substitute Equation (4.28) in Equation (4.27), and rearrange terms,

$$\frac{V_{in}}{R_1} = -\frac{V_o}{R_2} \tag{4.29}$$

Equivalently

$$V_o = -\left(\frac{R_2}{R_1}\right)V_{in} \tag{4.30}$$

which has a negative value indicating that the output signal is inverted and a transfer function is depicted by Figure 4.7c. Consequently, the *voltage gain* is estimated as

$$A = -\frac{R_2}{R_1} < 1 \tag{4.31}$$

Let us have a feel of amplification, i.e. what the output signal will be when the input signal is made sinusoidal, for example as depicted in Figure 4.8, with resistive input and output values as indicated.

Using Equation (4.31), $A = -5$. So, the output signal will be sinusoidal, inverted but amplified by a factor of 5 as shown in Figure 4.9; frequency of 0.1 GHz was used for this simulation where time is in microseconds (μs).

All amplifiers have a maximum permissible voltage swing at the output. For instance, an amplified output signal V_o will be *clipped*:

i. if the power rails (i.e. power supply $+V_c$ and $-V_c$) of the amplifier are not at the same amplitude, i.e. $+V_c \neq -V_c$;
ii. if the power rails are set too close to the maximum and minimum of the output signal, i.e. $V_c \approx V_o$. As a result, we can bound the output voltage thus: $-V_c \leq V_o \leq +V_c$; or
iii. if maximum permissible voltage swing exceeds the power supply peaks of the signal, i.e. $V_{peak} > V_c$; a situation for case of Figure 4.10b.

In conclusion, an inverting amplifier is a useful circuit allowing us to scale a signal to any voltage range we desire by adjusting the gain accordingly.

In *digital signal processing* (DSP), clipping occurs when the range of a chosen representation restricts the signal. For example, suppose a system uses 8-bit integers, and the amplitude of the

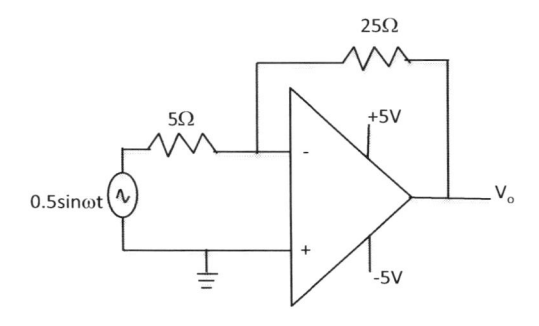

FIGURE 4.8 An inverting amplifier circuit.

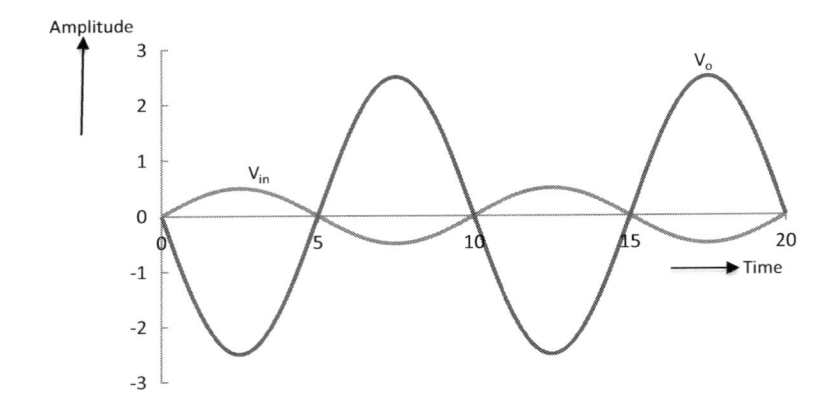

FIGURE 4.9   Input and output responses of an inverting amplifier.

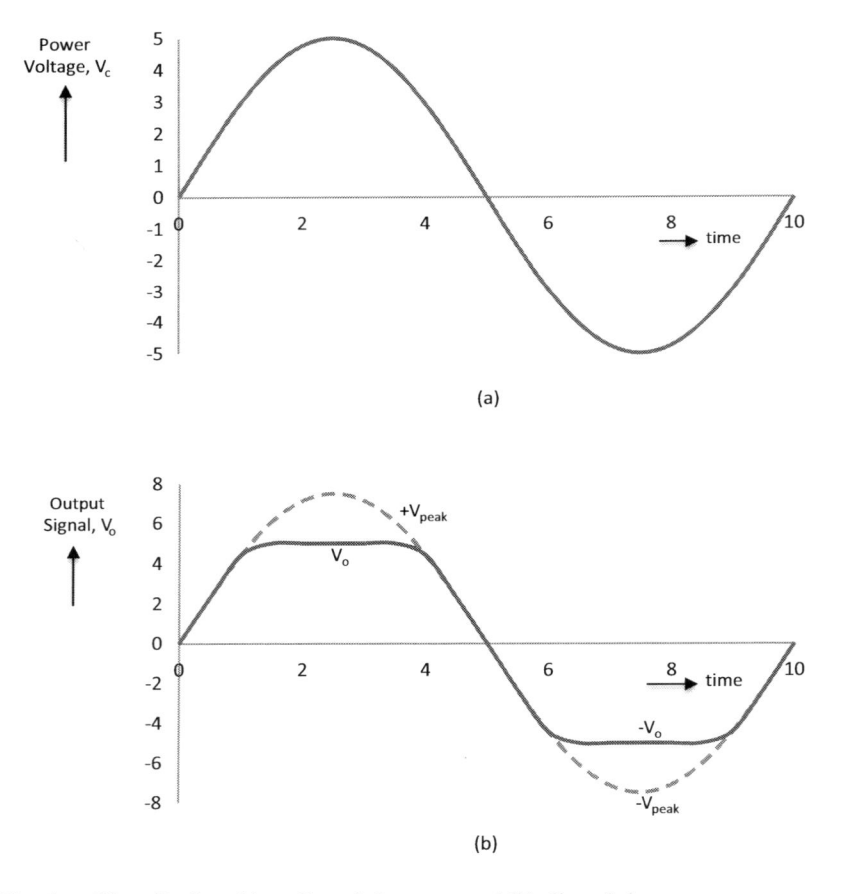

FIGURE 4.10   Amplifier clipping: (a) unclipped sine wave and (b) clipped sine wave.

sample signal is 120 bits. Fundamentally, the system will have the largest positive value of 127 (i.e. $2^7 - 1$) that can be represented. If during signal processing the amplitude of the signal doubled, the available sample values of 120 bits would become 240 bits. However, instead of the sample signals reaching 240 bits, they are truncated (clipped) to the system's maximum of 127. Clipping may be seen as a form of distortion that limits a signal once it exceeds a threshold, but it has been suggested as an effective peak-to-average power ratio reduction scheme for orthogonal frequency-division multiplexing (OFDM) system [3,4]. OFDM is a modulation technique used in many new and emerging broadband technologies including Internet, digital audio, and video broadcasting. It is a spectrally efficient method, which mitigates the severe problem of multipath propagation that causes massive data errors and loss of signal, particularly in the microwave and ultra-high frequency (UHF) spectrum. OFDM is not discussed in this book.

4.2.2 Examples of Inverting Amplifier Circuit Analyses

The following amplifier circuits would show the readers how basic analyses are done.

Example 4.1

Calculate the gain, output voltage, and power of the amplifier depicted by Figure 4.11a.

Solution

To facilitate application of current and voltage laws, and solve the problem, Figure 4.11a is relabeled and annotated as in Figure 4.11b. The assumptions made prior to solving this problem are (i) the amplifier operates within the linear range, i.e. not in saturation state; and the amplifier is *ideal*, i.e. the input resistances at (−) and (+) input pins are of infinite ohmic values (i.e. $R_{-/+} = \infty$).

First, observe if there is a negative feedback. Yes, there is; one arm is fed from the output to the negative pin of the amplifier through R_2 resistor. So, we can write one expression:

$V_- = V_+$ meaning the voltage at node A (V_A) is the same as that at node B (V_B).

$$V_A = V_B \tag{4.32}$$

Second, applying KCL to nodes A and B, we are able to derive the following equations. KCL states that *the sum of currents entering a node must equal those leaving the node*. Readers wanting to know more about KCL and others and their applications to electric circuits should consult [5] amongst other books. Note that the directions of currents I_i (where $i = -, +, o, 1, ..., 5$) are taken arbitrarily. Applying KCL to node A:

$$I_1 = I_- + I_2 \tag{4.33}$$

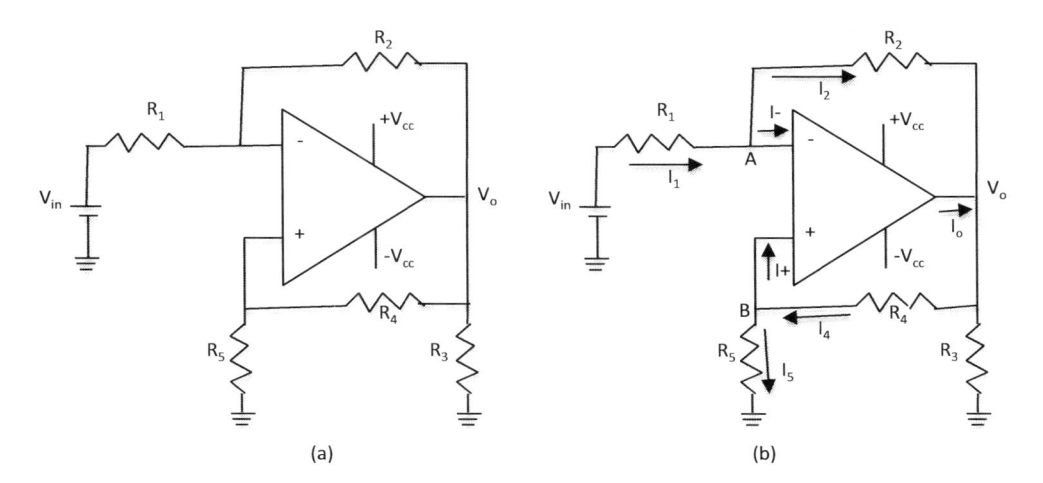

(a) (b)

FIGURE 4.11 An inverting amplifier circuit: (a) and (b) same as (a) with notations.

From the above assumptions, $I_- = I_+ = 0$, and the voltages at $(-)$ and $(+)$ input pins of the amplifier must be equal; i.e. $V_- = V_+$. So, the voltage at nodes A and B must be equal. Then write

$$V_A = V_B \tag{4.34}$$

Equation (4.33) can be rewritten in terms of voltages as

$$\frac{V_{in} - V_A}{R_1} = 0 + \frac{V_A - V_o}{R_2} \tag{4.35}$$

Applying KCL to node B:

$$I_4 = I_+ + I_5 \tag{4.36}$$

$$\frac{V_o - V_B}{R_4} = 0 + \frac{V_B}{R_5} \tag{4.37}$$

Third, write expression for the output current I_o:

$$I_o = I_2 + I_4 + I_3 = \frac{V_o - V_A}{R_2} + \frac{V_o - V_B}{R_4} + \frac{V_o}{R_3} \tag{4.38}$$

Since $V_A = V_B$, Equation (4.38) becomes

$$I_o = V_o \left[\frac{1}{R_2} + \frac{1}{R_3} + \frac{1}{R_4} \right] - V_A \left[\frac{1}{R_2} + \frac{1}{R_4} \right] \tag{4.39}$$

Rearranging Equation (4.35) in terms of V_o we write

$$V_o = V_A \left[1 + \frac{R_2}{R_1} \right] - V_{in} \left[\frac{R_2}{R_1} \right] \tag{4.40}$$

Similarly, for Equation (4.37)

$$V_o = V_B \left[1 + \frac{R_4}{R_5} \right] \tag{4.41}$$

Equating Equation (4.36) with Equation (4.37), and since $V_A = V_B$, we write V_A in terms of input voltage V_{in}:

$$V_A = V_{in} \left\{ \frac{\left(R_2 / R_1 \right)}{\left[\left(\dfrac{R_2}{R_1} - \dfrac{R_4}{R_5} \right) \right]} \right\} \tag{4.42}$$

Substitute Equation (4.40) in Equation (4.39) and the value of the amplifier's output voltage:

$$V_o = V_{in} \left\{ \frac{\left(R_2 / R_1 \right)}{\left[\left(\dfrac{R_2}{R_1} - \dfrac{R_4}{R_5} \right) \right]} \right\} \left[1 + \frac{R_4}{R_5} \right] \tag{4.43}$$

From Equation (4.43), the gain A of this amplifier is expressed thus

$$A = \frac{V_o}{V_{in}} = \left\{ \frac{\left(R_2 / R_1 \right)}{\left[\left(\dfrac{R_2}{R_1} - \dfrac{R_4}{R_5} \right) \right]} \right\} \left[1 + \frac{R_4}{R_5} \right] \tag{4.44}$$

Having obtained expressions for V_o (from Equation (4.43)), V_A (from Equation (4.42)), and I_o (from Equation (4.39)); and using Ohms law, the output power of the amplifier can be estimated, thus

$$P = I_o V_o \qquad (4.45)$$

Example 4.2

Consider an amplifier circuit depicted in Figure 4.12a where the non-inverting pin is grounded. Estimate the output voltage, current, and power of the amplifier for the following component values: the chip is powered by ±10 V, the signal input voltage is 7.5 V, the load resistor $R_l = 1$ kΩ, and other resistors: $R_1 = 5$ kΩ, $R_2 = 15$ kΩ, $R_3 = 10$ kΩ, and $R_4 = 2.5$ kΩ.

Solution

Use Figure 4.12b as a guide.

First: there is a negative feedback loop. So, $V_- = V_+$ and $I_- = I_+ = 0$. Given that V_- is grounded—meaning $V_+ = 0$; therefore, the voltage at *node A*

$$V_A = V_+ = 0 \qquad (4.46)$$

Second: applying KCL to node A:

$$I_1 = I_- + I_2 \qquad (4.47)$$

which is

$$\frac{V_{in} - V_A}{R_1} = 0 + \frac{V_A - V_B}{R_2} \qquad (4.48)$$

In view of Equation (4.46),

$$\frac{V_{in}}{R_1} = -\frac{V_B}{R_2}$$

Solving for V_B to have

$$V_B = -\frac{R_2}{R_1} V_{in}$$

$$= -\frac{15 \times 7.5}{5} = -22.5 \text{ V} \qquad (4.49)$$

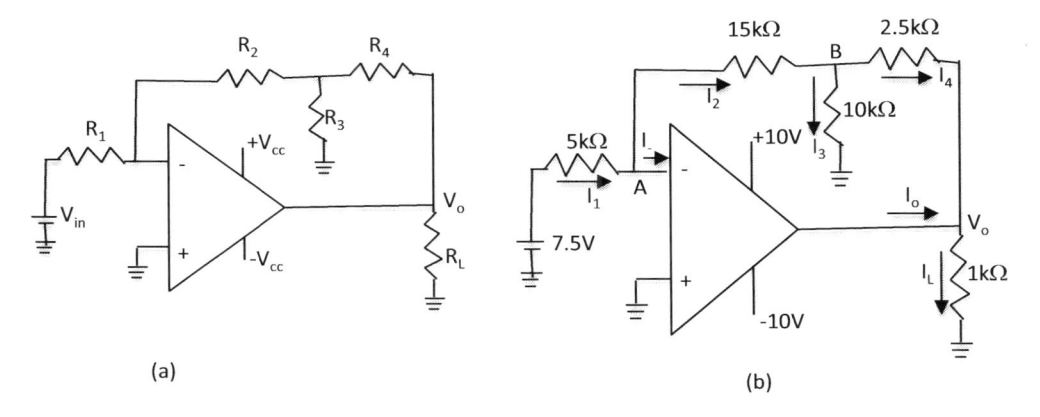

(a)

(b)

FIGURE 4.12 Another amplifier circuit arrangement: (a) and (b) same as (a) with notations and values.

Consider *node B* and apply KCL:

$$I_2 = I_3 + I_4 \tag{4.50}$$

which is

$$\frac{V_A - V_B}{R_2} = \frac{V_B}{R_3} + \frac{V_B - V_o}{R_4} \tag{4.51}$$

Expanding Equation (4.51) and solving

$$V_o = V_B \left[1 + \frac{R_4}{R_2} + \frac{R_4}{R_3} \right]$$

$$= -22.5 \left[1 + \frac{2.5}{15} + \frac{2.5}{10} \right]$$

$$= -31.875\,\text{V} \tag{4.52}$$

The output current I_o is

$$I_o = I_L - I_4$$

which is

$$I_o = \frac{V_o}{R_L} - \left(\frac{V_B - V_o}{R_4} \right) = V_o \left(\frac{1}{R_L} + \frac{1}{R_4} \right) - \frac{V_B}{R_4} = -35.625\,\text{mA} \tag{4.53}$$

The output power of the amplifier is

$$P_o = I_o V_o = \left(-35.625 \times 10^{-3} \right)(-31.875) = 1.1355\,\text{W} \tag{4.54}$$

4.2.3 VOLTAGE AND POWER AMPLIFIERS

Generally, amplifiers can be divided into two distinct types depending upon their power or voltage gain. Small signal amplifiers are generally referred to as *voltage amplifiers* because they convert a small input voltage into a much larger output voltage. Voltage amplifiers find applications as pre-amplifiers, instrumentation amplifiers, etc., which are designed to amplify very small signal voltage levels of only a few microvolts (μV) from sensors or audio signals.

Large signal amplifiers are generally referred to as power amplifiers because they amplify large input voltage signals. This type of amplifiers finds application where high switching currents are needed such as audio power amplifiers or switching amplifiers. Radio frequency (RF) power amplifiers are normally characterized in device physics by cutoff frequency, breakdown voltage, thermal conductivity, integration level, size, and cost [6].

4.2.3.1 Operation Circuit

An operation circuit combines the non-inverting and inverting amplifier, an example is that shown in Figure 4.13. We can derive the relationship between the input voltages (V_{1i} and V_{2i}, where $i = 0$, 1, 2, 3, 4) and the output voltage V_o by considering each triangular terminal at a time.

First, observe if there is a negative feedback. Yes, there is; one arm is fed from the output to the negative pin of the amplifier through R_o resistor. So, we can write one expression:

$$V_- = V_+ \tag{4.55}$$

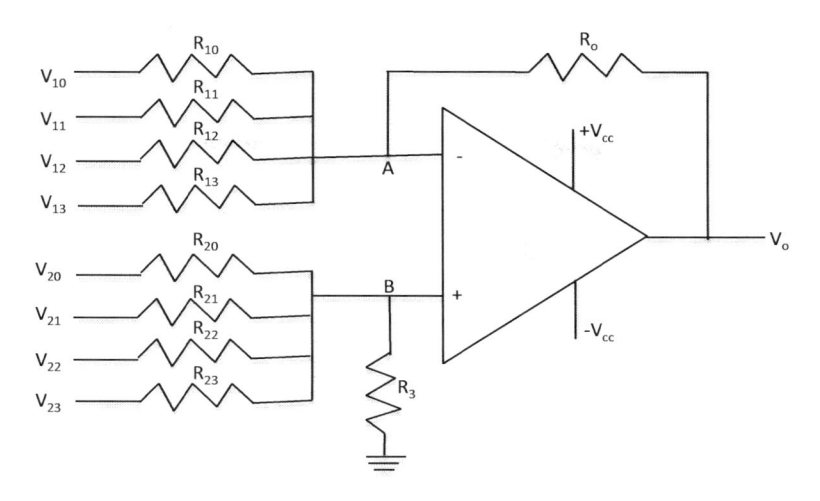

FIGURE 4.13 An operation circuit.

meaning the voltage at node A (V_A) is the same as that at node B (V_B).

$$V_A = V_B \tag{4.56}$$

Given an infinite resistance assumption at $(-)$ and $(+)$ input pins of the amplifier, the currents flowing into the pins are very negligible, i.e. $I_- = I_+ = 0$.

Application of KCL to each node, we obtain the following.

For the inverting node A:

$$\frac{V_{10} - V_A}{R_{10}} + \frac{V_{11} - V_A}{R_{11}} + \frac{V_{12} - V_A}{R_{12}} + \frac{V_{13} - V_A}{R_{13}} = \frac{V_A - V_o}{R_o} \tag{4.57}$$

Expand and collect terms:

$$V_o = V_A \left\{ R_o \left(\frac{1}{R_o} + \frac{1}{R_{10}} + \frac{1}{R_{11}} + \frac{1}{R_{12}} + \frac{1}{R_{13}} \right) \right\} - R_o \left(\frac{V_{10}}{R_{10}} + \frac{V_{11}}{R_{11}} + \frac{V_{12}}{R_{12}} + \frac{V_{13}}{R_{13}} \right) \tag{4.58}$$

For the non-inverting node B:

$$\frac{V_{20} - V_B}{R_{20}} + \frac{V_{21} - V_B}{R_{21}} + \frac{V_{22} - V_B}{R_{22}} + \frac{V_{23} - V_B}{R_{23}} = \frac{V_B}{R_3} \tag{4.59}$$

Expand and collect terms:

$$V_B = \frac{\left(\dfrac{V_{20}}{R_{20}} + \dfrac{V_{21}}{R_{21}} + \dfrac{V_{22}}{R_{22}} + \dfrac{V_{23}}{R_{23}} \right)}{\left[\dfrac{1}{R_3} + \dfrac{1}{R_{20}} + \dfrac{1}{R_{21}} + \dfrac{1}{R_{22}} + \dfrac{1}{R_{23}} \right]} \tag{4.60}$$

By abbreviating

$$R_p = \left[\frac{1}{R_3} + \frac{1}{R_{20}} + \frac{1}{R_{21}} + \frac{1}{R_{22}} + \frac{1}{R_{23}} \right] \tag{4.61}$$

$$R_n = \left(\frac{1}{R_o} + \frac{1}{R_{10}} + \frac{1}{R_{11}} + \frac{1}{R_{12}} + \frac{1}{R_{13}} \right) \tag{4.62}$$

We know from Equation (4.56) that $V_A = V_B$ and by substituting Equations (4.60)–(4.62) in Equation (4.58), we write the output voltage as

$$V_o = \left\{ \left(\frac{R_o R_n}{R_p} \right) \left(\frac{V_{20}}{R_{20}} + \frac{V_{21}}{R_{21}} + \frac{V_{22}}{R_{22}} + \frac{V_{23}}{R_{23}} \right) \right\} - R_o \left(\frac{V_{10}}{R_{10}} + \frac{V_{11}}{R_{11}} + \frac{V_{12}}{R_{12}} + \frac{V_{13}}{R_{13}} \right) \tag{4.63}$$

Hence, this operation circuit adds the source voltages of the non-inverting terminal (i.e. V_{20}, V_{21}, V_{22}, and V_{23}) and subtracts the source voltages of the inverting terminal (i.e. V_{10}, V_{11}, V_{12}, and V_{13}). Different coefficients can be applied to the input signals by adjusting the resistors. If all the resistors have the same value, then the output voltage is just the difference between the source voltages; i.e.

$$V_o = \left(V_{20} + V_{21} + V_{22} + V_{23} \right) - \left(V_{10} + V_{11} + V_{12} + V_{13} \right) \tag{4.64}$$

4.2.4 Summing and Averaging Amplifiers

The operational circuit in Figure 4.13 can be used to perform two other functions if the non-inverting terminal is grounded or earthed as shown in Figure 4.14.

This amplifier's operation can be "summed up" by

$$I_f + \sum_{k=1}^{n} I_k = 0 \tag{4.65}$$

indicating that the input and feedback currents must add up to zero, i.e. no current enters the op-amp's input terminals. We can rewrite Equation (4.65) as

$$\frac{V_o}{R_f} = -\left(\frac{V_1}{R_1} + \frac{V_2}{R_2} + \frac{V_3}{R_3} + \frac{V_4}{R_4} + \ldots + \frac{V_n}{R_n} \right) \tag{4.66}$$

In terms of output voltage, we rearrange

$$V_o = -\left(V_1 \frac{R_f}{R_1} + V_2 \frac{R_f}{R_2} + V_3 \frac{R_f}{R_3} + V_4 \frac{R_f}{R_4} + \cdots + V_n \frac{R_f}{R_n} \right) \tag{4.67}$$

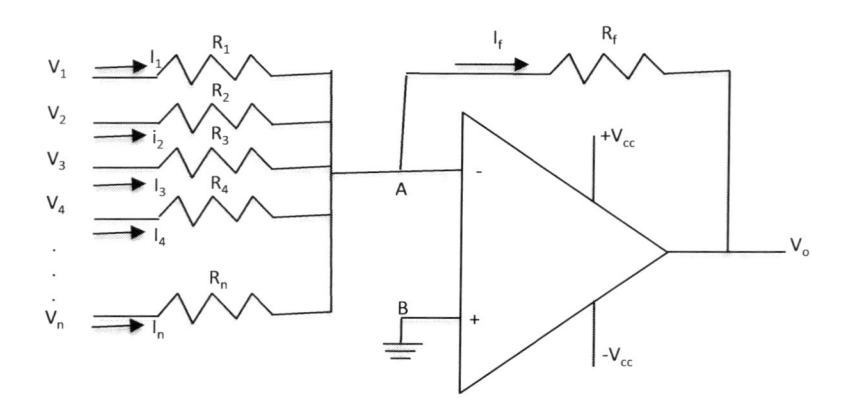

FIGURE 4.14 Summing amplifier and averager.

This expression, Equation (4.67), has the form of a series of "n" inverting amplifiers, whose outputs are summed together.

Figure 4.14 can also be used as an average, if $R_1 = R_2 = R_3 = R_4 = \cdots = R_n$ and $R_f = R_1/n$.

4.2.5 Comparator Amplifier and Schmitt Trigger

A comparator compares two analog signals and produces a 1-bit digital signal. For example, consider a simple comparator in Figure 4.15a, which behaves as follows:

- When V_+ is larger than V_- (i.e. $V_+ > V_-$), the output bit V_o is 1.
- When V_+ is smaller than V_- (i.e. $V_+ < V_-$), the output bit V_o is 0.

Consider the comparator circuit in Figure 4.15b where the comparator is used to compare the input to a standard voltage and switch the output if the input is above the threshold. The threshold V_2 can be obtained using voltage divider technique, as

$$V_2 = V_{\text{ref}} \left(\frac{R_2}{R_1 + R_2} \right) \tag{4.68}$$

If V_{in} goes more positive than the voltage set by V_2, then the output switches to $-V_{cc}$. However, when V_{in} drops below that value of V_2, the output switches back to $+V_{cc}$. The comparator output is shown in Figure 4.15c.

In essence, the basic comparator simply "compares" the input against a threshold and delivers a binary output that indicates whether the input is above or below the threshold. Primarily swinging its output to $\pm V_{cc}$ at a slightest difference between its inputs.

If the output is fed back via a resistor, say R_3, and connected to the R_2 resistor, as shown in Figure 4.16, a new comparator application called the *Schmitt trigger* is formed.

The threshold V_2, in Figure 4.16 of Schmitt trigger, differs from the ordinary comparator of Figure 4.15. In this instance, this threshold is expressed as

$$V_2 = V_{\text{ref}} \left(\frac{R_{123}}{R_1} \right) + V_o \left(\frac{R_{123}}{R_3} \right) \tag{4.69}$$

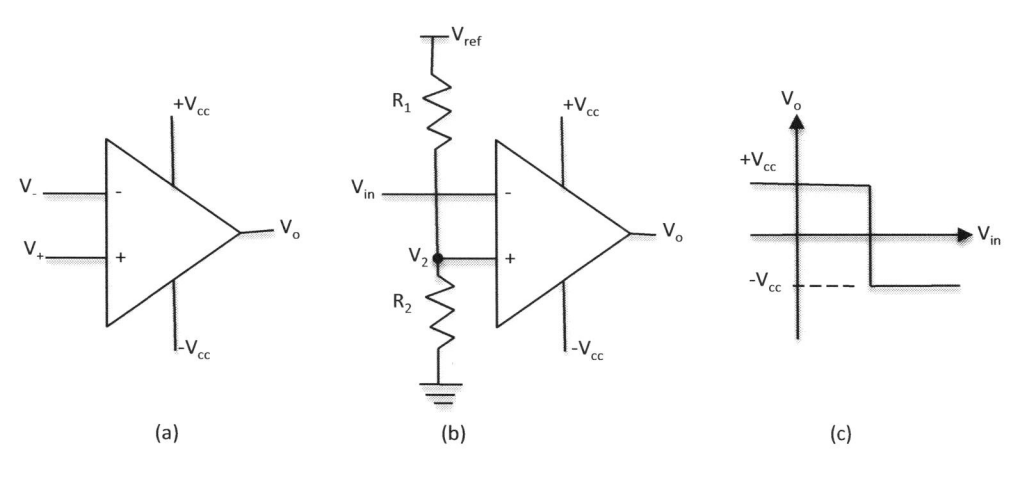

(a) (b) (c)

FIGURE 4.15 A simple comparator: (a) an op-amp comparator, (b) comparator circuit, and (c) comparator output.

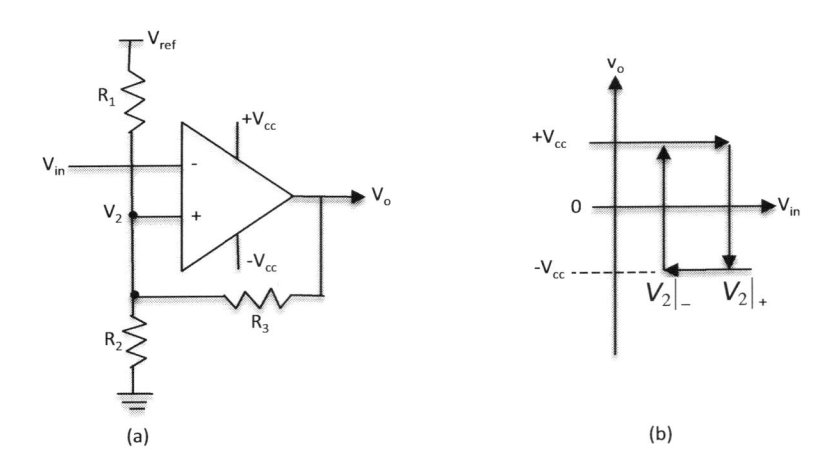

FIGURE 4.16 Schmitt trigger circuit: (a) inverting comparator with positive feedback and (b) output response.

where

$$R_{123} = R_1 \parallel R_2 \parallel R_3 = \frac{R_1 R_2 R_3}{R_1 + R_2 + R_3} \tag{4.70}$$

Because of the feedback, the output voltage would alternate between positive and negative V_{cc}, so, there would be dual thresholds or two stable states, i.e.

$$V_2\big|_+ = V_{ref}\left(\frac{R_{123}}{R_1}\right) + V_o\left(\frac{R_{123}}{R_3}\right) \tag{4.71}$$

$$V_2\big|_- = V_{ref}\left(\frac{R_{123}}{R_1}\right) - V_o\left(\frac{R_{123}}{R_3}\right) \tag{4.72}$$

As a consequence, if V_{in} is less than $V_2\big|_-$, the output voltage is $+V_{cc}$. However, if V_{in} is greater than $V_2\big|_+$, the output voltage is $-V_{cc}$. For $V_2\big|_- < V_{in} < V_2\big|_+$, the output voltage V_o can have two stable states, i.e. $V_o = \pm V_{cc}$. The effect of the second threshold voltage $V_2\big|_-$ is to prevent rapid cycling at the switching point, i.e. stabilizing the switching against rapid triggering by noise as it passes the trigger point. The graph of V_o versus V_{in} is given in Figure 4.16b. The path for the output voltage V_o on the graph in Figure 4.16b is indicated with arrows. The loop in the graph is commonly called *a hysteresis loop*.

Given that Figure 4.16 circuit has stable output states: positive saturation or negative saturation, it can be considered a *bistable multivibrator*. In essence, if a trigger of the correct polarity and amplitude is applied to the circuit, it will change states and remain there until triggered again. The trigger need not have a fixed pulse repetitive frequency; in fact, triggers from different sources, occurring at different times, can be used to switch this circuit.

Multivibrator is nothing but a two-stage amplifier, operating in two modes. The modes are called states of the multivibrator. Basically, the output of the first stage is fed to the input of the second stage while the output of the second stage is fed back to the input of the first stage. These input signals drive the active device of one stage to saturation while the other is cut off. The new set of signals, generating exactly opposite effects, follows. As a result, the cutoff stage now saturates while the saturated stage becomes cutoff. Primarily, there are three types of multivibrator circuits, namely (a) *bistable multivibrator*, which we have just discussed, (b) *astable multivibrator*, and (c) *monostable multivibrator*. The *astable* and *monostable* types have been discussed in Chapter 2.

It is appropriate to review the concept of transfer function, which we will use sparingly in the remainder of this chapter, as well as some aspects of this book.

4.2.6 TRANSFER FUNCTION: A REVIEW

The word "transfer function" becomes appropriate if the relationship between the input and output is expressed in terms of spatial or temporal frequency; for example, if the input and output signals are continuous-time variants, i.e. $v_{in}(t)$ and $v_o(t)$. For instance, the Laplace transform converts functions with a real dependent variable (such as time) into functions with a complex dependent variable (such as frequency, often represented by s). In its simplest form, the transfer function is the linear mapping of the Laplace transform of the input, $V_{in}(s)$, to the output $V_o(s)$; that is, taking the Laplace transforms of both input and output assuming zero conditions, and solve for the ratio of the output Laplace over the input Laplace. In this regard, the *transfer function* can be expressed as

$$H(s) = \frac{V_o(s)}{V_{in}(s)} = \frac{L\{v_o(t)\}}{L\{v_{in}(t)\}} \tag{4.73}$$

where $L\{.\}$ is the *Laplace transform* of $\{.\}$ and $H(s)$, or $G(s)$ in some books, is the transfer function of the *linear, time-invariant* (LTI) system.

In discrete-time systems, the transfer function is referred to as the *pulse-transfer function*, which, like Equation (4.73), is written in Z-transform instead of Laplace as

$$H(z) = \frac{V_o(z)}{V_{in}(z)} = \frac{Z\{v_o(t)\}}{Z\{v_{in}(t)\}} \tag{4.74}$$

The z-transform is the discrete-time counterpart of the Laplace transform. Like Laplace transform the z-transform allows insight into the transient behavior, the steady state behavior, and the stability of discrete-time systems. A working knowledge of the z-transform is essential to the study of digital filters and systems. This chapter does not attempt to delve deeply into the intricacies and application of z-transform beyond giving the reader a brief explanation of the relationship between Laplace and z-transform from the Laplace transform of a discrete-time signal. By definition, the Laplace transform $X(s)$, of a continuous-time signal $x(t)$, is given by the integral

$$X(s) = \int_0^\infty x(t)e^{-st}\, dt \tag{4.75}$$

where the complex variable $s = \sigma + j\omega$, and the lower limit of $t = 0$ allows the possibility that the signal $x(t)$ may include an impulse. If the continuous-time input signal $x(t)$ can be sampled as $x(m)$, assuming the sampling period *of unity* (i.e. sampling period $T_s = 1$), then the Laplace transform of Equation (4.75) becomes

$$X(e^s) = \sum_{m=0}^\infty x(m)e^{-sm} \tag{4.76}$$

If we substitute the variable e^s in Equation (4.76) with the variable z we obtain the one-sided z-transform equation

$$X(z) = \sum_{m=0}^\infty x(m)z^{-m} \tag{4.77}$$

Of course, we can write equivalent expression for two-sided signal (i.e. $-\infty \leq m \leq \infty$) as

$$X(z) = \sum_{m=-\infty}^{\infty} x(m)z^{-m} \tag{4.78}$$

In essence, transfer function is frequently used to refer to LTI systems. In practice, most real systems have non-linear input/output characteristics, but when systems operate within nominal, non-saturated parameters, they have the behavior close to that of LTI systems.

Example 4.3

Express the transfer function of a two-stage, inverting op-amps described by Figure 4.17a.

Consider "Stage 1" of the op-amp involving *nodes A* and *B*. By considering the amplifiers as ideal, the current I_- entering the (−) port is zero, (i.e. $I_- = 0$), and the voltages at (+) port and (−) port (i.e. at node A (V_A)) to be the same. We then write

$$V_+ = V_- = V_A \tag{4.79}$$

Since the (+) port of the amplifier is grounded, then $V_- = 0$. So,

$$V_+ = V_- = V_A = 0 \tag{4.80}$$

Apply KCL to node A:

$$I_1 = I_2 \left(\text{since } I_- = 0\right) \tag{4.81}$$

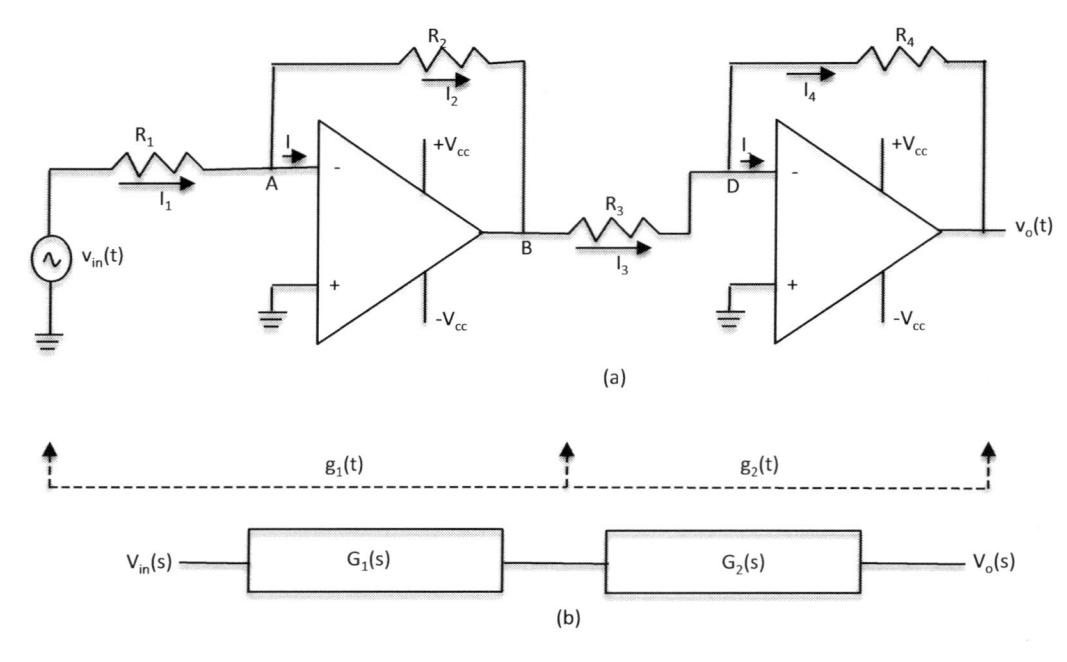

(a)

(b)

FIGURE 4.17 Analysis of two-stage op-amps circuit: (a) a two-stage op-amps system and (b) transfer function of a two-stage system.

where

$$I_1 = \frac{V_{in} - V_A}{R_1} \tag{4.82}$$

$$I_2 = \frac{V_A - V_B}{R_2} \tag{4.83}$$

The gain of "Stage 1" can be expressed, from Equation (4.83), as

$$g_1(t) = \frac{I_2}{I_1} = \frac{V_A - V_B}{V_{in} - V_A} = -\frac{R_2}{R_1} \tag{4.84}$$

Consider "Stage 2": the output voltage of "Stage 1" becomes the input voltage of "Stage 2." Again, the current I_- entering the (−) port is zero (i.e. $I_- = 0$), and the voltages at (+) port and (−) port (i.e. at node D (V_D)) to be the same. We then write

$$V_+ = V_- = V_D = 0 \tag{4.85}$$

Apply KCL to node D:

$$I_3 = I_4 \,(\text{since } I_- = 0) \tag{4.86}$$

where

$$I_3 = \frac{V_B - V_D}{R_3} \tag{4.87}$$

$$I_4 = \frac{V_D - V_o}{R_4} \tag{4.88}$$

The gain of stage 2 can be expressed from Equations (4.82) and (4.83) as

$$g_2(t) = \frac{I_4}{I_3} = \frac{V_D - V_o}{V_B - V_D} = -\frac{R_4}{R_3} \tag{4.89}$$

The overall gain of the two-stage amplifier is the product of the two stages' gains, i.e.

$$A = \frac{v_o(t)}{v_{in}(t)} = g_1(t)g_2(t) = \left(\frac{R_2}{R_1}\right)\left(\frac{R_4}{R_3}\right) \tag{4.90}$$

The above two-stage op-amp diagram, Figure 4.17a, can be modeled as a two-stage system, as shown in Figure 4.17b, in the time domain. Some analyses are easier to perform in the frequency domain. In order to convert time to the frequency domain, apply the Laplace transform to determine the transfer function of the system. In this instance, the Laplace transform converts time functions of each stage into functions with a complex dependent variable (such as frequency, represented by s). Consequently, the transfer function is the ratio of the output Laplace transform to the input Laplace transform assuming zero initial conditions. From Figure 4.17b and Equation (4.90), the two-stage op-amp can be written as

$$H(s) = \frac{V_o(s)}{V_{in}(s)} = G_1(s)G_2(s) = \left(\frac{R_2}{R_1}\right)\left(\frac{R_4}{R_3}\right) \tag{4.91}$$

Many important characteristics of dynamic or control systems are determined from their *transfer function*.

4.2.7 Differentiator Amplifier

The differentiator generates an output signal proportional to the first derivative of the input with respect to time. An ideal differentiator circuit is shown in Figure 4.18a. At node A, in time domain, $v_- = v_+ = 0$, and $i_R + i_C = 0$, we can write

$$\underbrace{\frac{v_o(t)}{R}}_{i_R} + \underbrace{C\frac{dv_i(t)}{dt}}_{i_C} = 0 \tag{4.92}$$

Rearranging in terms of output voltage

$$v_o(t) = -RC\frac{dv_i(t)}{dt} = -\tau\frac{dv_i(t)}{dt} \tag{4.93}$$

where τ is time constant (in seconds). Integrating each term with respect to time and solving for the output voltage:

$$v_o(t) = -RC\frac{dv_i(t)}{dt} \tag{4.94}$$

The ideal differentiator is inherently unstable in the presence of high frequency noise because the differentiator would amplify the noise no matter small noise that might present itself. To have a feel for this problem, suppose a small noise with response given by $v_{noise}(t) = \delta \sin(2\pi ft)$, where δ is

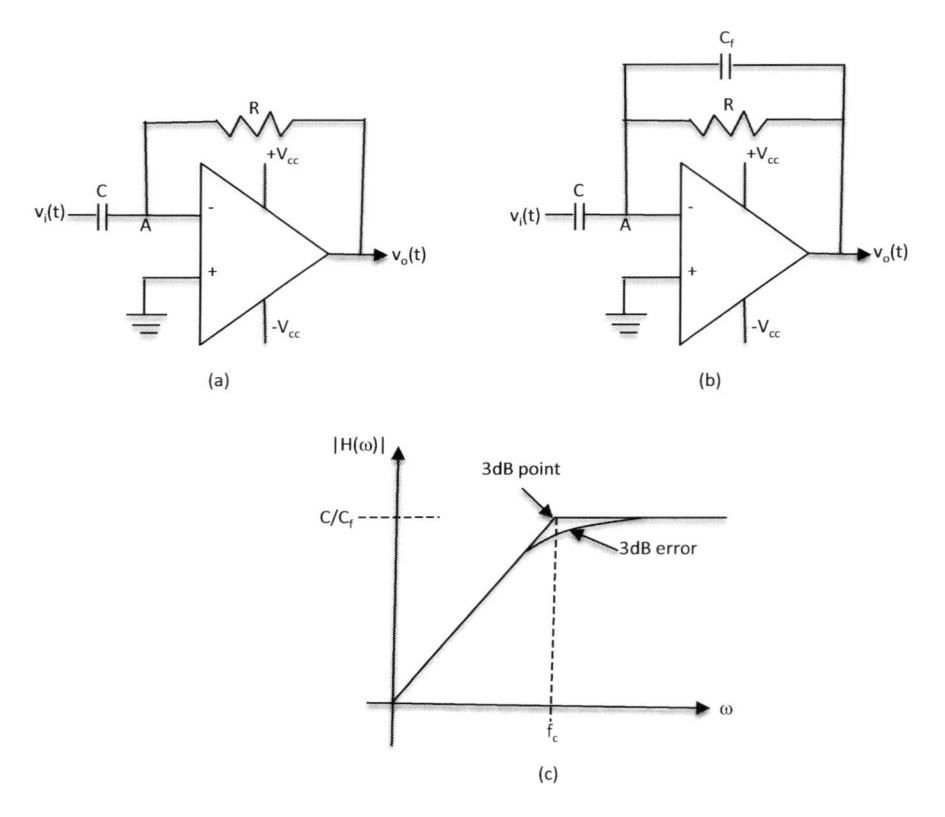

(a)   (b)

(c)

FIGURE 4.18 Differentiator: (a) ideal differentiator, (b) modified differentiator, and (c) Bode plot of the transfer function of (b).

amplitude. The output would be $v_o(t) = 2\pi f \delta \cos(2\pi ft)$. Let us put some values to these variables, say $\delta = 1.0\,\mu V$, and $f = 10\,MHz$. Substituting these values in the output voltage expression, we have

$$v_o(t) = 2\pi f \delta \cos(2\pi ft) = 2 \times \pi \times 10^7 \times 10^{-6} \cos(2\pi ft) = 62.83 \cos(2\pi ft) \tag{4.95}$$

This means that the output voltage amplitude would magnify to 62.83 V from the noise small magnitude of 1.0 μV. To circumvent this problem, a parallel capacitor C_f is inserted across the feedback resistor as shown in Figure 4.18b. In fact, the introduction of the capacitor compromises the internal compensation of the op-amp. Placing a capacitor C_f in the feedback network restores the compensation, making the overall circuit stable. The choice of C_f is not trivial if we want to preserve the differentiator circuit characteristics.

Figure 4.18b, in essence, is an active first-order, *high-pass filter* (HPF). Its transfer function can be computed as

$$H(s) = \frac{V_o(s)}{V_i(s)} = -\frac{sCR}{sRC_f + 1} \tag{4.96}$$

with the filter having a high frequency gain of $\left(\dfrac{C}{C_f}\right)$, and $s = j\omega$. The Bode magnitude plot of Equation (4.96) is shown in Figure 4.18c, and f_c is the cutoff frequency, $\left(\text{i.e. } f_c = \dfrac{1}{2\pi RC_f}\right)$. More is said about the transfer function in the next section. Obviously, a constant input (regardless of its magnitude) generates a zero-output signal.

A good use for an op-amp differentiator is in the field of chemical instrumentation in obtaining the first derivative of a potentiometric titration curve for the easier location of the titration final points (i.e. points of maximum slope).

4.2.8 INTEGRATOR AMPLIFIER

By swapping the resistor R with capacitor C in the differentiator, as shown in Figure 4.19a, an integrator is obtained. In this configuration, the integrator generates an output signal proportional to the time integral of the input signal. The output is proportional to the charge accumulated in capacitor C, which serves as the integrating device.

From Figure 4.19a and at node B, in time domain, $v_- = v_+ = 0$, and $i_R + i_C = 0$, we can write

$$\underbrace{\frac{v_i(t)}{R}}_{i_R} + \underbrace{C\frac{dv_o(t)}{dt}}_{i_C} = 0 \tag{4.97}$$

Rearranging in terms of output voltage:

$$dv_o(t) = \frac{1}{CR}v_i(t)dt \tag{4.98}$$

Integrating each term with respect to time, the exact output voltage level is given by

$$v_o(t) = \frac{1}{CR}\int_{t_o}^{t} v_i(\tau)d\tau \tag{4.99}$$

where t_o is the initial (switch on) time.

The circuit in Figure 4.19a works quite well, unfortunately it integrates small errors like random noise, and can soon float out of the output range. In addition, real op-amps or signals

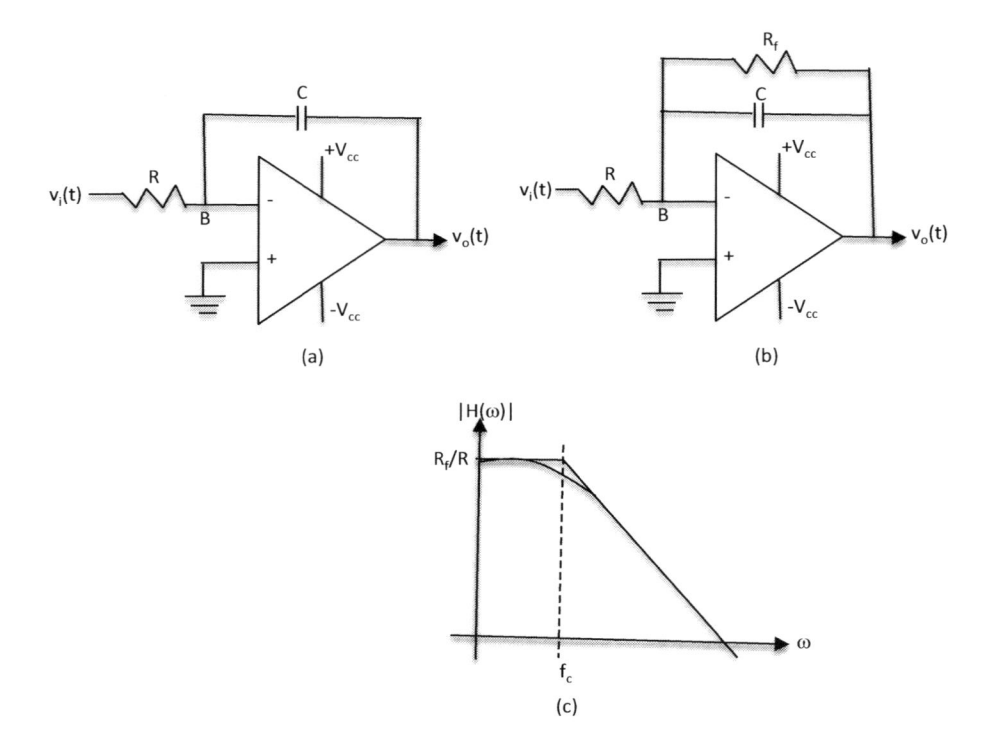

FIGURE 4.19   Integrator amplifier: (a) ideal integrator, (b) active integrator, and (c) Bode plot of the transfer function of (b).

connected to the input have always a dc offset. This offset is indeed integrated and after a given time will saturate the amplifier output. This saturation is essentially a manifestation of the instability of the circuit at low frequency. Given that the initial charge of the capacitor is undefined, it thus makes the initial output state unpredictable. A way of solving this problem is to insert a very large resistor R_f across the capacitor, as shown in Figure 4.19b, to drain off the errors accumulated by random noise, as well as reducing the amplifier dc gain. An intuitive way to understand the effect of this feedback resistance R_f is that it does not allow the capacitor to be charged at its own pleasure.

Like the differentiator, the choice of the feedback resistor R_f is not so trivial if we want to preserve the characteristic of a good integrator. Like the condition specified in Chapter 3, Section 3.5.4.1, a necessary good integrator condition is

$$\omega \gg \frac{1}{CR_f} \tag{4.100}$$

If, on the other hand, the dc current must be integrated, we can place a *switch* in parallel with the capacitor (instead of R_f) to be opened when the integration is started. In this way, the capacitor state will have been completely defined. This switch could be a transistor switch allowing for control from somewhere else in the circuit.

Figure 4.19b, in essence, is an active first-order, *low-pass filter* (LPF). Its transfer function can be computed as

$$H(s) = \frac{V_o(s)}{V_i(s)} = -\left(\frac{R_f}{R}\right)\frac{1}{CR_f s + 1} \tag{4.101}$$

where the term $\left(\dfrac{R_f}{R}\right)$ is the filter's dc gain. The Bode magnitude plot of (4.101) is shown in Figure 4.18c, and f_c is the cutoff frequency $\left(\text{i.e. } f_c = \dfrac{1}{2\pi CR_f}\right)$. For $\omega \gg \dfrac{1}{CR_f}$, the plot exhibits zero slope. However, for $\omega \gg \dfrac{1}{CR_f}$, the filter will have a 6 dB per octave roll-off, which is the proper slope for an integrator. It follows that the circuit with R_f acts as an integrator only for frequencies such that $\omega \gg \dfrac{1}{CR_f}$.

A good use for an op-amp integrator could be as part of a position sensing system.

4.3 TIME-BASE GENERATORS

Time-base generators are the circuits that provide an output waveform, a part of which is characterized by a linear variation of voltage or current with respect to time. Op-amp integrator and summer and oscillator (multivibrators) circuits are employed in constructing time-base generators. In this section, the operations of oscillator circuits that generate time-dependent single pulse or a train of pulses, sine and triangular waves, will be discussed separately.

4.3.1 OSCILLATORS AND MULTIVIBRATORS

A further application of positive feedback in op-amp circuits is in the construction of oscillator circuits. Oscillator circuits are designed to put out periodic waveforms like sine waves or square waves. Oscillators play a lot of roles in our modern electronic devices and appliances. For example, oscillators are at the heart of the laboratory waveform generators, and they form a basic part of any radio or TV tuner, and they generate the clock signals that characterize and control all computers' *central processing units* (CPUs).

4.3.1.1 Basic Oscillator

An oscillator can be thought of as an amplifier that provides itself (through feedback) with an input signal. The primary purpose of an oscillator is to generate a given waveform at a constant peak amplitude and specific frequency and to maintain this waveform within certain limits of amplitude and frequency. Having considered transfer function in the previous section, we can utilize the same concept to describe the basic and necessary condition of oscillation.

An oscillator must provide amplification. Amplification of signal power occurs from input to output. In an oscillator, a portion of the output is fed back to sustain the input, as shown in Figure 4.20a. Enough power must be fed back to the input circuit for the oscillator to drive itself. In this case, the gain of the amplifier

$$A(s) = \frac{V_o(s)}{V_{\text{in}}(s)} \tag{4.102}$$

As we have seen in the op-amps analyses thus far, there are resistors and capacitors forming the feedback circuits. A convenient starting point for discussing self-sustaining feedback is to consider the feedback configuration shown in Figure 4.20b, which comprises of an amplifier $A(s)$, a feedback system $g(s)$, and a switch that toggles between two positions x and y. This feedback amplifier network is connected to the signal source V_{in}. The amplifier can be connected from its output to its input by flipping the switch from position yy to position xx. In position yy, the switch connects the feedback network $g(s)$ to the signal source V_{in}.

The transfer function of the system is

$$H(s) = \frac{V_o(s)}{V_{\text{in}}(s)} = g(s)A(s) \tag{4.103}$$

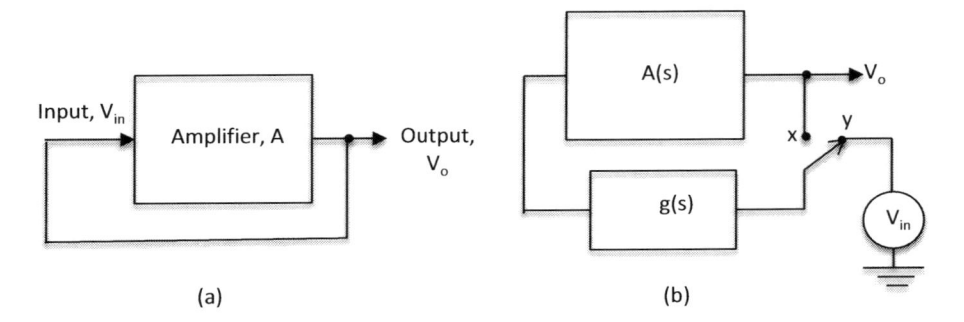

(a)   (b)

FIGURE 4.20 Block diagrams of a basic oscillator: (a) basic oscillator block diagram and (b) improved oscillator block diagram.

Suppose $s = j\omega$ and at some frequency, say f_o, we can achieve unity loop gain (i.e. $|H(s)| = 1.0$), with zero phase angle (i.e. $0°$). If the signal source V_{in} puts out a sine wave at this frequency f_o, the amplifier output V_o will be a sine wave of the same amplitude and phase. If, however, the switch is changed to position xx, the signal input to the feedback network will not change. Instead, the output signal V_o from the amplifier will drive its input. The circuit is thus a stable oscillator.

If the loop gain is greater than unity (i.e. $|H(s)| > 1$), at frequency f_o, the circuit will oscillate when the switch is closed and the amplitude of the output voltage V_o will increase with time until the amplifier overloads or clips.

However, if the loop gain is less than unity (i.e. $|H(s)| < 1$), at frequency f_o, the amplitude of the output voltage V_o will decrease or damp out with time until the output becomes zero.

Thus, the necessary condition for a stable output sine wave is that the loop gain $|H(s)|$ must be unity at the frequency f_o for which its phase is $0°$. This condition is called *Barkhausen criterion*.

4.3.2 Relaxation Oscillator (or Square Wave Generator)

A relaxation oscillator is a circuit that repeatedly alternates between two states. The length of its stay at each state depends on the charging of a capacitor. Figure 4.21a shows the circuit of a square wave generator, also called a relaxation oscillator. This square wave generator is like the Schmitt trigger circuit in that the reference voltage (V_{ref}) for the comparator action depends on the output voltage. For this oscillator, like the comparator amplifier, the reference voltage V_{ref} is determined at node xx of Figure 4.21a using voltage divider rule:

$$V_{ref} = \pm V_{cc}\left(\frac{R_2}{R_1 + R_2}\right) \tag{4.104}$$

The charge–discharge curve must follow the R_3C universal dc transient expression, which includes the variables: time t and time constant τ. That is,

$$v_{+ref} = v_{cc} + \left[v_{-ref} - v_{cc}\right]e^{-t/\tau} \tag{4.105}$$

Rearranging in terms of the exponent term to have

$$e^{-t/\tau} = \frac{\left(v_{+ref} - v_{cc}\right)}{\left(v_{-ref} - v_{cc}\right)} \tag{4.106}$$

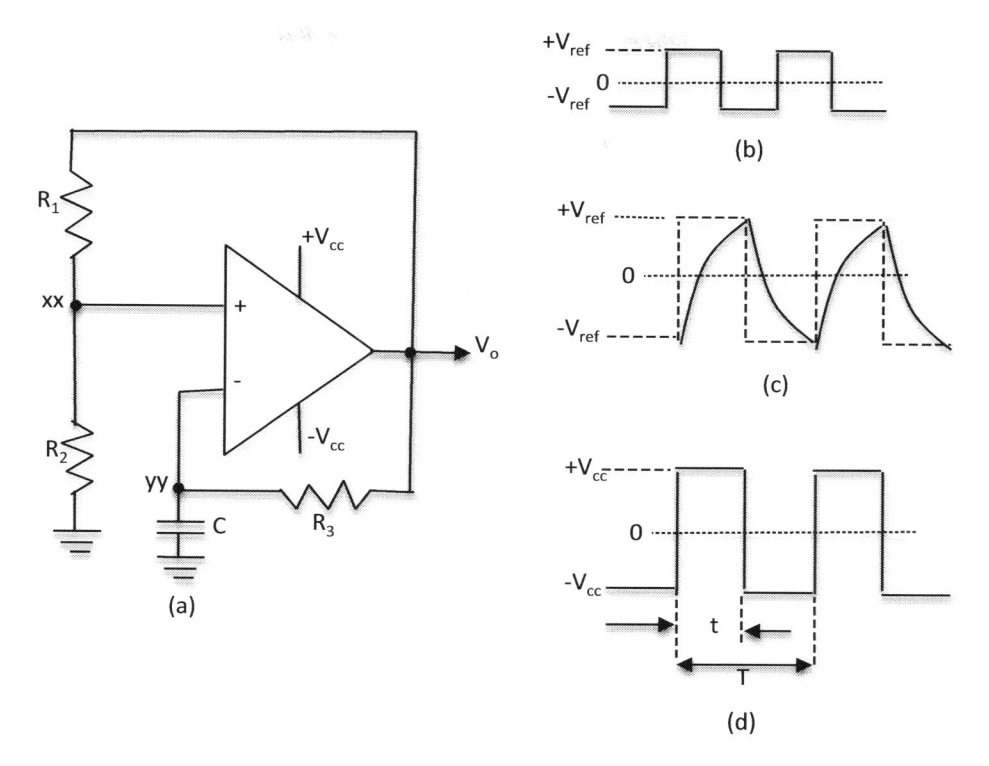

(b)

(c)

(a)

(d)

FIGURE 4.21   Oscillator circuit with nodes' waveforms: (a) square wave oscillator, (b) waveform at *xx*, (c) waveform at *yy*, and (d) output V_o square waveform.

Take log-to-base-*e* of both sides of the equation and rearrange in terms of *t* to have

$$t = -\tau \log_e \left[\frac{(v_{+\text{ref}} - v_{cc})}{(v_{-\text{ref}} - v_{cc})} \right] \tag{4.107}$$

The time required to discharge the capacitor from $v_{+\text{ref}}$ is the same to $v_{-\text{ref}}$. So, a complete charge and discharge cycle creates one cycle of the waveform. Therefore, the period *T* of the waveform is expressed as

$$T = 2t = -2\tau \log_e \left[\frac{(v_{+\text{ref}} - v_{cc})}{(v_{-\text{ref}} - v_{cc})} \right] \tag{4.108}$$

From this expression (i.e. Equation (4.108)), the circuit's frequency *f* of oscillation can be estimated using

$$f = \frac{1}{T} = \frac{1}{2t} \tag{4.109}$$

Since the circuit's frequency of oscillation is dependent on the charge and discharge of the capacitor *C* through a feedback resistor R_3, the frequency of this oscillator may be adjusted by varying the size of any of these components. Technically, this relaxation oscillator circuit can also be classified as an *astable multivibrator* because the circuit has *no stable* state. With no external signal applied, the op-amp alternately switches from cutoff to saturation at a frequency determined by the time constants of the coupling circuits. An astable multivibrator is also called a free running relaxation oscillator.

Example 4.4

Calculate the frequency of oscillation of Figure 4.21a if $R_1 = 8.6$ kΩ, $R_2 = 10$ kΩ, $R_3 = 100$ kΩ, and $C = 0.01$ µF. Suppose the op-amp is rail sourced at ± 10 V.

Solution:

Using (4.104), the threshold voltage is

$$V_{ref} = \pm 10 \left(\frac{10}{8.6 + 10} \right) = \pm 5.38 \text{ V}$$

Time constant $\tau = R_3 C = 100 \times 10^3 \times 0.01 \times 10^{-6} = 1$ ms.

By substituting τ in (4.108), and using other parameters, the pulse period

$$T = -2 \times 10^{-3} \log_e \left[\frac{5.38 - 10}{-5.38 - 10} \right] = 2.4 \text{ ms}$$

Frequency, f, is inversely proportional to period, so the frequency of oscillation is

$$f = \frac{1}{T} = \frac{1}{2.4 \times 10^{-3}} = 416.7 \text{ Hz}$$

Example 4.5

Design a square wave generator using op-amp to have an output frequency of 2.5 kHz with variable duty cycle from 25% to 75%.

Solution:

A circuit for non-symmetrical square wave generator is shown in Figure 4.22a. Let us now design the circuit components capable of operating at 2.5 kHz frequency. Assume $R_1 = R_2 = 10$ kΩ and $V_o = \pm V_{sat} = \pm V_{cc} = \pm 15$ V.

When $V_o = +V_{sat} = +15$ V, then capacitor C will charge through R_3 due to forward-biased diode D_1. However, when $V_o = -V_{sat} = -15$ V, capacitor C will discharge through R_4 due to forward-biased diode D_2. Selecting different combinations of R_3 and R_4, charging and discharging time can be varied and, consequently, a non-symmetrical square wave can be obtained.

Duty cycle (or duty factor) is the time that the generator spends in an active state as a fraction of the total time under consideration. Figure 4.22b, for instance, time t is the active state or pulse duration and the total time or pulse period of the wave is T. Hence, the *duty cycle, d_c,* is

$$d_c = \frac{t}{T} \tag{4.110}$$

For a symmetrical wave or pulse, as shown in Figure 4.22b, $t_1 = t_2 = t$, as such the period $T = 2t$. For non-symmetrical wave or pulse, as shown in Figure 4.22c, $t_1 \neq t_2$, as such, period $T = (t_1 + t_2)$. Duty cycle is important because it relates to peak and average power in the determination of total energy output of any system, particularly radar.

We know that frequency f is inversely proportional to period T, so the square wave period is

$$T = \frac{1}{f} = \frac{1}{2.5} = 0.4 \text{ ms}$$

Since $R_1 = R_2 = 10$ kΩ, it is obvious that $V_{ref} = \pm 0.5 \ V_{cc}$. If we take the capacitor as having a value of 1.0 µF, and using Equation (4.108), we can evaluate R_3 due to forward-biased diode D_1. For instance,

$$R_3 = \frac{T}{-2C \log_e \left[V_{ref} - V_{cc} / -V_{ref} - V_{cc} \right]} = \frac{0.4 \times 10^{-3}}{-2 \times 10^{-6} \log_e \left[7.5 / 22.5 \right]} = \frac{0.4 \times 10^{-3}}{-2 \times 10^{-6} \times -1.0986} = 182 \ \Omega$$

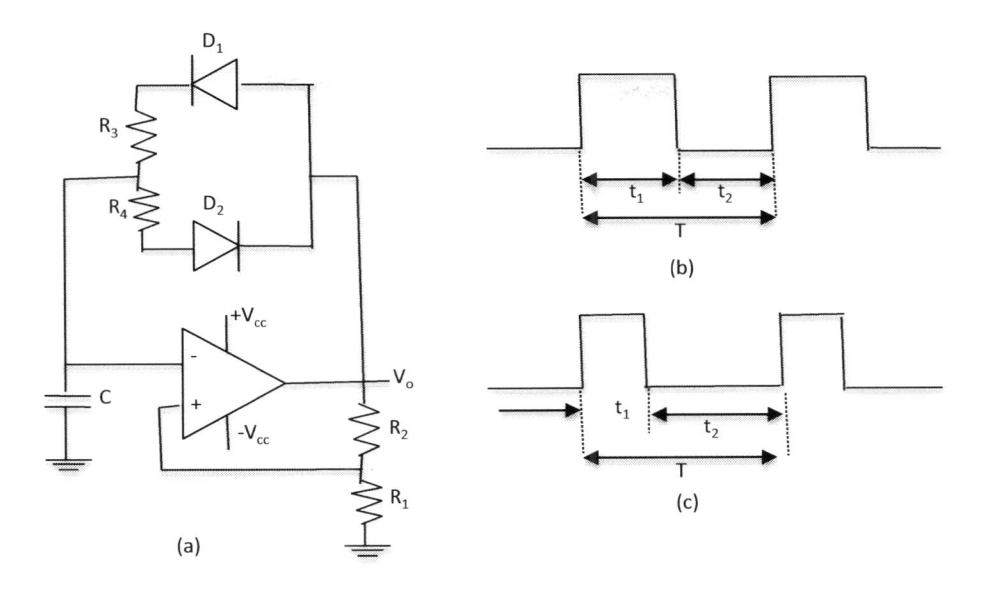

FIGURE 4.22 Non-symmetrical wave generator of period T: (a) wave generator, (b) symmetrical pulse: $t_1 = t_2$, and (c) nonsymmetrical pulse: $t_1 \neq t_2$.

For a symmetrical wave, i.e. square wave, $R_4 = R_3 = 182\ \Omega$. For the case of this example, we need to vary the duty cycle from 25% to 75%, meaning that we modify R_4 and R_3 accordingly in the same proportion. For instance, $R_4 = 4R_3$ and $R_4 = 1.33R_3$. The modified circuit will give frequency $f = 2.5\,\text{kHz}$ with variable duty cycle from 25% to 75%.

4.3.3 MONOSTABLE MULTIVIBRATOR

In some applications, the need arises for a pulse of known height and width generated in response to a trigger signal. A monostable multivibrator can generate such a standardized pulse. A monostable multivibrator has a stable state in which it can remain indefinitely if not disturbed. It can be made to change to its other state by the application of a suitable triggering pulse. After the time interval is determined by component values, the circuit returns to its stable state. At times, a monostable multivibrator is also called a *one shot*. An electronic circuit of a monostable multivibrator can be developed by modifying the astable circuit of Figure 4.21 and including a trigger circuit as well as a diode D_1 connected in parallel to the timing capacitor C_1, as shown in Figure 4.23a. This diode is inserted to prevent the phase inverting input terminal of the amplifier from going positive; this gives a monostable circuit. The operation of the circuit is described as follows.

In the stable state, the output V_o is high at positive saturation, i.e. $+V_{osat}$, which is $+V_{cc}$, and D_1 conducts thereby clamping the op-amp at $-v_{xx}$, which is the reference voltage at node xx, given by

$$\pm v_{xx} = \pm V_{cc}\left[\frac{R_2 \parallel R_4}{R_2 \parallel (R_1 + R_4)}\right]V_{cc}$$

$$= \pm V_{cc}\left[\frac{R_4(R_1 + R_2 + R_4)}{(R_2 + R_4)(R_1 + R_4)}\right] \tag{4.111}$$

It is assumed that the resistor R_4 is much greater than R_2 so that its loading effect may be neglected. This voltage $\pm v_{xx}$ is larger than $v_{\pm ref}$ (of the astable situation) and therefore V_o stays at $+V_{cc}$. Note that, from Equation (4.104),

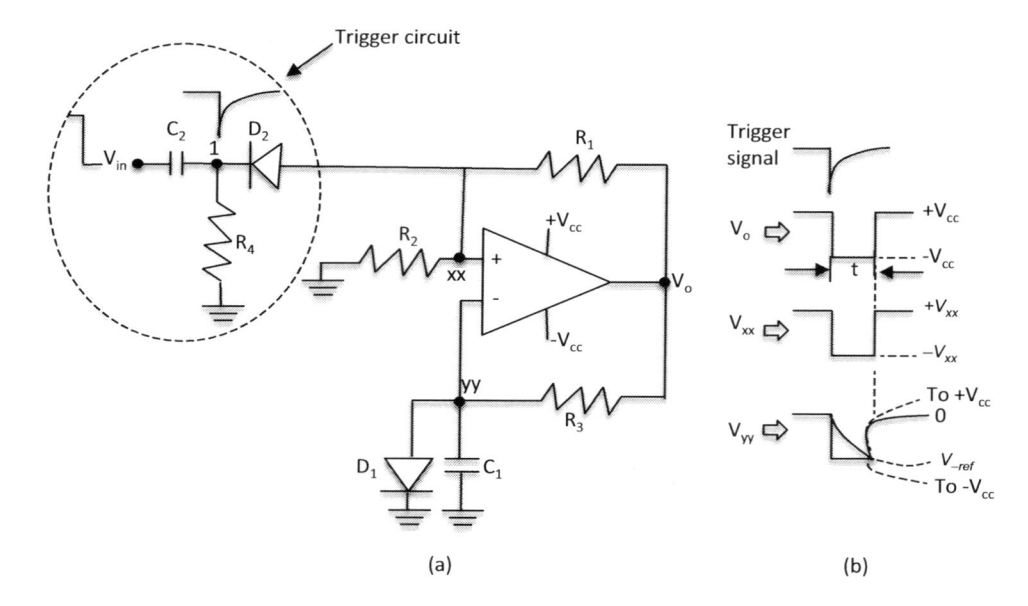

(a)   (b)

FIGURE 4.23   Monostable circuit and waveforms: (a) op-amp monostable circuit and (b) node's waveforms.

$$V_{\pm\text{ref}} = \pm V_{cc}\left(\frac{R_2}{R_1 + R_2}\right) \qquad (4.112)$$

When a negative-going transition appears at the input V_{in}, the $R_4 C_2$ differentiator produces a negative pulse. Node xx is pulled low, and the output changes to a negative saturation, i.e. $-V_{cc}$. The diode D_2 is now reverse biased and isolates the op-amp circuit from the differentiator circuit. The capacitor now starts discharging toward $-V_{cc}$ through R_3. When it crosses $v_{-\text{ref}}$, the output changes back to $+V_{cc}$. The circuit switches back to its permanently stable state, i.e. $+V_{o\text{sat}}$ ($+V_{cc}$) when the voltage at node xx is equal to that at node yy. Otherwise, the circuit can switch to a quasi-permanent state at either $+V_{cc}$, zero, $v_{-\text{ref}}$, or $-V_{cc}$. The waveforms of the output voltage, and voltages at nodes xx and yy are illustrated in Figure 4.23b.

The timing period for this monostable circuit is

$$t = C_1 R_3 \ln\left[1 + \frac{R_2}{R_1}\right] \qquad (4.113)$$

Multivibrators have several usages, depending on their configuration. They could be used as sawtooth generator, as square wave and pulse generator, as standers frequency source, as well as for many specialized uses in radar and TV circuits, and as memory element in computers.

4.3.4   TRIANGLE WAVE GENERATOR

The square wave output of a relaxation oscillator (of Figure 4.21) can be used to drive the active integrator circuit (of Figure 4.19b) to produce a triangle wave generator. The triangle wave generation circuit is shown in Figure 4.24a, which is now a two-stage system with two distinct circuits, and its output waveform shown in Figure 4.24b. When the voltage is at the $-V_{cc}$ level, the triangle wave output rises with the slope, Δm, which is

$$\Delta m = \frac{V_{cc}}{R_3 C_1} \qquad (4.114)$$

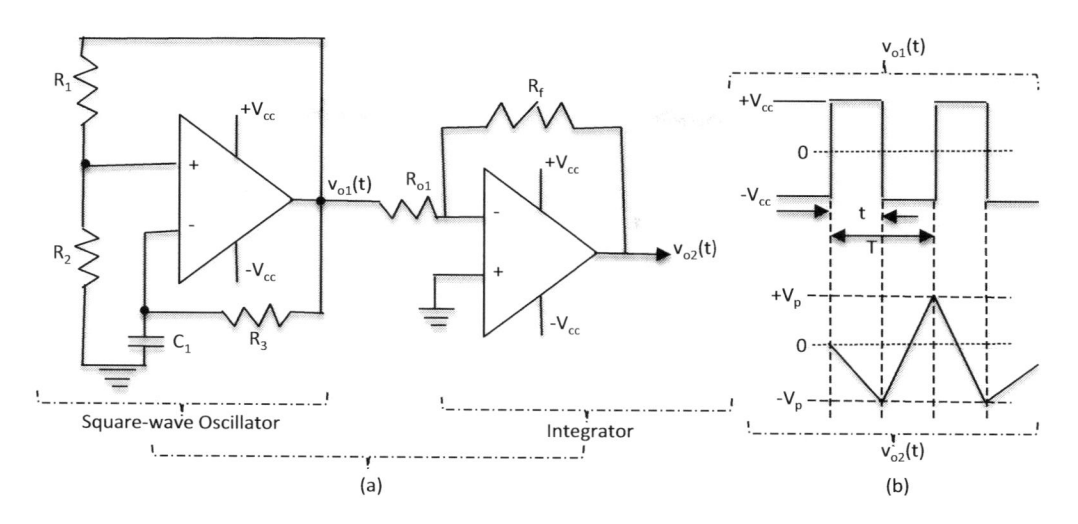

FIGURE 4.24 Triangle wave oscillator: (a) triangle wave generation circuit and (b) output waveform $v_{o2}(t)$.

If we denote the peak values of the triangle wave by $+V_p$ and $-V_p$, it follows that

$$2V_p = \frac{\Delta m}{2}T \tag{4.115}$$

where T is the period of the triangle wave. We can estimate the peak values of the triangle wave by

$$V_p = \pm\frac{R_{ol}}{R_f}V_{cc} \tag{4.116}$$

By substituting (4.114) and (4.116) in (4.115), and noting that the period of the triangle wave T is the reciprocal of the frequency f_T of triangle oscillator, we can write this oscillator frequency as

$$f_T = \frac{1}{4}\frac{R_f}{R_{ol}}\frac{1}{C_1 R_3} \tag{4.117}$$

4.3.5 SINE WAVE OSCILLATOR CIRCUITS

There are many types of sine wave oscillator circuits and variants. In an application, the choice of which circuits to use depends on the frequency and the desired monotonicity of the output waveform. Two prominent oscillator circuits, namely the Wien bridge oscillator and phase-shift oscillator, are considered.

4.3.5.1 Wein Bridge Oscillator

The Wien bridge oscillator is a unique circuit because it generates an oscillatory output signal without having a sinusoidal input source. Instead, the amplifier supplied its own input through the two arms of the bridge, namely Z_1 and Z_2, as shown in Figure 4.25a, which are connected to the positive terminal to form a frequency selective feedback network. This network comprises of capacitors and resistors, as seen in Figure 4.25b and c. Instead of direct sinusoidal input source, the Wien bridge oscillator uses the capacitors in the feedback network with initial voltages to create the output. The oscillator amplifies the signal with the two negative feedback passive elements (Z_3 and Z_4). To analyze the circuit in Figure 4.25a, we may artificially separate the oscillating feedback network and the amplifier itself.

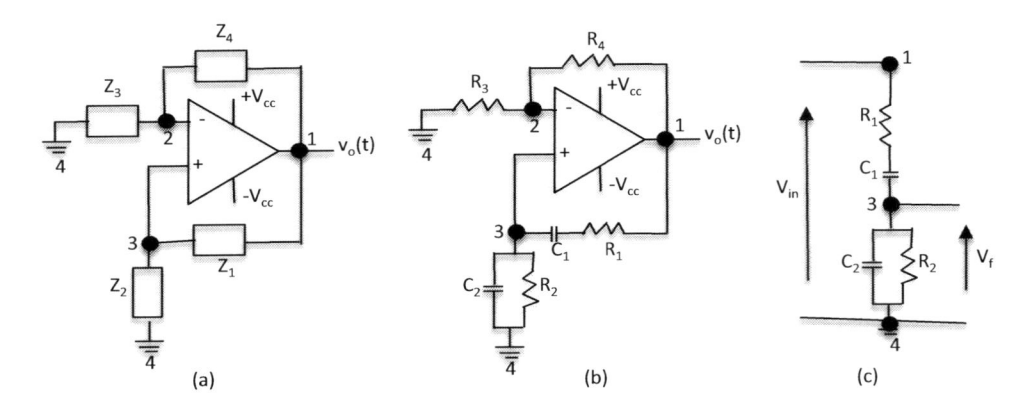

FIGURE 4.25 Sine wave oscillator (a–c).

For an ideal op-amp, the amplifier gain, K, is

$$K = 1 + \frac{Z_4}{Z_3} \tag{4.118}$$

The two arms of the bridge determine the gain β of the feedback network, i.e. from Figure 4.28c:

$$\beta = \frac{V_f}{V_{in}} \tag{4.119}$$

For oscillation to occur, the product of these gains must be equal to unity, i.e. $|K\beta| = 1$. This relationship becomes clearer later in the next few paragraphs.

Let us arrange the passive elements with resistors and capacitors as shown in Figure 4.25b. The two arms of the bridge, namely; R_1, C_1 in series, and R_2 and C_2 in parallel are called *frequency sensitive arms* because the components of these arms determine the frequency of the oscillator. We can arrange the input V_{in} to the feedback network as being between nodes 1 and 4, and the output V_f of the feedback network as between nodes 3 and 4, as depicted in Figure 4.25c. Such a feedback network is called *lead-lag network* because at very low frequencies it acts like a lead while at high frequencies it acts like a lag network.

Comparing Figure 4.25a with Figure 4.25b, we write the series arm of the bridge as

$$Z_1 = R_1 + \frac{1}{j\omega C_1} = \frac{1 + j\omega R_1 C_1}{j\omega C_1} \tag{4.120}$$

and the parallel arm of the bridge as

$$Z_2 = R_2 \left\| \frac{1}{j\omega C_2} = \frac{R_2 \times \dfrac{1}{j\omega C_2}}{R_2 + \dfrac{1}{j\omega C_2}} = \frac{R_2}{1 + j\omega R_2 C_2} \tag{4.121}$$

Replacing $j\omega = s$, we write

$$Z_1 = R_1 + \frac{1}{sC_1} = \frac{1 + sR_1 C_1}{sC_1} \tag{4.122}$$

$$Z_2 = \frac{R_2}{1 + sR_2C_2} \tag{4.123}$$

From Figure 4.25c, the current I that flows from node 1 to node 4 is

$$I = \frac{V_{in}}{Z_1 + Z_2} \tag{4.124}$$

And, the feedback voltage is

$$V_f = IZ_2 \tag{4.125}$$

which, in view of Equation (4.124) in Equation (4.125), the feedback voltage becomes

$$V_f = IZ_2 = \frac{Z_2 V_{in}}{Z_1 + Z_2} \tag{4.126}$$

We can now express this equation in terms of the feedback network gain, thus

$$\beta = \frac{V_f}{V_{in}} = \frac{Z_2}{Z_1 + Z_2} \tag{4.127}$$

Rearranging

$$\beta = \frac{Z_2}{Z_1 + Z_2} = \frac{\left(\dfrac{R_2}{1 + sR_2C_2}\right)}{\left(\dfrac{1 + sR_1C_1}{sC_1}\right) + \left(\dfrac{R_2}{1 + sR_2C_2}\right)} \tag{4.128}$$

Expanding Equation (4.128) to have

$$\beta = \frac{sR_2C_1}{1 + s\left(R_1C_1 + R_2C_2 + R_2C_1\right) + s^2 R_1 R_2 C_1 C_2} \tag{4.129}$$

Replacing s with $j\omega$, and s^2 with $-\omega^2$ in Equation (4.129), the gain expresses as

$$\beta = \frac{j\omega R_2 C_1}{1 + j\omega\left(R_1C_1 + R_2C_2 + R_2C_1\right) - \omega^2 R_1 R_2 C_1 C_2}$$

$$= \frac{j\omega R_2 C_1}{\left(1 - \omega^2 R_1 R_2 C_1 C_2\right) + j\omega\left(R_1C_1 + R_2C_2 + R_2C_1\right)} \tag{4.130}$$

Rationalizing Equation (4.130) to have

$$\beta = \frac{\omega^2 R_2 C_1\left(R_1C_1 + R_2C_2 + R_2C_1\right) + j\omega R_2 C_1\left(1 - \omega^2 R_1 R_2 C_1 C_2\right)}{\left(1 - \omega^2 R_1 R_2 C_1 C_2\right) + \omega^2\left(R_1C_1 + R_2C_2 + R_2C_1\right)^2} \tag{4.131}$$

In order to have zero phase shift of the feedback network, the imaginary part of Equation (4.131) must equal to zero, i.e.

$$\omega R_2 C_1\left(1 - \omega^2 R_1 R_2 C_1 C_2\right) = 0 \tag{4.132}$$

where $\omega R_2 C_1 \neq 0$. So

$$1 - \omega^2 R_1 R_2 C_1 C_2 = 0 \tag{4.133}$$

Expressing ω in terms of other variables:

$$\omega^2 = \frac{1}{R_1 R_2 C_1 C_2} \tag{4.134}$$

Or equivalently

$$\omega = \sqrt{\frac{1}{R_1 R_2 C_1 C_2}} \tag{4.135}$$

Noting that $\omega = 2\pi f$, then, in terms of frequency of the oscillator, we write

$$f = \frac{1}{2\pi\sqrt{R_1 R_2 C_1 C_2}} \tag{4.136}$$

By changing the resistor and capacitor values in the positive feedback network, the output frequency can be changed.

At this frequency, i.e. ω in Equation (4.135), and substituting it into the transfer function (β) of Equation (4.131), the feedback network gain is given by

$$\beta = \frac{\omega^2 R_2 C_1 (R_1 C_1 + R_2 C_2 + R_2 C_1) + j\omega R_2 C_1 (1 - \omega^2 R_1 R_2 C_1 C_2)}{(1 - \omega^2 R_1 R_2 C_1 C_2) + \omega^2 (R_1 C_1 + R_2 C_2 + R_2 C_1)^2} = \frac{R_1 C_2}{R_1 C_1 + R_2 C_2 + R_2 C_1} \tag{4.137}$$

If this expression (Equation (4.137)) equals to unity, the circuit will oscillate with a *stable* output when the loop is closed.

In practice, the circuit is often designed with $R_1 = R_2 = R$ and $C_1 = C_2 = C$. In this case, the frequency would become

$$f = \frac{1}{2\pi\sqrt{R^2 C^2}} = \frac{1}{2\pi RC} \tag{4.138}$$

The oscillation frequency can be continuously tuned using coupled variable resistors. At this oscillation frequency, the loop gain is

$$\beta = \frac{R_1 C_2}{R_1 C_1 + R_2 C_2 + R_2 C_1} = \frac{RC}{3RC} = \frac{1}{3} \tag{4.139}$$

Because the attenuation (or loop gain, β) at the resonant frequency is 1/3, the negative feedback must have a theoretical gain K must be 3, i.e. ensuring that $|K\beta| = 1$. This condition (given by Equation (4.118) when $K = 3$) is the necessary condition for stable oscillations at frequency, f, defined by Equation (4.138). From Equation (4.118), if $K = 3$, we write

$$K = 1 + \frac{Z_4}{Z_3} = 3 \tag{4.140}$$

or

$$Z_4 = 2Z_3 \tag{4.141}$$

This means that the feedback resistor R_4 (= Z_4) must be set to twice the value of R_3 (= Z_3) to satisfy this condition. Figure 4.26a depicts this condition; i.e. when $K = 3$.

Other conditions can be deduced from Equation (4.140), such as

- If K is less than three (i.e. $K < 3$), then oscillations attenuate. Figure 4.26b depicts this condition.
- If K is greater than 3 (i.e. $K > 3$), then oscillations amplify. Figure 4.26c depicts this condition.

The op-amp in the Wien bridge circuit oscillator is not ideal; as such the oscillator is susceptible to distortion, which might make the oscillator unstable. A solution to making the oscillations steady is to include a diode network in series with the feedback resistor R_4 as shown in Figure 4.27. Naturally the resistor R_5 in the diode network is the same as the feedback resistor R_4.

To minimize distortions due to the op-amp saturation when the gain is larger than one, it is required to provide a circuit with variable gain. Essentially, we need an overall gain larger than one

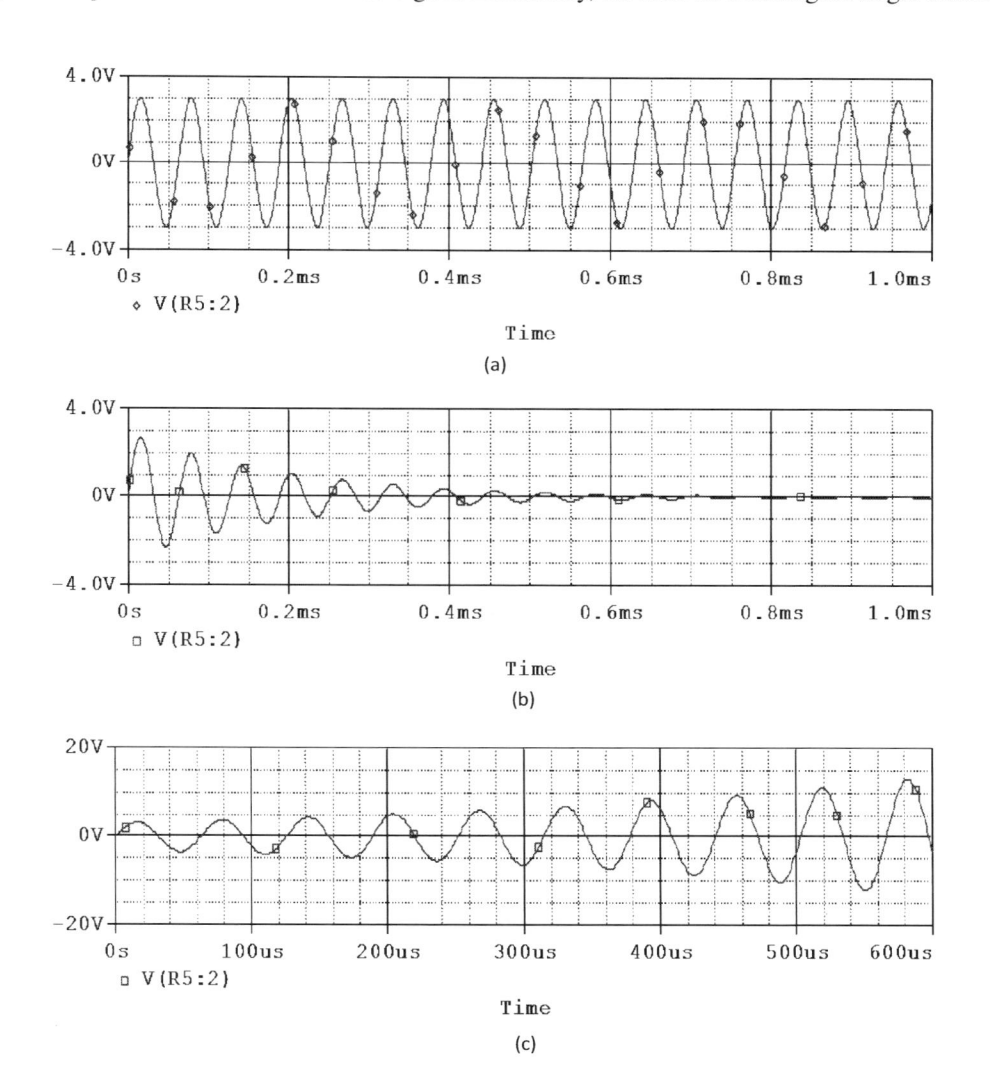

FIGURE 4.26 Oscillation conditions: (a) $K = 3.0$, (b) $K = 2.9$, and (c) $K = 3.05$.

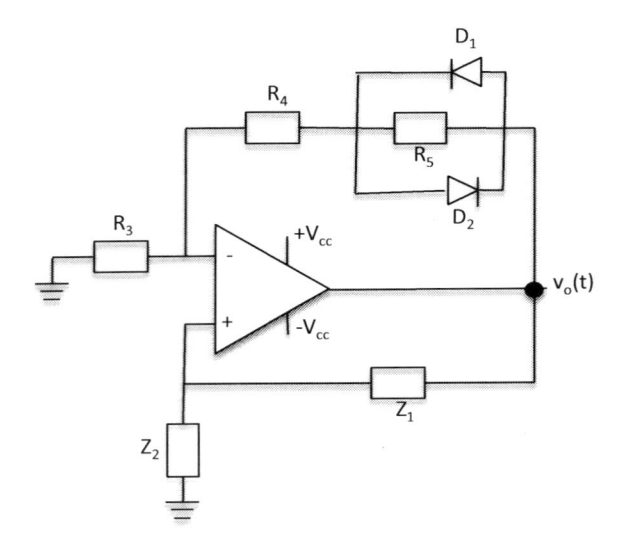

FIGURE 4.27   Stable sine wave oscillator.

for small signal to sustain the oscillation and gain of about 1 or less for large signal to avoid distortion. The negative feedback path shown in Figure 4.27 does the job. For large signals one of the diodes becomes forward biased reducing the feedback resistance and the op-amp gain. For smaller signal the gain is not affected by the diodes.

In practice though, due to the fact that op-amps are not ideal, it is necessary to make the value of K a little higher than 3 with the diode configuration so that the diodes will work properly.

The Wien bridge oscillator has few parts and good frequency stability, but the basic circuit has high output distortion.

Example 4.6

Design the components of the feedback network of the Wein oscillator necessary to achieve the 10.5 kHz frequency of oscillation.

Solution:

As noted in the above discussion, the elements in the feedback network circuit are identical, i.e. $R_1 = R_2 = R$ and $C_1 = C_2 = C$. In that case, from Equation (4.138), we write the frequency of oscillation as

$$f = \frac{1}{2\pi\sqrt{R^2 C^2}} = \frac{1}{2\pi RC}$$

If we select $R = 15$ kΩ, then we can write C (capacitor) in terms of frequency f and resistor R to obtain the capacitor's value, thus

$$C = \frac{1}{2\pi f R} = \frac{1}{2 \times \pi \times 10.5 \times 10^3 \times 15 \times 10^3} = 1.01\,\text{nF}$$

Nothing stops you from choosing different values for R (to get C), or for C (to get R).

4.3.5.2   Phase-Shift Oscillator

A phase-shift oscillator is another sine wave oscillator. It contains an inverting op-amp (with an inverting gain, i.e. its gain is negative; a negative gain is equivalent to a phase shift of 180°), and a feedback filter network that shifts the phase of the amplifier output by 180° at the oscillation frequency.

A simple phase-shift oscillator circuit is shown in Figure 4.28a. For the circuit to be an oscillator, the feedback network must introduce an additional phase of $\pm 180°$ so that the total phase is $0°$ (or $\pm 360°$). This is achieved by using an RC network that introduces a phase shift of $180°$ between input and output, where each sector of the RC filter produces a $60°$ phase shift. By cascading three RC HPFs, there is a frequency at which the phase shift is $+180°$ and the gain is non-zero. The output of the amplifier is given to feedback network driving the amplifier. The basic assumption is that the phase-shift filters are independent of each other. The total phase shift around a loop is $180°$ of op-amp and $180°$ due to the three RC sections, giving a total of $360°$ phase shift. This satisfies the required condition for positive feedback and the circuit works as an oscillator. The phase-shift network of Figure 4.28a is a *lead network*. However, if the resistors are interchanged with capacitors, it becomes a *lag network*.

To analyze the circuit in Figure 4.28a, suppose we break the feedback loop at point xx, to separate the oscillating feedback network and the amplifier itself. As a result, we can decouple the amplifier gain from the gain of the phase-shift network. For an ideal op-amp, the amplifier gain, K, is

$$K = \frac{R_f}{R_i} \tag{4.142}$$

The next stage is to obtain the loop transfer function β of the phase-shift network. By applying the Kirchhoff voltage law (KVL) to the phase-shift network, as depicted by Figure 4.28b, the loop transfer function β can be developed in terms of the loop's passive components. For instance:

Loop 1:

$$V_{xx} = I_1 \left(R + jX_c \right) - I_2 R \tag{4.143}$$

where X_c is capacitive reactance, i.e. $X_c = -\dfrac{1}{\omega C}$

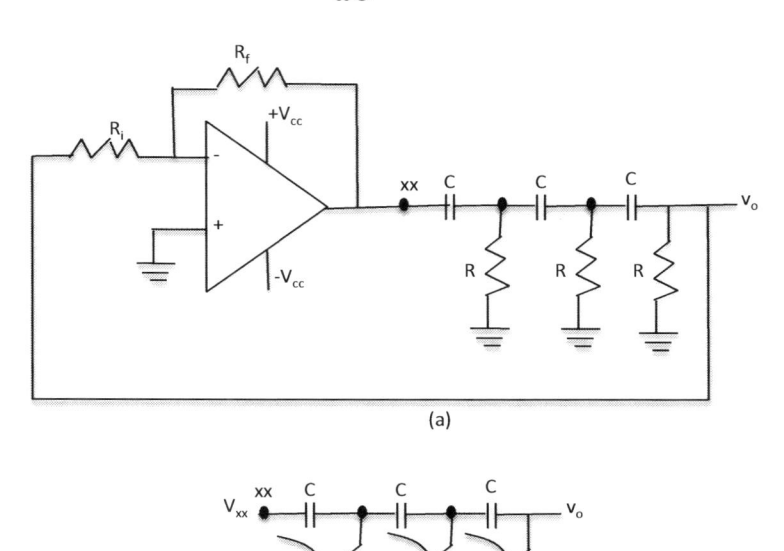

(a)

(b)

FIGURE 4.28 A phase-shift modulator: (a) simple phase-shift modulator and (b) phase-shift network.

Loop 2:

$$0 = -I_1 R + I_2 \left(2R - jX_c\right) - I_3 R \tag{4.144}$$

Loop 3:

$$0 = -I_2 R + I_3 \left(2R - jX_c\right) \tag{4.145}$$

Solving for I_3, to have

$$I_3 = \frac{\begin{vmatrix} R - jX_c & -R & V_{xx} \\ -R & 2R - jX_c & 0 \\ 0 & -R & 0 \end{vmatrix}}{\begin{vmatrix} R - jX_c & -R & 0 \\ -R & 2R - jX_c & -R \\ 0 & -R & 2R - jX_c \end{vmatrix}} \tag{4.146}$$

or

$$I_3 = \frac{V_{xx} R^2}{\left(R - jX_c\right)\left[\left(2R - jX_c\right)^2 - R^2\right] - R^2 \left(2R - jX_c\right)} \tag{4.147}$$

Again, from Figure 4.28b, the output voltage, V_o, is

$$V_o = I_3 R \tag{4.148}$$

Substitute Equation (4.147) in Equation (4.148) to have

$$V_o = \frac{V_{xx} R^3}{\left(R - jX_c\right)\left[\left(2R - jX_c\right)^2 - R^2\right] - R^2 \left(2R - jX_c\right)}$$

$$= \frac{V_{xx} R^3}{\left(R^5 - 5RX_c^2\right) + j\left(X_c^3 - 6R^2 X_c\right)} \tag{4.149}$$

From Equation (4.149), the transfer function of the phase-shift network is given by

$$\beta = \frac{V_o}{V_{xx}} = \frac{R^3}{\left(R^5 - 5RX_c^2\right) + j\left(X_c^3 - 6R^2 X_c\right)} \tag{4.150}$$

For 180° phase shift, the imaginary part of Equation (4.150) must be equal to zero, i.e.

$$X_c^3 - 6R^2 X_c = 0 \tag{4.151}$$

or

$$\left(X_c^2 - 6R^2\right) X_c = 0 \quad \text{where} \quad X_c \neq 0 \tag{4.152}$$

So,

$$X_c^2 = \left(\frac{1}{\omega C}\right)^2 = 6R^2 \tag{4.153}$$

Alternately

$$\omega = \frac{1}{RC\sqrt{6}} \tag{4.154}$$

Noting that $\omega = 2\pi f$, then, in terms of frequency of the oscillator, we write

$$f = \frac{1}{2\pi RC\sqrt{6}} \tag{4.155}$$

This frequency would sustain oscillations generated. Of course, the frequency depends on the values of R and C.

If we substitute this frequency of oscillation given by Equation (4.154) in Equation (4.150), the loop transfer function (β) is given by

$$\beta = -\frac{1}{29} \tag{4.156}$$

The negative sign means that the phase inversion is from the voltage.

At this frequency, the gain of the op-amp must be, at least, 29 to satisfy $|K\beta| = 1$. As a result, for oscillations to occur the gain of the op-amp inverting amplifier must be

$$|K| \geq \frac{R_f}{R_i} \geq 29 \tag{4.157}$$

So,

$$R_f \geq 29R_i \tag{4.158}$$

To prevent the overloading of the amplifier because of RC networks, it is necessary that $R_i \geq 10R$ and the capacitor value can be calculated using Equation (4.154). Also, to prevent the RC sections from loading each other, buffers are placed between each section, as shown in Figure 4.29. In this case, the buffered phase-shift oscillator will perform closer to the calculated frequency and gain.

4.3.5.3 Crystal Oscillator

Crystal oscillators are based on the property of piezoelectricity exhibited by some crystals and ceramic materials. Piezoelectric materials change size when an electric field is applied between two of its faces. Conversely, if we apply a mechanical stress, piezoelectric materials generate an electric field. Some crystals have internal mechanical resonances with very high quality factors; for example, quartz can reach quality factors of 10^4 and can be indeed used to generate very stable oscillators. It should be noted that mechanical resonance stability depends mainly on the fact that the resonance value is determined by the crystal geometry. And, a slight change in the crystal's temperature actually improves the resonator's stability. Active temperature stabilization can clearly improve frequency stability. Figure 4.30a shows the circuit symbol of a crystal oscillator, while Figure 4.30b is the equivalent circuit modeled using ideal passive components. As seen in Figure 4.30a, a crystal has two parallel faces. To apply an electric field to a crystal is necessary to make a conductive coating on two parallel faces, and this process creates a capacitor with an

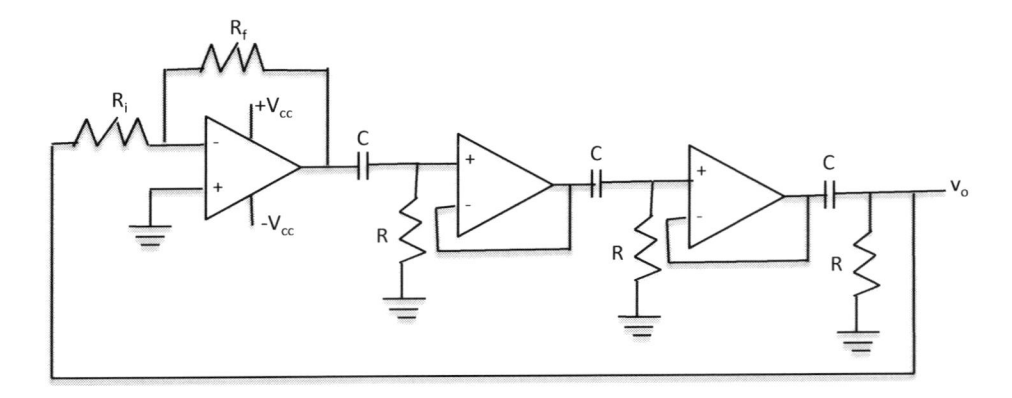

FIGURE 4.29 Buffered phase-shift modulator.

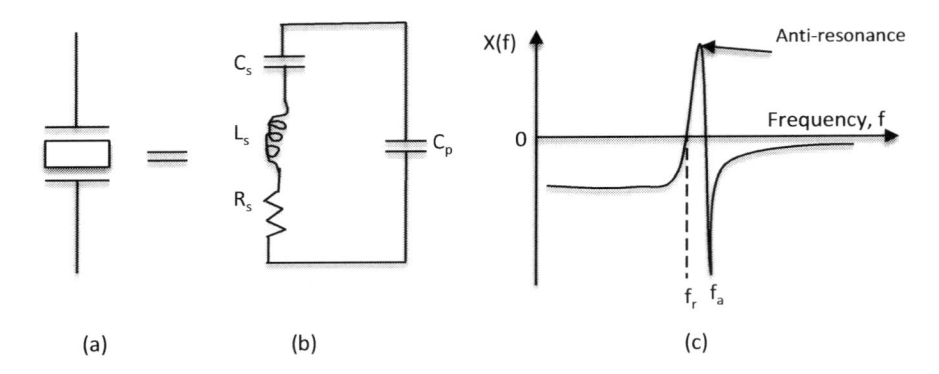

(a)   (b)   (c)

FIGURE 4.30 Crystal oscillator: (a) symbol, (b) equivalent circuit, and (c) reactance of the crystal.

interposed dielectric. This explains the presence of the capacitor of capacitance C_p in the model. The motional parameters of the crystal, e.g. inertia, stiffness, and internal losses, are represented by series inductance, capacitance, and resistance ($L_sC_sR_s$ series circuit) of the second parallel arm; this series circuit accounts for the particular mechanical resonance we want to use to build the oscillator. The reactance of the crystal approaches zero at the point of series resonance f_r and reaches a maximum at the anti-resonant frequency f_a (as seen in Figure 4.30c).

To understand the frequency-dependent impedance of the quartz crystal, it is convenient to analyze its equivalent electrical circuit using conventional filter theory.

The equivalent impedance

$$Z_{eq} = \frac{Z_p Z_s}{Z_p + Z_s} \tag{4.159}$$

where

$$Z_p = -j\frac{1}{\omega C_p} \tag{4.160}$$

$$Z_s = R_s + j\left(\omega L_s - \frac{1}{\omega C_s}\right) \tag{4.161}$$

By substituting Equations (4.160) and (4.161) in Equation (4.159), we write

$$Z_{\text{eq}} = \frac{\dfrac{-j}{\omega C_p}\left[R_s + j\left(\omega L_s - \dfrac{1}{\omega C_s}\right)\right]}{R_s + j\left(\omega L_s - \dfrac{1}{\omega C_s} - \dfrac{1}{\omega C_p}\right)} \tag{4.162}$$

Expanding (4.162) we write

$$Z_{\text{eq}} = \frac{\left(\dfrac{\omega L_s - \dfrac{1}{\omega C_s}}{\omega C_p}\right)\left[\dfrac{-jR_s}{\omega C_p}\right]}{R_s\left[1 + j\left(\dfrac{\omega L_s}{R_s} - \dfrac{1}{\omega C_s R_s} - \dfrac{1}{\omega C_p R_s}\right)\right]} \tag{4.163}$$

which we can rewrite as

$$Z_{\text{eq}} = \left(\frac{\omega L_s - \dfrac{1}{\omega C_s}}{\omega C_p R_s}\right)\frac{\left\{1 + j\left[\left(\dfrac{-R_s}{\omega C_p}\right)\left(\dfrac{\omega C_p}{\omega L_s - \dfrac{1}{\omega C_s}}\right)\right]\right\}}{\left[1 + j\left(\dfrac{\omega L_s}{R_s} - \dfrac{1}{\omega C_s R_s} - \dfrac{1}{\omega C_p R_s}\right)\right]} \tag{4.164}$$

Resonance occurs at a frequency f when Z_{eq} is purely real, i.e. when the imaginary terms in the numerator and denominator are equal. Implying that

$$\left(\frac{-R_s}{\omega C_p}\right)\left(\frac{\omega C_p}{\omega L_s - \dfrac{1}{\omega C_s}}\right) = \frac{\omega L_s}{R_s} - \frac{1}{\omega C_s R_s} - \frac{1}{\omega C_p R_s} \tag{4.165}$$

or

$$\frac{-R_s}{\omega L_s - \dfrac{1}{\omega C_s}} = \frac{\omega L_s}{R_s} - \frac{1}{\omega C_s R_s} - \frac{1}{\omega C_p R_s} \tag{4.166}$$

alternately

$$\frac{\omega C_s R_s^2}{1 - \omega^2 L_s C_s} = \omega L_s - \frac{1}{\omega C_s} - \frac{1}{\omega C_p} \tag{4.167}$$

Rearranging Equation (4.167) to have

$$\omega^4 - \alpha\omega^2 + \kappa = 0 \tag{4.168}$$

where

$$\alpha = \frac{1}{C_p}\left[1 + \frac{2C_p}{L_sC_s} - \frac{R_s^2 C_p}{L_s^2}\right] \tag{4.169}$$

$$\kappa = \frac{C_p + C_s}{L_s^2 C_s^2 C_p} \tag{4.170}$$

Solving Equation (4.168) quadratically to have

$$\omega = \sqrt{\left(\frac{1}{L_sC_s} + \frac{1}{2L_sC_p} - \frac{R_s^2}{2L_s^2}\right) \pm \sqrt{\left(\frac{1}{2L_sC_p} - \frac{R_s^2}{2L_s^2}\right)^2 - \frac{R_s^2}{2L_s^3C_s}}} \tag{4.171}$$

If we assume that

$$\left(\frac{1}{2L_sC_p} - \frac{R_s^2}{2L_s^2}\right)^2 \gg \frac{R_s^2}{2L_s^3C_s}$$

Equation (4.171) naturally reduces to

$$\omega \approx \sqrt{\left(\frac{1}{L_sC_s} + \frac{1}{2L_sC_p} - \frac{R_s^2}{2L_s^2}\right) \pm \left(\frac{1}{2L_sC_p} - \frac{R_s^2}{2L_s^2}\right)} \tag{4.172}$$

Equation (4.172) gives two solutions:

a. the resonant frequency f_r, which is obtained using the negative sign, i.e.

$$\omega \approx \sqrt{\left(\frac{1}{L_sC_s} + \frac{1}{2L_sC_p} - \frac{R_s^2}{2L_s^2}\right) - \left(\frac{1}{2L_sC_p} - \frac{R_s^2}{2L_s^2}\right)} \approx \sqrt{\frac{1}{L_sC_s}} \tag{4.173}$$

or

$$f_r \cong \frac{1}{2\pi}\sqrt{\frac{1}{L_sC_s}} \tag{4.174}$$

b. the anti-resonant (or parallel resonant) frequency f_a, which is obtained using the positive sign, i.e.

$$\omega \approx \sqrt{\left(\frac{1}{L_sC_s} + \frac{1}{2L_sC_p} - \frac{R_s^2}{2L_s^2}\right) + \left(\frac{1}{2L_sC_p} - \frac{R_s^2}{2L_s^2}\right)}$$

$$\approx \sqrt{\frac{1}{L_sC_s} + \frac{1}{L_sC_p} - \frac{R_s^2}{L_s^2}} \tag{4.175}$$

or

$$f_a \cong \frac{1}{2\pi}\sqrt{\frac{1}{L_sC_s} + \frac{1}{L_sC_p} - \frac{R_s^2}{L_s^2}} \tag{4.176}$$

Practically, the quartz crystal equivalent circuit parameters have typical values in the range: R_1 less than 100 Ω, C_1 in the order of few femto-Farads (fF ≈ 10^{-15} F), L_1 is of the order of 1 mH, and C_0 is typically a few pico-Farads (pF ≈ 10^{-12} F). So, we can further approximate that $\dfrac{1}{L_sC_p} \gg \dfrac{R_s^2}{L_s^2}$. As such, the anti-resonant frequency f_a approximates to

$$f_a \cong \frac{1}{2\pi}\sqrt{\frac{1}{L_sC_s}+\frac{1}{L_sC_p}} \cong \frac{1}{2\pi}\sqrt{\frac{1}{L_sC_s}}\sqrt{1+\frac{C_s}{C_p}} \tag{4.177}$$

Using the typical values of C_0 and C_1, we found that the ratio C_s/C_p is far less than one, i.e. $\left(C_s/C_p < 1\right)$. As a consequence, upon application of the binomial approximation, we can write

$$\sqrt{\left(1+\frac{C_s}{C_p}\right)} \approx \left(1+\frac{C_s}{2C_p}\right) \tag{4.178}$$

which further simplifies the anti-resonant frequency f_a expression to

$$f_a \cong \frac{1}{2\pi}\sqrt{\frac{1}{L_sC_s}}\left(1+\frac{C_s}{2C_p}\right) \tag{4.179}$$

A comparison of Equation (4.179) with Equation (4.174) shows that

$$f_a \cong f_r\left(1+\frac{C_s}{2C_p}\right) \tag{4.180}$$

An area typically chosen for operation of the oscillator is either near series resonance, or at the more inductive area of parallel resonance. At frequencies between fr and fa, the impedance of the crystal appears inductive. The frequency separation Δf is the difference between fr and fa.

$$\Delta f = f_a - f_r \tag{4.181}$$

Example 4.7

Estimate the series resonant frequency, anti-resonant frequency, and the separation frequency for a crystal having the following resistive and reactive values: $C_s = 6.0$ fF, $L_s = 25.8$ mH, $R_s = 24.5$ Ω, $C_p = 2.2$ pF.

Solution:

Note that $C_s = 6.0 \times 10^{-15}$ F, $L_s = 0.0258H$, $R_s = 24.5$ Ω, $C_p = 2.2 \times 10^{-12}$ F.

   i. The series resonant frequency fr is evaluated using Equation (4.174):
        $fr = 12.792$ MHz
  ii. The anti-resonant frequency fa is evaluated using Equation (4.176):
        $fr = 12.809$ MHz
 iii. The frequency separation Δf is obtained using Equation (4.181):
        $\Delta f = 0.017$ MHz or 17 kHz.

4.3.5.4 Pierce Oscillator

The section closely follows the description of Pierce oscillator in [7]. Figure 4.31a shows an ideal series-resonant circuit of a Pierce oscillator. The total phase shift around the loop is 360°: where ideally the amplifier provides 180°; the R_1C_1 acts as an integration network and provides a 90° phase

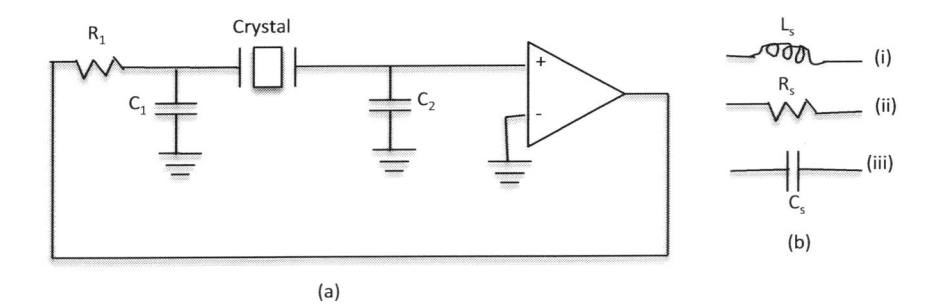

FIGURE 4.31   Pierce oscillator: (a) pierce circuit and (b) equivalent crystal impedances at: (i) above series resonance, (ii) at series resonance, and (iii) below series resonance.

lag; and the crystal, together with C_2, acts as a second integration network and provides a second $90°$ phase lag.

At series resonance, the crystal's impedance is a pure resistance, and together with C_2, it acts like a R_sC_2 integrating network, providing a $90°$ phase lag, as shown in Figure 4.31b, case (ii). Below series resonance, the crystal's impedance is capacitive, C_s, as shown in Figure 4.31b, case (iii), and together with C_2, it acts like a capacitive voltage divider with $0°$ phase shift. Above series resonance, the crystal's impedance is inductive L_s, as shown in Figure 4.31b, case (i), and together with C_1, it provides a $180°$ phase lag. Thus, the crystal can provide anything from a $0°$ to $180°$ phase lag by just a small increase or decrease in frequency from series resonance.

In reality, of course, the amplifier provides slightly more than a $180°$ phase shift, due to the transistor's internal capacitance and storage time, and the R_1C_1 integrating network provides something less than a $90°$ phase shift. Naturally, there is a close correlation in the Pierce between the circuit's short-term frequency stability and the crystal's internal series resistance R_s. The lower the crystal's resistance at series resonance, the smaller the frequency shift needed to change the crystal's impedance from capacitive to inductive (or vice versa) and correct any phase errors around the loop.

The Pierce oscillator has many desirable characteristics, namely:

- It works in the frequency range of 1 kHz to 200 MHz.
- Large phase shifts in RC networks and large shunt capacitances to ground on both sides of the crystal make the oscillation frequency relatively insensitive to small changes in the series resistances or shunt capacitances.
- The RC roll-off networks and shunt capacitances to ground minimize any transient noise spikes, which give the circuit a high immunity to noise.

Despite these advantages, the Pierce oscillator needs a relatively high amplifier gain to compensate for relatively high gain losses in the circuitry surrounding the crystal.

Technically, the Pierce and Colpitts oscillators have identical circuit layouts except for the location of the ground point. The basic difference between the circuits is that the crystal in the Pierce is designed to look into the lowest possible impedance across its terminals, whereas the crystal in the Colpitts is designed to look into a high impedance across its terminals. This basic difference results in other differences.

4.3.5.5   Miller Effect

Before we finish discussion on op-amps, it is important to know that in real op-amps, there is generally an internal (sometimes external) capacitor that sets the op-amp's dominant pole and forces it to roll off before other less well-controlled parasitic capacitances would cause that.

In practical IC implementations, the capacitances available are very small (a few *picoFaraday*, pF). The question is: How can such small capacitances give rise to such low cutoff frequencies (a few hundred Hz)? The answer is provided by an understanding of the Miller Effect, which basically explains how an impedance sitting across an amplifier is effectively converted into a smaller impedance at the input and roughly an equal impedance at the output of the amplifier: the amount that the effective input impedance is smaller than the actual one across the amplifier is a function of the amplifier's gain. This explains how an op-amp, with a huge gain and a very small internal capacitance, can act like it has a huge capacitance at its input (a smaller impedance corresponds to a larger capacitance). In essence, the Miller Effect effectively takes impedance across an amplifier and replaces it with smaller impedance at the input and a roughly equivalent one at the output. It should be noted that this effect works only for amplifiers that share a common terminal between the input and output (e.g. ground). The effect works for op-amps with negative and positive gains. However, for op-amps with positive gains greater than one ($A > 1$), the effect gives rise to negative impedances at the input (meaning, reverse polarity of current that flows for an applied voltage, but Ohm's Law still applies). The output impedance is still positive.

4.4 SUMMARY

This chapter has discussed the basic principles of core electronic devices starting from basic electronic op-amp characteristics and applications, including simple comparators, multivibrators, and oscillators.

PROBLEMS

4.1   What kind of an amplifier is depicted in Figure 4Q.1? Find its output voltage.

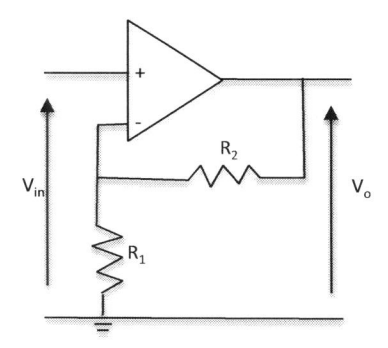

FIGURE 4Q.1   An op-amp.

4.2   What function does a monostable circuit perform?

4.3   Multivibrator circuits are extensively used by systems for generating and processing. How many states do these circuits have? How many types of multivibrators are available and what are the ways of implementing each type?

4.4   Design a summing amplifier to produce the average of three-input voltages. The amplifier must be designed such that each input signal sees the maximum possible input resistance under the condition that the maximum allowed resistance in the circuit is 1.25 MΩ.

4.5   Suppose the output voltage of the amplifier in Figure 4Q.2 is given as $v_o = -(7v_{11} + 14v_{12} + 3.5v_{13} + 10v_{14})$. Suppose $I_- = 0$, and the allowable resistance value is 280 kΩ, design a summing amplifier that produces the output voltage v_o.

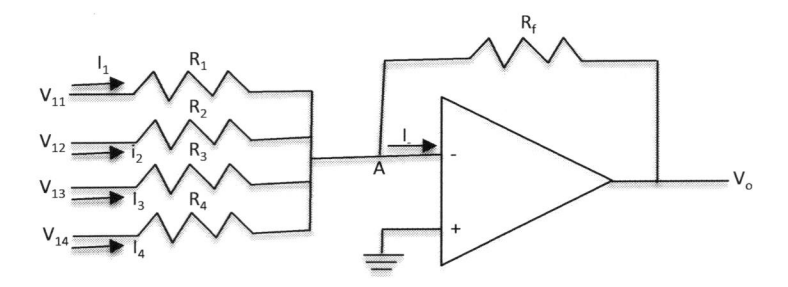

FIGURE 4Q.2   A summing amplifier.

4.6   When is a circuit an oscillator?

4.7   Design a Wien bridge oscillator with a frequency of 1 and 10 kHz.

REFERENCES

1. Boylestad, R. and Nashelsky, L. (1996). *Electronic Devices and Circuit Theory*. Englewood Cliffs, NJ: Prentice Hall College Division. 2. Carter, B. and Huelsman, L.P. (1963). *Handbook of Operational Amplifier Active RC Networks*. Dallas: Texas Instruments Application Report, SBOA093A.

3. Thompson, S.C., Proakis, J.G. and Zeidler, J.R. (2005). The effectiveness of signal clipping for PAPR and total degradation reduction in OFDM systems. *Proceedings of the IEEE GLOBECOM*, St. Louis, MT, pp. 2807–2811.

4. Guel, D. and Palicot, J. (2009). Clipping formulated as an adding signal technique for OFDM Peak Power Reduction. *Proceedings of the IEEE Vehicular Technology Conference*, Barcelona, pp. 26–29.

5. Kolawole, M.O., Adegboyega, G.A. and Temikotan, K.O. (2012). *Basic Electrical Engineering*. Akure: Aoge Publishers.

6. Tasić, A., Serdijn, W.A., Larson, L.E. and Setti, G. (eds) (2009). *Circuits and Systems for Future Generations of Wireless Communications*. New York: Springer.

7. Matthys, R.J. (1992). *Crystal Oscillator Circuits*. Malabar, FL: Krieger Publishing.

5 MOS Field-Effect Transistor (MOSFET) Circuits

As discussed in Chapter 3, an FET operates as a conducting semiconductor channel with two ohmic contacts; namely, the *source* and the *drain*, where the number of charge carriers in the channel is controlled by a third contact called the *gate*. In the vertical direction, the gate junction (i.e. the gate channel-substrate structure) may be regarded as an orthogonal two-terminal device that is either a metal oxide semiconductor (MOS) structure, or a reverse-biased rectifying device, that controls the mobile charge in the channel by capacitive coupling (field effect). An example of FETs based on these principles is *metal-oxide-semiconductor* FET (MOSFET; or simply MOS). As briefly noted in Chapter 3, Section 3.1.1, there occur several variants of MOSFET (or MOS). In the MOS, "metal" is more commonly a heavily doped polysilicon layer: n^+ or p^+ layer. NMOS implies n-type substrate and PMOS implies p-type substrate. The variants of MOSFETs (MOSs) are due to improved technology in devices' fabrication. Prominent variants are gallium arsenide (GaAs) substrate-based *metal-semiconductor FET* (MESFET), *high-electron mobility transistor* (HEMT), and *pseudomorphic* HEMT (or pHEMT). The main difference between HEMTs and MESFETs is the epitaxial layer structure. In the HEMT structure, compositionally different layers are grown in order to optimize and to extend the performance of the FET. These different layers form heterojunctions since each layer has a different bandgap. These structures are shown in Figure 3.4, which is reproduced here, as seen in Figure 5.1.

Structures grown with the same lattice constant but different bandgaps are referred to as lattice-matched HEMTs. Those structures grown with slightly different lattice constants are called pHEMTs. The MESFETs and HEMTs are grown on a semi-insulating GaAs substrate using molecular beam epitaxy (MBE), or metal–organic chemical vapor deposition (MOCVD), which currently is less common. Epitaxy is the natural or artificial growth of crystals on a crystalline substrate determining their orientation. Other commonly used names for HEMTs include MODFET (modulation-doped FET), TEGFET (two-dimensional electron gas FET), and SDHT (selectively doped heterojunction transistor).

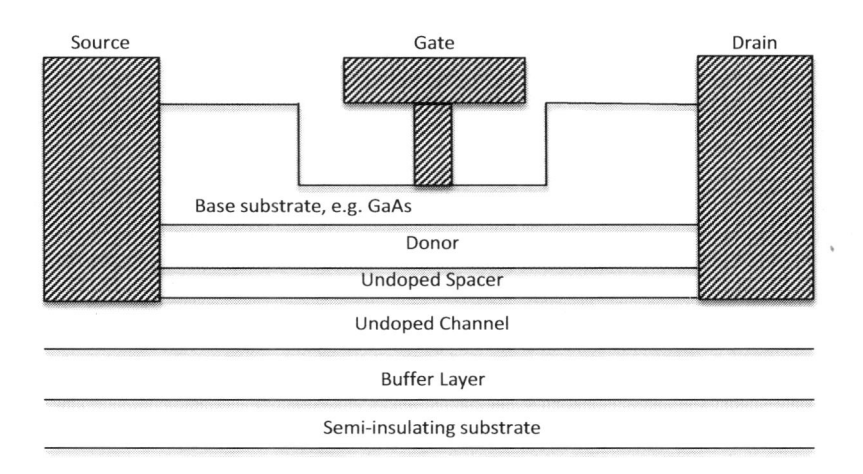

FIGURE 5.1 Structure of a basic HEMT (same as Figure 3.4).

In this chapter, we present the basic structure and characteristics of the MOSFETs (or MOSs), and their arrangement to forming logic gates and digital circuits. A summary of small-signal (incremental) models of two major types of MOSFET devices used in analog, digital, and mixed-mode integrated circuits is also presented. A very important application of MOSFETs (MOSs) is in the arrangement known as a *complementary MOS* (CMOS) *system*, which we will consider in the next chapter (Chapter 6).

5.1 BASIC STRUCTURE OF MOSFET

The MOS transistors are the basic building blocks of most computer chips, as well as of chips that include analog and digital circuits.

Typically, a MOSFET has four nodes: the *source* (S), *drain* (D), the poly-silicon called the *gate* (G), and *bulk* (B). The bulk (B) is also called the body (B); in addition, the bulk is also called the bulk substrate or substrate. An n-type MOS is called NMOS, while the p-type MOS is called PMOS. The structures and notational symbols of NMOS and PMOS are shown in Figures 5.2 and 5.3, respectively. As seen in the MOSFET structural arrangements of n-channel and p-channel of Figures 5.2 and 5.3, respectively, the gate electrode is placed on top of a very thin insulating layer, and there are a pair of small n-type and p-type regions just under the drain and source electrodes. The gate-channel insulator is made of a very thin metal-oxide layer, typically silicon dioxide (SiO_2) of dielectric ε_{ox}, ($\sim 3.9\varepsilon_0$, where ε_0 is the permittivity of vacuum or free space, or simply electric constant). The quantity ε_0 is defined as

$$\varepsilon_o = \frac{1}{\mu_o c^2} = 8.8542 \times 10^{-12} \, \mathrm{F \, m^{-1}} \tag{5.1}$$

where c ($\sim 3 \times 10^8 \mathrm{m \, s^{-1}}$) is the speed of light and μ_0 ($= 4\pi \times 10^{-7} \, \mathrm{H \, m^{-1}}$) is the permeability of free space. The unit F is Farads.

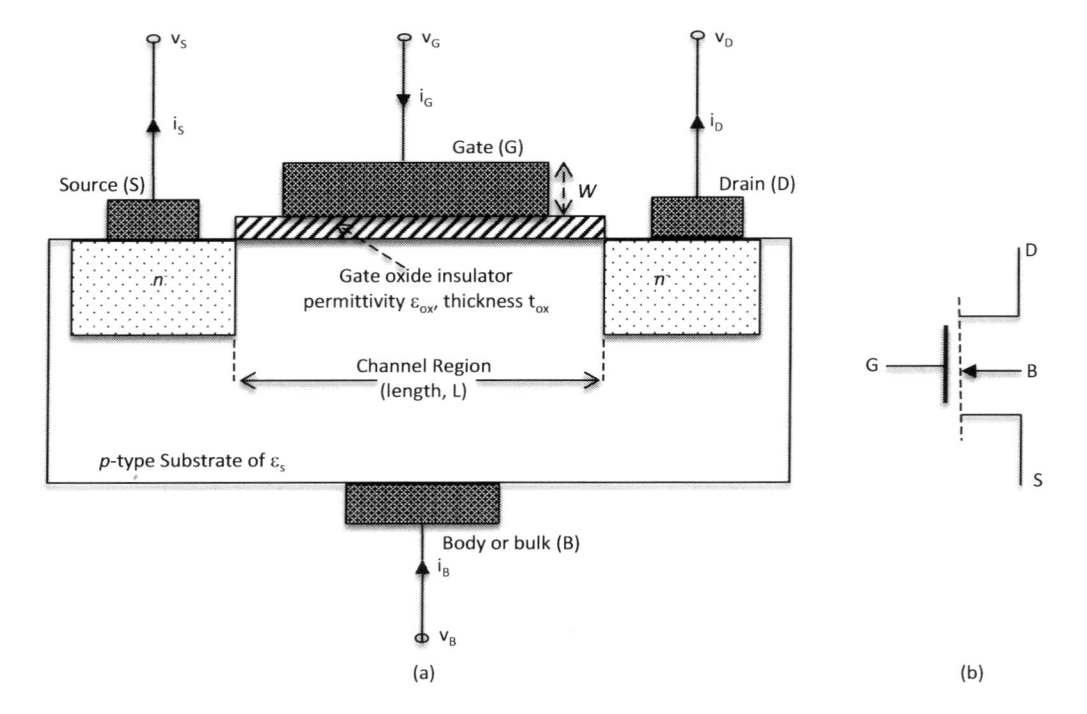

(a)   (b)

FIGURE 5.2   Structure and symbol of n-channel MOSFET: (a) structure of NMOS and (b) symbol.

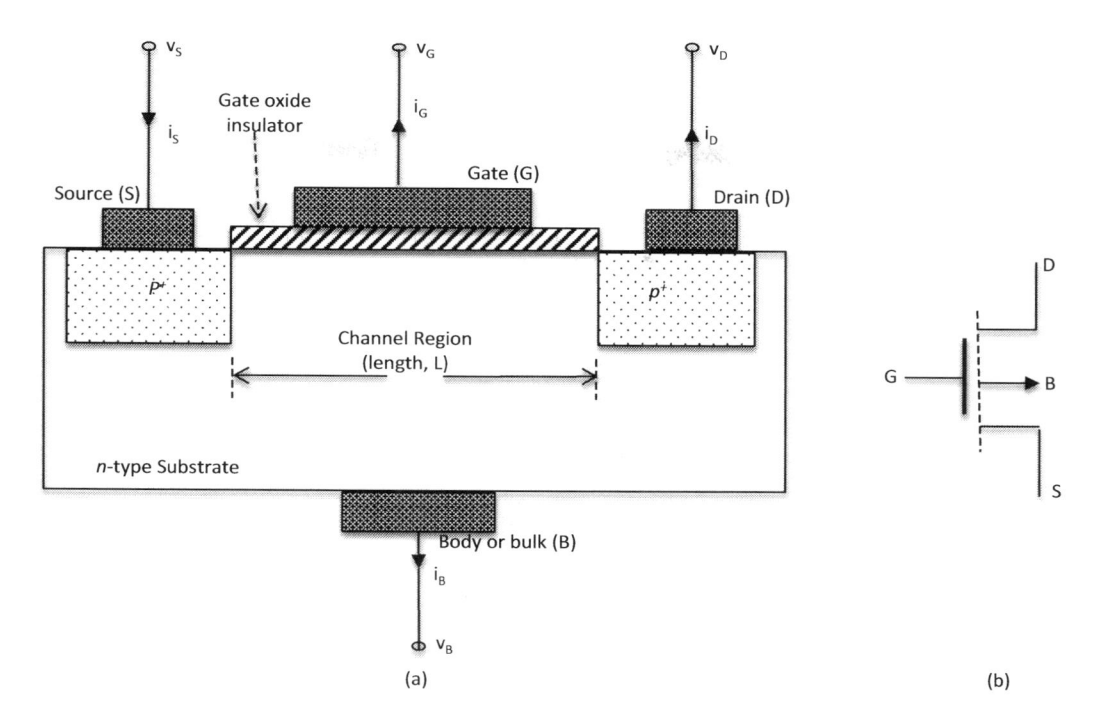

FIGURE 5.3 Structure and symbol of p-channel MOSFET: (a) structure of PMOS and (b) symbol.

There is a difference in the permittivity values of silicon dioxide wafer and the body substrate's ε_s. Typically, $\varepsilon_s \sim 11.7\varepsilon_0$.

The current at two electrodes—drain and source—is controlled by the action of an electric field at the gate electrode. The MOSFET operation depends on its ability to induce and modulate a conducting sheet of minority carriers at the semiconductor–oxide interface. The gate voltage V_G controls the flow of electrons from the source to the drain. When a positive voltage is applied to the gate (i.e. for $+ V_G$), it attracts electrons to the interface between the gate dielectric and the semiconductor. These electrons form a conducting channel between the source and the drain, called the *inversion layer*. No gate current (I_G) is required to maintain the inversion layer at the interface since the gate oxide blocks any carrier flow. The net result is that the gate voltage V_G controls the current between the drain and the source. This process makes MOS to act as a voltage-controlled device.

When the voltage applied between the metal and the semiconductor is more negative $V_G < 0$ (i.e. than the flatband voltage $V_{FB} < 0$), a MOSFET structure with a p-type semiconductor will enter the *accumulation* regime of operation.

The capacitance of the gate-channel-substrate MOSFET structure depends on the bias voltage on the gate and is determined by oxide thickness. The dependence is shown in Figure 5.4.

There are roughly three regimes of operation, namely:

i. *Accumulation*: in this region, carriers of the same type as the substrate (body) accumulates at the surface;

ii. *Depletion*: in this region, the surface is devoid of any carriers leaving only a space charge or depletion layer; and

iii. *Inversion*: in this region, carriers of the opposite type from the body aggregate at the surface to "invert" the conductivity type.

Two voltages: threshold voltage V_{th} and flatband voltage V_{FB} separate the three regimes. In the Figure 5.4, Q_G represents the metal gate charge. (More is said of threshold voltage later in this chapter.)

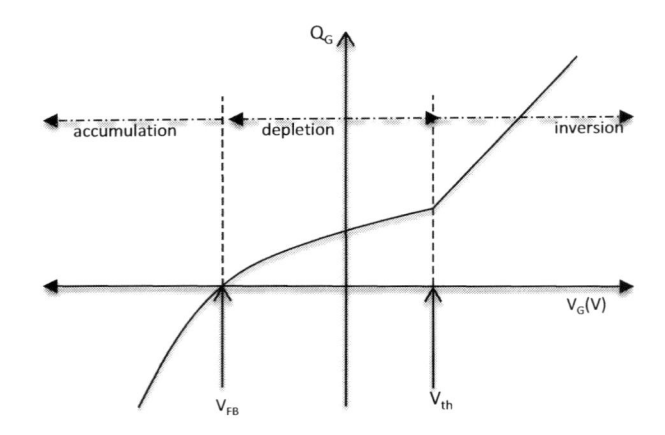

FIGURE 5.4 MOSFET capacitance Q–V curve.

Due to SiO_2 layering, MOSFET has a very high input impedance compared to even reverse-biased p-n junction depletion region input impedance in junction field-effect transistor (JFET).

There are two modes of MOSFET transistors: enhancement and depletion.

If a positive voltage is applied to the gate ($+V_G$), this will push away the "holes" inside the p-type substrate and attracts the moveable electrons in the n-type regions under the source and drain electrodes. When the positive gate voltage is increased, it pushes the p-type holes further away and enlarges the thickness of the created channel. This process increases the amount of current that can go from source to the drain: this is the reason transistor of this kind is called an *enhancement-mode* MOSFET. This is the opposite way around to a JFET, where increasing the gate voltage tends to reduce the drain-source current which will flow when we do not apply a specific gate voltage.

Suppose a narrow channel is embedded into the substrate between the n-type regions for drain and source, as shown in Figure 5.5. When V_{DS} is 0 V, negative gate voltage ($-V_G$) induces a positive charge into the channel near the interface with the oxide layer. The recombination of induced positive charge with the existing negative charge in the channel causes a depletion of majority carriers. This action is the reason transistor of this kind is called *depletion-mode* MOSFET. If the gate voltage V_G is made more negative, majority carriers may be virtually depleted, which in effect the channel can be eliminated. Under these circumstances, the drain current I_D is zero ($I_D = 0$). In essence, the term *depletion mode* means that a channel exists even with zero gate voltage. The least negative value of gate-source voltage (V_{GS}) for which the channel is depleted of majority carriers is called the *threshold voltage* V_{th}. This *voltage* is positive for the *enhancement-mode* MOSFETs.

The threshold voltage V_{th} is the minimum gate-to-source voltage differential that is needed to create a conducting path between the source and drain terminals, and is one of the most important parameters characterizing metal-insulator-semiconductor devices. V_{th} is dependent on the doping density and some physical parameters that characterize the MOSFET structure, including the gate material and the substrate bias voltage V_{sb}.

The physical construction of an enhancement MOSFET is identical to the depletion MOSFET, with one exception: the conduction channel is physically implanted rather than induced—as in enhancement type. Both types—enhancement (Figure 5.6a and b) and depletion (Figure 5.6c and d) MOSFETs—are precisely identical in nearly every way: same modes, same equations, and same terminal names. Engineers control whether a device is an enhancement or depletion type by adding carefully controlled amounts of dopants (impurities) in the semiconductor.

In essence, the current through the channel region is controlled with gate voltage, V_G. In practice though, the bulk B must be properly connected before power is applied. However, when the

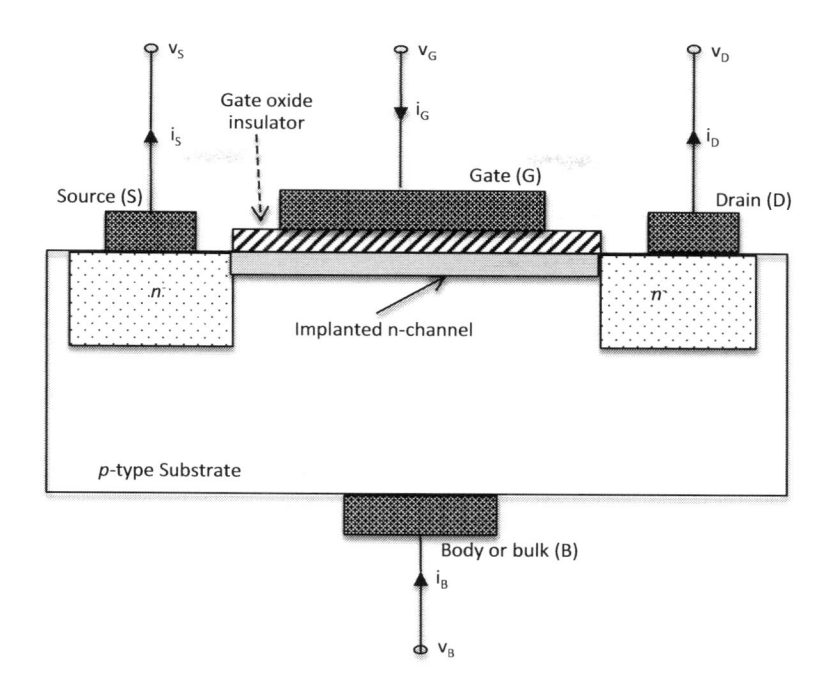

FIGURE 5.5 Structure of depleted MOSFET.

source-bulk connection is shorted—internally connected to the source—i.e. $V_{SB} = 0.0$ V, new symbols are formed, as shown in Figure 5.6e–j. Figure 5.6g and i are the most commonly used symbols in *Very Large-Scale Integration* (VLSI) logic design. More is said about VLSI in Chapter 7. By using complementary MOS transistors, CMOS circuit is formed. (More is said about CMOS in Chapter 6.)

It is worth noting that the source and drain regions of the FET are interchangeable, in contrast to the bipolar transistors whose emitter and collector regions are not interchangeable, as their emitter is much more heavily doped than the collector. Almost all logic circuits, microprocessor and memory chips contain exclusively MOSFETs and their variants. The basic characteristics of MOSFET circuits are discussed in the next section.

5.2 MOSFET CHARACTERISTICS

Suppose we connect two external voltage sources V_{s1} and V_{s2} to MOSFET terminals, as shown in Figure 5.7a. These two external voltage sources (V_{s1}, V_{s2}) provide the drain-source voltage V_{DS} and the gate-source voltage V_{GS}. The voltage V_{DS} may cause a drain-to-source current I_D, provided that there is a path for this current from the drain (through the device) to the source. Whether such a path exists or not depends on the value of V_{GS}. For small values of V_{GS}, no such path is established within the device, and the drain current I_D is zero. The MOSFET device then looks like an open circuit between the drain and the source. For a sufficiently high V_{GS}, an internal current path, called a channel, is established between the drain and the source. Now the drain current I_D can flow. The precise value of V_{GS} determines how easy it is for the channel to conduct the current; the higher the V_{GS} value, the easier such conduction is and the larger the value of I_D, other things being equal. For a given V_{GS}, the value of I_D will also depend on V_{DS} and will tend to increase with the latter. Figure 5.7b and c show typical current versus voltage characteristic curves of a MOSFET for (I_D–V_{DS}) and (I_D–V_{GS}), respectively.

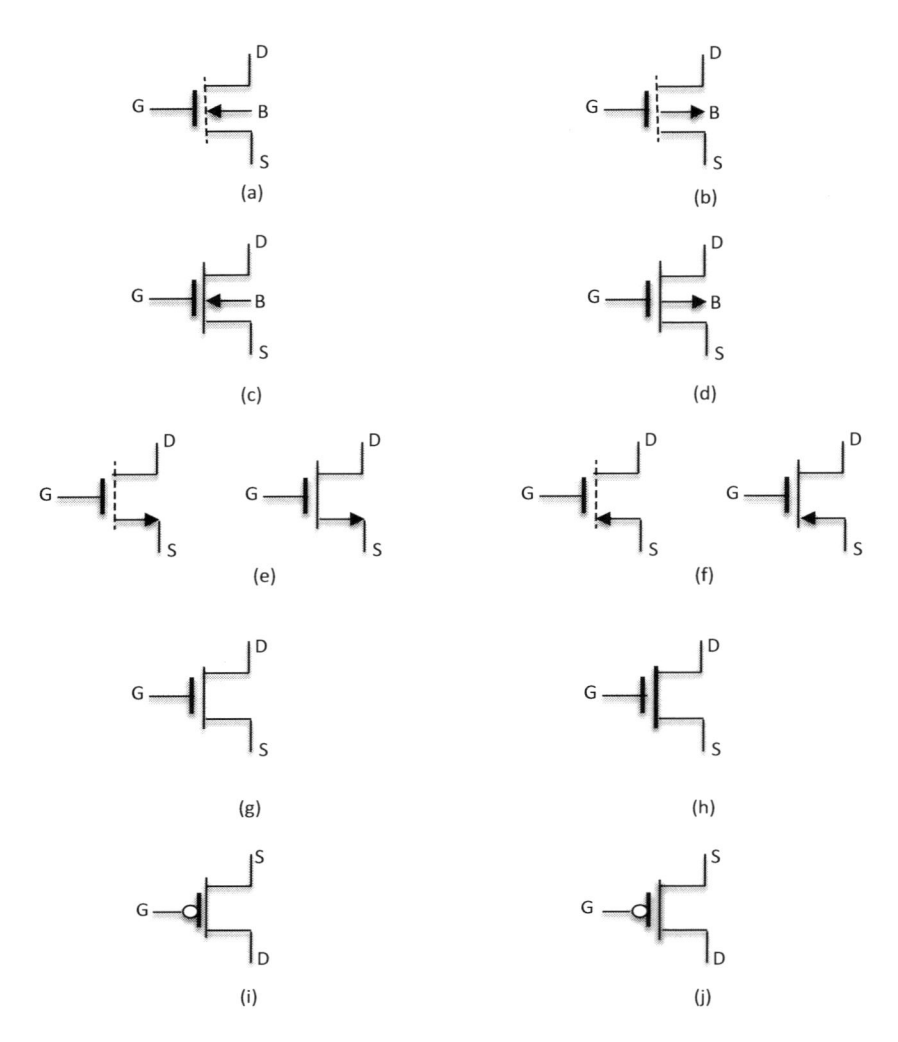

FIGURE 5.6 Symbols representing *MOSFET* circuits: (a) *NMOS* enhancement-mode transistor, (b) *PMOS* enhancement-mode transistor, (c) *NMOS* depletion-mode transistor, (d) *PMOS* depletion-mode transistor, (e) three-terminal *NMOS* transistor, (f) three-terminal *PMOS* transistor, (g) shorthand notation of *NMOS* enhancement-mode transistor, (h) shorthand notation of *NMOS* depletion-mode transistor, (i) shorthand notation of *PMOS* enhancement-mode transistor, and (j) shorthand notation of *PMOS* depletion-mode transistor.

As depicted in Figure 5.7b, two regions (linear and saturation) are depicted by the behavior of I_D and V_{DS}. In the *linear region,* for small values of V_{DS} (i.e. $V_{DS} < V_{GS} - V_{th}$), an increase in V_{DS} is accompanied by an increase in drain current I_D. MOSFET behaves like a resistor in this region: the reason it is called *ohmic region (also called non-saturation region* or *triode region).* However, as V_{DS} increases, the voltage drops across the gate oxide at the drain side of the channel V_{GD} (= $V_{DS} - V_{GS}$) decreases. This reduced voltage lowers the field across the channel. The channel is seen as being "pinched off," and the drain current I_D increases much more slowly with respect to increases in V_{DS} than in the ohmic region near the origin. Ideally, once pinch-off is achieved, further increases in V_{DS} produces no change in I_D and the current droops. The region where this occurs is called *saturation region.* Technically, the saturation drain voltage V_{SAT} equates to threshold voltage (i.e. $V_{SAT} = V_{th}$). The saturation region is in fact similar in nature to velocity saturation in the JFET. Increases in V_{GS} (i.e. $V_{GS} > V_{th}$) result in increasing saturation values of I_D.

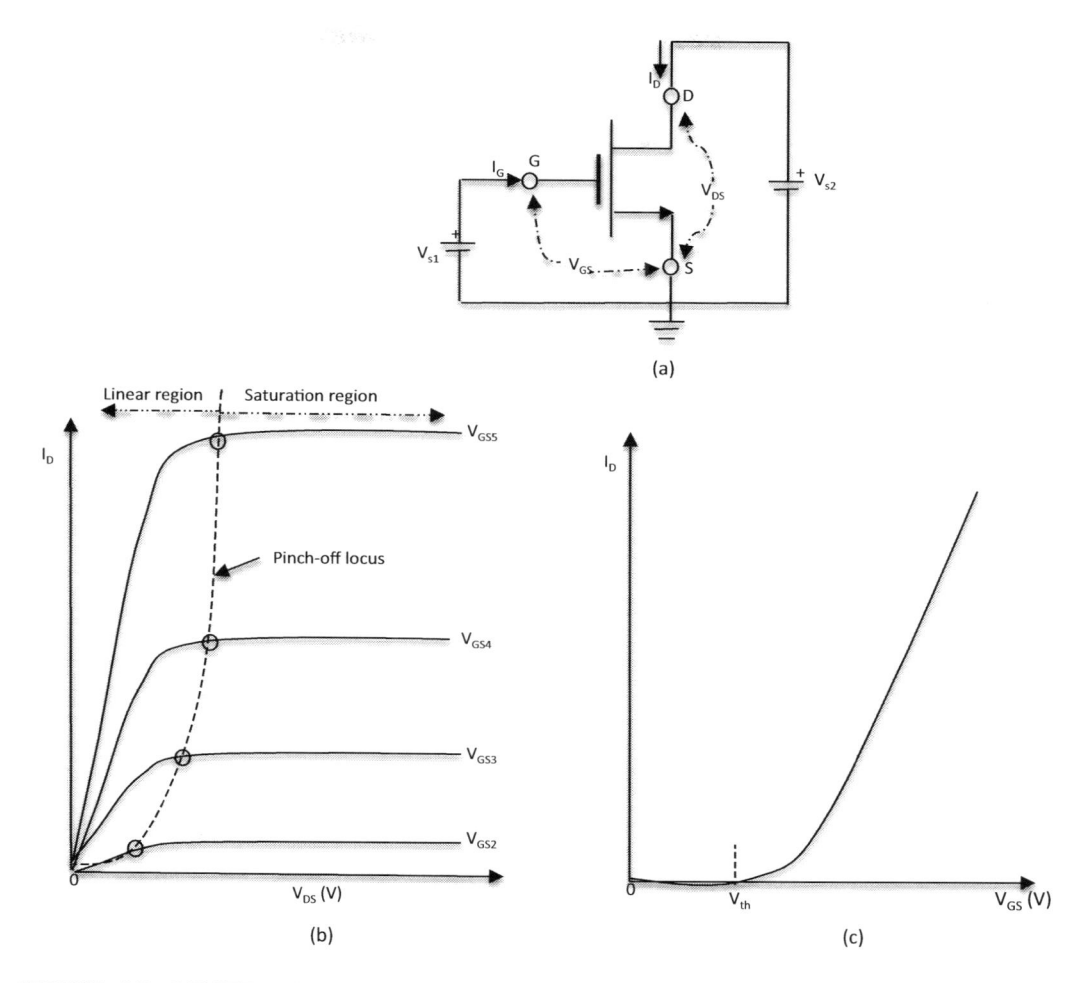

FIGURE 5.7 MOSFET characteristics: (a) MOSFET bias, (b) I_D–V_{DS} characteristics, and (c) I_D–V_{GS} characteristic.

A good engineering approximation of the currents in the two regions—*ohmic* and *saturation*—can be expressed thus

a. *Linear region*:

$$I_D = \kappa \left(\frac{W}{L} \right) \left[2(V_{GS} - V_{th})V_{DS} - V_{DS}^2 \right] \tag{5.2}$$

The ratio (W/L) is called the *aspect ratio*, which serves as a scale factor for the drain current.

b. *Saturation region*:

$$I_{D(\text{SAT})} = \kappa \left(\frac{W}{L} \right) (V_{GS} - V_{th})^2 (1 \pm \gamma V_{DS}) \tag{5.3}$$

where the sign "±" changes the relation for NMOS and PMOS. When current flows from drain-to-source as in NMOS "+" in Equation (5.3) is used while "−" is used for PMOS where current flows from source-to-drain. In reality, in the saturation region, I_D increases slightly with V_{DS}. To accommodate for this slight increase, the *body-effect* factor γ is introduced. (γ is also called *channel length modulation factor*.) If the actual MOS characteristics are extended back into the second quadrant (just as in *bipolar junction transistor* (BJT), Chapter 3, Section 3.5.2), they meet at $V_{DS} = -\dfrac{1}{\gamma}$, which is referred to as the *Early-Effect* factor. Also, quantifiable using

$$\gamma = \frac{\sqrt{2q\varepsilon_s N_A}}{C_{\text{ox}}}\left(V^{-1}\right) \tag{5.4}$$

where N_A is the density of doped carriers (given)—a negative quantity for *p*-type and a positive quantity for *n*-type, ε_s is the body substrate's permittivity, q is the electric charge carried by a single electron or proton ($= 1.6 \times 10^{-19}$ Coulombs), and C_{ox} is the thin-oxide field capacitance under the gate region (in Farads, F), which is defined as

$$C_{\text{ox}} = \frac{C_o}{A_c} = \frac{\varepsilon_{\text{ox}}}{t_{\text{ox}}} \tag{5.5}$$

where A_c, ε_{ox}, and t_{ox} correspond to the capacitor area, permittivity, and thickness, respectively, of the silicon-oxide wafer. C_{ox} is a very important parameter in MOSFET technology.

The values of γ depend both on the process technology used to fabricate the device and on the channel length L that the circuit designer selects. Typical values of γ are in the range 0.01–0.03 V^{-1}. For completeness, the term $\left(1 \pm \gamma V_{DS}\right)$ in Equation (5.3) is included to account for the channel-length modulation effect—an affect analogous to the base-width modulation in BJT—which is usually negligible in digital circuits but not analog circuits.

Note that the way the channel length L is manipulated affects the operation speed and the number of components that can be fitted onto a chip. Increasing or reducing channel length L arises another concept called *long-channel* or *short-channel* effects, respectively. Both channel effects are attributed to two physical phenomena: the limitation imposed on electron drift characteristics in the channel and the modification of the threshold voltage V_{th} due to the shortening channel length. The designer goal therefore is to design a MOSFET that is of small length as possible, or make a channel length much greater than the sum of the drain and source depletion widths, ensuring no leakage paths from drain to source, without channel effects compromising the device performance much.

c. *Pinch-off*:

$$I_{D(\text{pinch})} = \kappa\left(\frac{W}{L}\right)V_{DS}^2 \tag{5.6}$$

where W, L, and κ are channel width, channel length, and process parameter, respectively. The common term $\left\{\kappa\left(\dfrac{W}{L}\right)\right\}$ that appears in Equations (5.2)–(5.6) is called *transconductance parameter*, and is denoted as

$$\zeta = \kappa\left(\frac{W}{L}\right) \tag{5.7a}$$

and the process parameter κ is defined as

$$\kappa = \mu_n C_o / 2 \tag{5.7b}$$

where μ_n is electron mobility. It is worth noting that V_{th} also depends on C_{ox} as well as the doping densities of the n-type drain and source and p-substrate, as will be seen in the next few paragraphs.

Any two conductors separated by an insulator have capacitance: as a consequence, a series of capacitors or parasitics arise in MOSFET structures or topologies. More is discussed of these parasitics later in this chapter, as well as in Chapter 6. We can deduce from the preceding equations that

i.   Two (or more) MOSFETs having the same value of V_{th} but with different current capabilities can be fabricated on the same chip by using different aspect ratios.
ii. Given process parameter range κ, high values of I_D are obtainable with high aspect ratio.

d. *Cut-off*: MOSFET devices are in cutoff mode when the "input" voltage is below a threshold value for NMOS; that is, when $I_D = 0$; $V_{GS} < V_{th}$, or when the "input" voltage is greater than a threshold value for PMOS; that is, when $I_D = 0$; $V_{GS} > V_{th}$. Note that V_{th} is negative for a PMOS.

Important device parameters are the channel conductance and transconductance, which are defined as follows.

e. *Channel conductance*:

$$g_d = \left.\frac{dI}{dV_{DS}}\right|_{V_{GS}} = \begin{cases} \zeta(V_{th} - V_{DS}) \\ 0 \end{cases} \quad \text{for} \quad \begin{matrix} V_{DS} \leq V_{SAT} \\ V_{DS} > V_{SAT} \end{matrix} \tag{5.8}$$

f. *Transconductance*: defines the gain of the MOSFET—can be expressed as

$$g_m = \left.\frac{dI}{dV_{GS}}\right|_{V_{DS}} = \begin{cases} \zeta V_{DS} \\ \zeta(V_{GS} - V_{th}) \end{cases} \quad \text{for} \quad \begin{matrix} V_{DS} \ll (V_{GS} - V_{th}) \\ V_{DS} > (V_{GS} - V_{th}) \end{matrix} \tag{5.9}$$

Transconductance is usually measured at saturation region with fixed V_{DS}. As can be seen from Equations (5.8) and (5.9), high values of channel conductance g_d and transconductance g_m are obtained for large electron mobilities, large gate insulator capacitances C_{ox} (i.e. thin gate insulator layers), and large aspect ratios.

The threshold voltage's base formula V_{th} can be expressed as

$$V_{th} = V_{t\text{-mos}} + V_{FB} \tag{5.10}$$

where $V_{t\text{-mos}}$ is the ideal threshold voltage—meaning that there is no work function difference between the gate and substrate materials and V_{FB} is the flatband voltage that corresponds to the voltage which when applied to the gate electrode yields a flat energy band in the semiconductor. Naturally, the charge in the oxide or at the interface changes V_{FB}.

$$V_{t\text{-mos}} = 2\Phi_b + \frac{\sqrt{2q\varepsilon_s N_A (2\Phi_b + |V_{sb}|)}}{C_{ox}} \tag{5.11}$$

where Φ_b is the semiconductor *work function*, defined by

$$\Phi_b = \frac{kT}{q} \ln\left(\frac{N_A}{n_i}\right)$$

$$= V_T \ln\left(\frac{N_A}{n_i}\right) \tag{5.12}$$

The work function varies with the type of metal deposited onto silicon dioxide (SiO_2). This metal may be Magnesium (Mg), Aluminum (Al), Copper (Cu), Silver (Ag), Nickel (Ni), or Gold (Au). Note that Φ_b for SiO_2 is not exactly the same as that of the metal in vacuum.

The substrate bias voltage V_{sb} is often given; if not, it can be taken as zero.

T = absolute temperature (in kelvins, K)

k = Boltzmann's constant = 1.38×10^{-23} J K^{-1}

n_i = intrinsic carrier density (in silicon) = 1.45×10^{10} cm^3

V_T = thermal voltage.

Other terms are as previously defined.

$$V_{FB} = -\left(\frac{E_g}{2} \pm \Phi_b\right) - \frac{Q_{fc}}{C_{ox}} \tag{5.13}$$

where Q_{fc} is a fixed charge for surface states (often given). Again, like in Equation (5.3), the $\pm$ sign in bracket (.) of Equation (5.13) determines the MOS type: plus "+" if NMOS device, and minus "−" if PMOS device. In some texts, the bracket (.) is called the work function difference between the gate and the wafer. E_g is the bandgap energy of silicon, which is temperature dependent, given as

$$E_g = 1.16 - \frac{0.704 \times 10^{-3} T^2}{1108 + T} \tag{5.14}$$

E_g is the same as the energy required to break the *covalent bond* of doping materials. For instance, at room temperature E_g for silicon is about 1.1 eV and that for germanium is about 0.7 eV.

Since any additional charge affects the flatband voltage V_{FB} and thereby also the threshold voltage V_{th}, great care has to be taken during fabrication to avoid the incorporation of charged ions as well as creation of surface states.

In view of Equations (5.11)–(5.13) in Equation (5.10), the threshold voltage can be expressed thus

$$V_{th} = 2\Phi_b + \frac{\sqrt{2q\varepsilon_{ox}N_A\left(2\Phi_b + |V_{sb}|\right)}}{C_{ox}} - \left(\frac{E_g}{2} \pm \Phi_b\right) - \frac{Q_{fc}}{C_{ox}}$$

$$= 2\Phi_b + \gamma\sqrt{\left(2\Phi_b + |V_{sb}|\right)} - \left(\frac{E_g}{2} \pm \Phi_b\right) - \frac{Q_{fc}}{C_{ox}} \tag{5.15}$$

Advances in technology have enabled fabrication of many MOFETs (MOSs) into a chip. As an example, CD4007 contains six MOSFETs, three *PMOS* and three *NMOS* transistors, which includes an inverter pair. The transistors are accessible via the 14-pin terminals. Figure 5.8 shows the 14-pin connection diagram and the corresponding schematic of the transistor array package. Fabrication process changes and improves with technology.

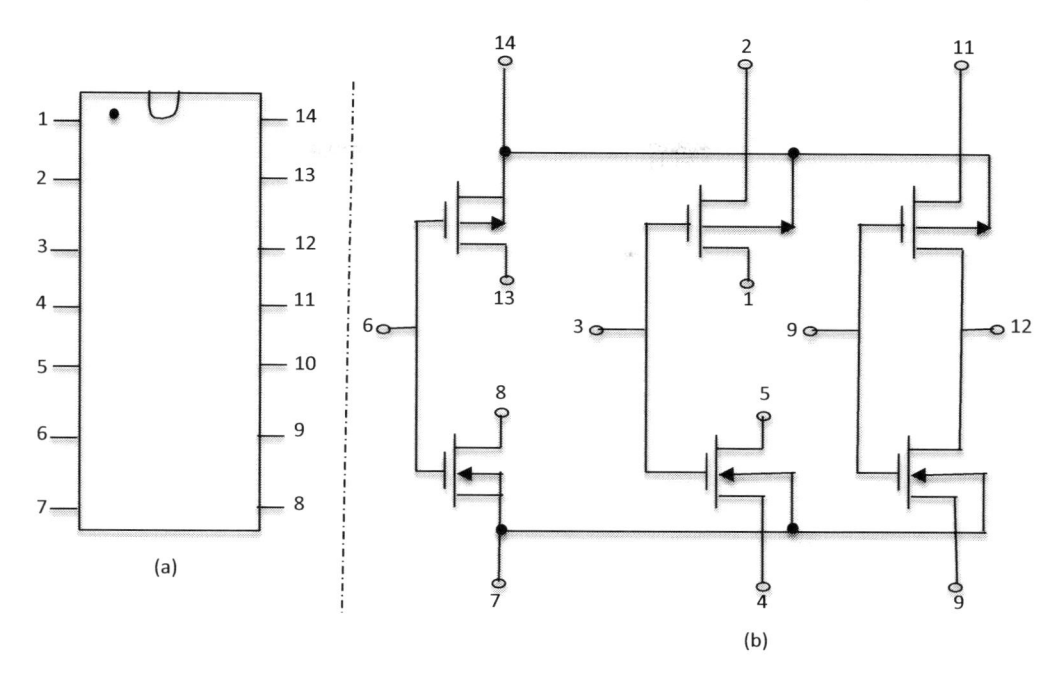

FIGURE 5.8 MOSFET array package in CD4007 chip: (a) CD4007 pin numbering and (b) MOSFETs arrangement in CD4007 chip.

Example 5.1

a. Calculate the oxide capacitance per unit area for $t_{ox} = 1.5$ nm and 10.0 nm assuming $\varepsilon_{ox} = 3.9\varepsilon_0$, where the permittivity of free space $\varepsilon_0 = 8.8542 \times 10^{-14}$ F cm^{-1}.

b. Determine the area of a 1.8 µF metal-oxide-metal capacitor for the two oxide thicknesses given in (a).

c. Estimate the body-effect factor for $t_{ox} = 1.5$ nm, $\varepsilon_{ox} = 3.9\varepsilon_0$, body substrate's permittivity $\varepsilon_s = 11.7\varepsilon_0$, the electric charge carried by a single electron or proton is 1.6×10^{-19} (Coulombs), and $N_A = 10^{16}$ atoms cm^{-2}.

Solution:

a. From Equation (5.5), the oxide capacitance per unit area can be expressed as

$$C_{ox} = \frac{\varepsilon_{ox}}{t_{ox}}$$

and substituting values and noting the units of each term, we have for

i. $t_{ox} = 1.5$ nm, $C_{ox} = \dfrac{3.9 \times 8.8542 \times 10^{-14}\left(F\,cm^{-1}\right)}{1.5 \times 10^{-7}\,(cm)} = 2.302\ \mu F\,cm^{-2}$

ii. $t_{ox} = 10.0$ nm, $C_{ox} = \dfrac{3.9 \times 8.8542 \times 10^{-14}\left(F\,cm^{-1}\right)}{1.0 \times 10^{-6}\,(cm)} = 345.3\ nF\,cm^{-2}$

b. Rearrange Equation (5.5) in terms of A_c, and noting the units of each term, the capacitor areas are

i. $t_{ox} = 1.5\,nm,$

$$A_c = \frac{C_{ox}t_{ox}}{\varepsilon_{ox}} = \frac{1.8\times10^{-6}\times1.5\times10^{-7}\,(Fcm)}{3.9\times8.8542\times10^{-14}\,\left(Fcm^{-1}\right)}$$

$$= \frac{1.8\times15\times10^{-14}}{3.9\times8.8542\times10^{-14}}\,cm^2 = 0.7819\,cm^2\;or\;78.19\,mm^2$$

ii. $t_{ox} = 10.0\,nm$, following same process as (a)(i), the capacitor area
    $A_c = 5.213\,cm^2$ or $521.3\,mm^2$

c. Given:
    $\varepsilon_s = 11.7\varepsilon_{0'} = 11.7 \times 8.8542 \times 10^{-15}$ F m^{-1} = $11.7 \times 8.8542 \times 10^{-14}$ F cm^{-1}
    $q = 1.6 \times 10^{-19}$ (Coulombs)
    $N_A = 10^{16}$ atoms cm^{-2}.
    and for $t_{ox} = 1.5\,nm$, from (i) of item (a),
    $C_{ox} = 2.302\ \mu F\ cm^{-2} = 2.302 \times 10^{-6}$ F cm^{-2}
    By substituting these values in Equation (5.4), the *body-effect* factor is estimated as

$$\gamma = \frac{\sqrt{2q\varepsilon_s N_A}}{C_{ox}} = \frac{10^{-8}}{2.302\times10^{-6}}\sqrt{2\times1.6\times1.17\times8.8542} = 0.02501\,V^{-1}$$

Example 5.2

Consider the NMOS circuit in Figure 5.9 with $\kappa = 0.5$ mA V^{-2}, threshold voltage, $V_{th} = 2$ V, supplied voltage, $V_{DD} = 12$ V, resistor $R_D = 0.5$ kΩ, and unity *aspect ratio*. Find v_o and i_D when $v_{in} = 1.6$ and 9 V.

Solution:

Using Kirchhoff's voltage law for the loops (a) node GS and (b) node DS in Figure 5.9, we have

$$(a)\ GS : v_{GS} = v_{in} \tag{5.16}$$

$$(b)\ DS : V_{DD} = R_D i_D + v_{DS} \tag{5.17}$$

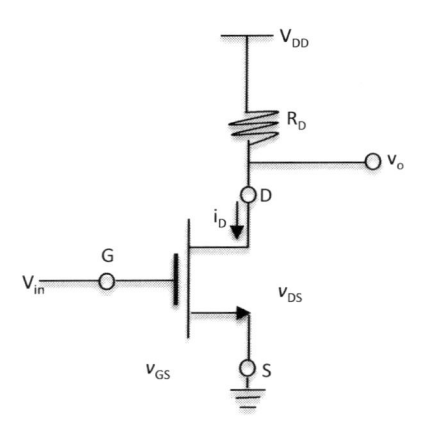

FIGURE 5.9   A simple NMOS circuit.

i. As $v_{in} = 1.6\,V$, from (5.16) we get $v_{GS} = v_{in} = 1.6\,V$. Since $v_{GS} < V_{th}$, then the circuit is in cutoff state and $i_D = 0$. The output voltage v_o can be obtained from Equation (5.17) as

$$v_{DS} = V_{DD} = 12\,V$$

ii. As $v_{in} = 9\,V$, from Equation (5.16) we get $v_{in} = v_{GS} = 9\,V$. Since $v_{GS} > V_{th}$ then the circuit is considered not in the cutoff state. Assume the NMOS is in the active state. Then, from Equation (5.2), the current is

$$i_D = \kappa \left(\frac{W}{L}\right)\left[2(v_{GS} - V_{th})v_{DS} - v_{DS}^2\right]$$

$$= 0.5 \times 10^{-3}(1)\left[2(9-2)v_{DS} - v_{DS}^2\right]$$

$$= 0.5 \times 10^{-3}\left[14 v_{DS} - v_{DS}^2\right] \tag{5.18}$$

Substituting i_D in Equation (5.17) to get

$$12 = 0.5 \times 10^3 \left\{0.5 \times 10^{-3}\left[14 v_{DS} - v_{DS}^2\right]\right\} + v_{DS} \tag{5.19}$$

Rearranging Equation (5.19) to have

$$v_{DS}^2 - 18 v_{DS} + 48 = 0 \tag{5.20}$$

which is a quadratic equation in v_{DS}. The next thing is to have the roots of v_{DS}.
   Recall: the roots of a quadratic equation of a single variable x:

$$ax^2 + bx + c = 0$$

can be solved using the coefficients from

$$x = \frac{-b \pm \sqrt{b^2 - 4ac}}{2a}$$

provided coefficient $a \neq 0$.
   Consequently, the two roots of v_{DS} are: 3.255 and 14.745 V. The second root is not realistic as the circuit is supplied by a 12 V. Therefore, $v_{ds} = v_o = 3.255\,V$. The current i_D is obtained either by substituting $v_{DS} = 3.255\,V$ in Equation (5.18);

$$i_D = 0.5 \times 10^{-3}\left[14 v_{DS} - v_{DS}^2\right] = 0.5\left[(14 \times 3.255) - 3.255^2\right] = 17.49\,mA$$

Or, by substituting $v_{DS}\,(= 3.255\,V)$ in Equation (5.17), to have

$$i_D = \frac{V_{DD} - v_{DS}}{R_D} = \frac{12 - 3.255}{0.5 \times 10^3} = 17.49\,mA$$

An inquiring reader might ask why choose Equation (5.2) not Equation (5.3) to solving i_D? The answer becomes obvious soon. Suppose we use Equation (5.3) instead of Equation (5.2). We know that $0.01 \leq \gamma \leq 0.03\,V^{-1}$. Suppose we choose $\gamma = 0.03\,V^{-1}$, from Equation (5.3) we can write for NMOS

$$i_D = 0.5 \times 10^{-3}(9-2)^2(1 + 0.03 V_{DS})$$

$$= 0.245(1 + 0.03 V_{DS})$$

Substituting this in Equation (5.17), and noting that $V_{DD} = 12V$, to have

$$12 = 0.5 \times 10^3 \times 0.245(1 + 0.03v_{DS}) + v_{DS}$$

$$= 12.25 + 0.3675v_{DS} + v_{DS}$$

$$-0.25 = 1.3675v_{DS}$$

which makes $v_{DS} = -0.1828\,\text{V}$

Since $v_{DS} = -0.1828\,\text{V} < \{v_{GS} - V_{th}\}\,(= 7\,\text{V})$ then NMOS is NOT in the active region. The previous assumption of using Equation (5.2) as being in active state is valid.

5.3 MOSFET'S SMALL-SIGNAL AND LARGE-SIGNAL MODELS

In real circuit operation, the MOSFET devices operate under time-varying terminal voltages. Depending on the magnitude of the time-varying voltages, the dynamic operation of the devices can be classified as either small-signal or large-signal operations. If the variation in voltages is sufficiently small, then a small-signal model can be used to represent the device with its linear elements such as resistors, capacitors, and current source. However, when the variation in voltages is large, i.e. changes that are of the same magnitude as the device operating parameters, an analytical, nonlinear "large-signal" model can represent the device. The device's capacitive effects influence the dynamic operations of both small- and large-signal models. Thus, a capacitance model describing the intrinsic and extrinsic components of the device capacitance is another essential part of a MOSFET circuit model. In most cases, the same capacitance model can be used for both large-signal transient analysis and small-signal ac analysis [2].

5.3.1 SMALL-SIGNAL MODEL FOR MOSFET

Like in BJTs, discussed in Chapter 2, Section 2.5, small-signal and large-signal models can be developed for MOSFETs (or MOSs). The small-signal regime is a very important mode of operation of MOSs, as well as for other active devices. As in most semiconductor devices, MOS transistor has a number of operating modes described by nonlinear equations. As in the analysis of signal processing circuits, a convenient approach to describing the nonlinearity is to find the dc operating point (also called the *quiescent* point or *Q*-point, as shown in Figure 5.10) and then linearize the

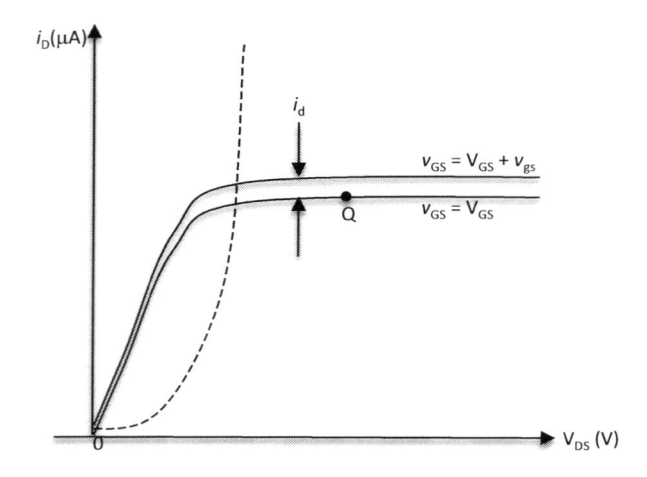

FIGURE 5.10   Illustration of nonlinearity of MOS I_D–V_{DS} characteristics.

characteristics around the dc bias operating point. In doing so, we would be able to examine what happens with the signals that vary around the dc bias point.

Linearization of the device characteristics amounts to taking a Taylor expansion of the device nonlinear characteristic around the Q-point. Suppose the dc bias voltages and currents at the Q-point are V_{GS}, I_D, and V_{DS}. The increment in drain current i_d is due to an increment in gate-source voltage $\Delta V_{GS} = v_{gs}$ when the MOSFET is saturated—with all other voltages held constant. So, the total voltage v_{GS} has the dc component V_{GS} and ac signal component $v_{gs} = \Delta V_{GS}$,

$$v_{GS} = V_{GS} + \Delta V_{GS} = \underbrace{V_{GS}}_{dc} + \underbrace{v_{gs}}_{ac} \tag{5.21}$$

And, the total drain current i_D can be written as a linear summation of dc component I_D and the incremental signal ac component i_d:

$$i_D = \underbrace{I_D}_{DC} + \underbrace{i_d}_{AC} \tag{5.22}$$

The next task is to find what constitute signal components v_{gs} and i_d?

By doing a Taylor expansion around the dc operating point—Q-point, for instance

$$i_D = I_D + \left.\frac{\partial i_D}{\partial v_{GS}}\right|_Q (v_{gs}) + \frac{1}{2}\left.\frac{\partial^2 i_D}{\partial v_{GS}^2}\right|_Q (v_{gs})^2 + \frac{1}{6}\left.\frac{\partial^3 i_D}{\partial v_{GS}^3}\right|_Q (v_{gs})^3 + \cdots \tag{5.23}$$

If the small-signal voltage is really *small*, then we can neglect all terms beyond the linear term; that is, neglecting all higher-order terms in the Taylor expansion beyond the first partial derivative. Hence Equation (5.23) reduces to

$$i_D = I_D + \left.\frac{\partial i_D}{\partial v_{GS}}\right|_Q (v_{gs}) = I_D + \underbrace{g_m v_{gs}}_{i_d} \tag{5.24}$$

where the partial derivative becomes the *transconductance*, g_m. We have seen in Equation (5.3) the functional dependence of the total drain current in saturation, and by evaluating its first partial derivative we have the equivalent expression in Equation (5.9) for the case of $V_{DS} > (V_{GS} - V_{th})$.

The small-signal drain current i_d due to v_{gs} is therefore given by

$$i_d = g_m v_{gs} \tag{5.25}$$

leading to

$$i_D = \frac{1}{r_o} v_{ds} + \underbrace{g_m v_{gs}}_{i_d} \tag{5.26}$$

Another important effect is that of *backgate bias* (also called the "body effect"). As explained in Section 5.1, every MOSFET is actually a four-terminal device, and one must recognize that variations in the potential of the bulk relative to the other device terminals will influence device characteristics. Although the source and bulk terminals are usually tied together, there are important instances when they are not. As such, Equation (5.26) is valid for the case where the effect of substrate potential is neglected. The substrate potential effect changes threshold voltage, which changes the drain current, and as a consequence the substrate acts like a *backgate*. This effect must be included in Equation (5.26), thus

$$i_D = \frac{1}{r_o} v_{ds} + \underbrace{g_m v_{gs}}_{i_d} + \underbrace{g_{mb} v_{bs}}_{\text{backgate}} \tag{5.27}$$

where the backgate transconductance, at the quiescent-point Q, is defined as

$$g_{mb} = \frac{di_D}{dv_{BS}}\bigg|_Q = -\frac{\gamma g_m}{2\sqrt{2\psi_s + V_{BS}}} \tag{5.28}$$

The term g_{mb} is dependent on v_{BS} in both saturation and non-saturation regions, where ψ_s is the surface potential, which is the change in potential from bulk to surface, and defined in terms of gate-substrate voltage V_G as

$$\psi_s = V_G + \frac{Q_G}{C_{ox}} \tag{5.29}$$

Note that the surface potential determines the carrier concentration. In practice, $g_m/5 \leq g_{mb} \leq g_m/3$

For circuit analysis, it is convenient to use equivalent small-signal models for MOS devices, as in Figure 5.11. This model is exactly the same for both NMOS and PMOS devices. The terms missing are the output resistance r_o—the inverse of the output conductance g_o—which we find next.

In the saturation/active mode, the steady-state characteristics of a MOS transistor are described using Equation (5.3) with Equation (5.7) to have

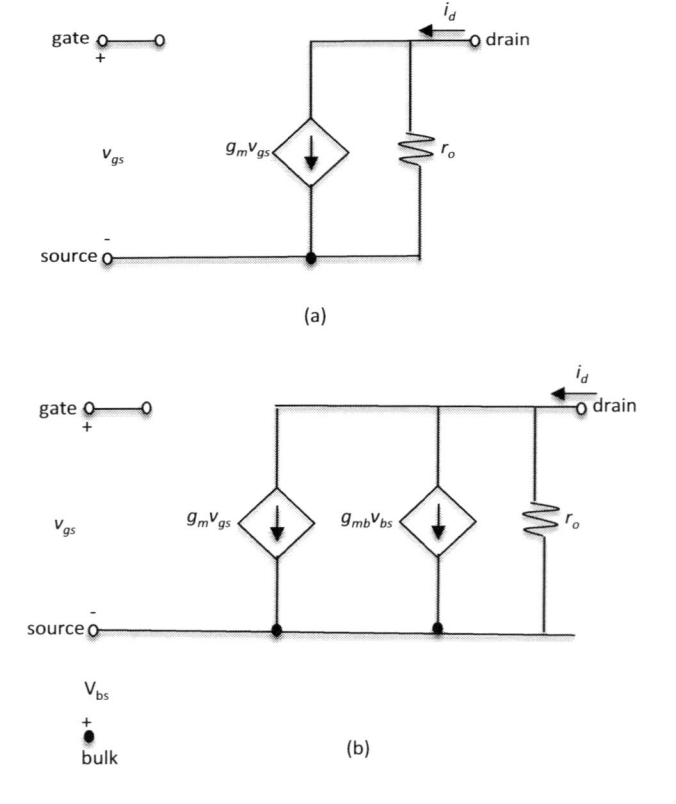

FIGURE 5.11 Signal model of MOSFET (MOS) for 3-terminal and 4-terminal configurations: (a) hybrid-π model for 3-terminal MOS and (b) hybrid-π model for 4-terminal MOS.

$$i_D = \kappa \left(\frac{W}{L} \right) (v_{GS} - V_{th})^2 (1 \pm \gamma v_{DS}) \tag{5.30}$$

From Equation (5.30), the output conductance g_o can be obtained if we find the change in drain current δi_D due to an increment in the drain-source voltage δv_{DS}; that is,

$$g_o = \left. \frac{\partial i_D}{\partial v_{DS}} \right|_Q = \pm \kappa \left(\frac{W}{L} \right) (v_{GS} - V_{th})^2 \gamma \cong \pm \gamma I_D \tag{5.31}$$

Consequently, the output resistance

$$r_o = \left| \frac{1}{g_o} \right| = \frac{1}{\gamma I_D} = \frac{L}{\kappa W (v_{GS} - V_{th})^2 \gamma} \, \Omega \tag{5.32}$$

Amplification factor (also called maximum voltage gain)

$$A_f = g_m r_o \tag{5.33}$$

where g_m and r_o are as previously defined.

We can argue that, for similar operating point, MOSFET has higher transconductance g_m and lower output resistance r_o than BJT.

As noted earlier in this section, MOSFET device's capacitive effects influence the dynamic operations of both small- and large-signal models. Thus, a capacitance model describing the intrinsic and extrinsic components of the device capacitance is another essential part of a MOSFET circuit model that needs to be considered.

5.3.2 MOSFET Device Capacitances

The techniques employed in Chapter 3, Section 3.5 for the analysis of the small- and large-signal properties of junction-gate FETs, can equally be applied to MOSFET transistors. To examine the variation of charge in the substrate, gate, and channel with electrode potential, it is useful to identify these charges as seen in Figure 5.12.

In Figure 5.12, Q_G is the total gate charge, Q_N is the total inversion or channel charge, Q_B is the total substrate charge, and Q_{IS} is the equivalent interface charge. These charges are related by the charge neutrality condition; i.e.

$$Q_G + Q_B + Q_N + Q_{IS} = 0 \tag{5.34}$$

where

$$Q_G = W \int_0^L Q_g \, dy \tag{5.35}$$

$$Q_N = W \int_0^L Q_n(y) \, dy \tag{5.36}$$

$$Q_B = W \int_0^L Q_d \, dy \tag{5.37}$$

$$Q_{IS} = W \int_0^L Q_{is} \, dy = WLQ_{is} \tag{5.38}$$

L and W are the channel length and width, respectively. Figure 5.13 shows the small-signal capacitance curves. The capacitance of the MOSFET structure is linear in accumulation and inversion regions, but nonlinear in the depletion region and presumably smallest, as seen in Figure 5.13b.

In the accumulation mode the capacitance C_{acc} $(= C_{ox})$ is due to the voltage drop across t_{ox}. In the depletion region, the capacitance C_{dep} $(\neq C_{ox})$ is just due to the voltage drop across the oxide and the

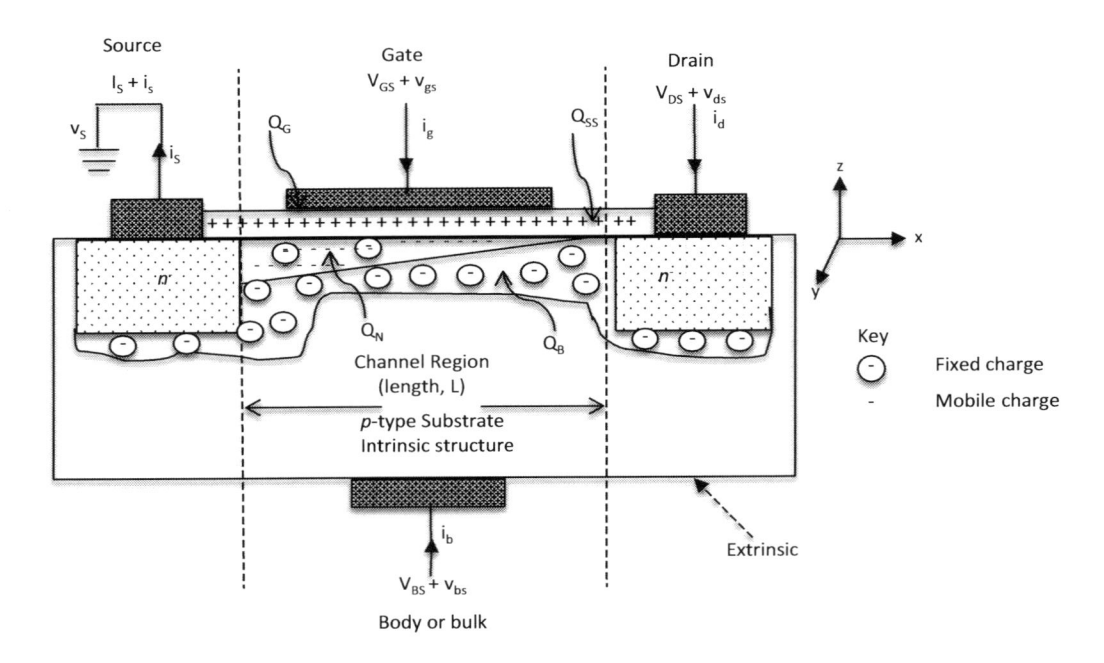

FIGURE 5.12 Notations of n-channel MOSFET's structure and charge components.

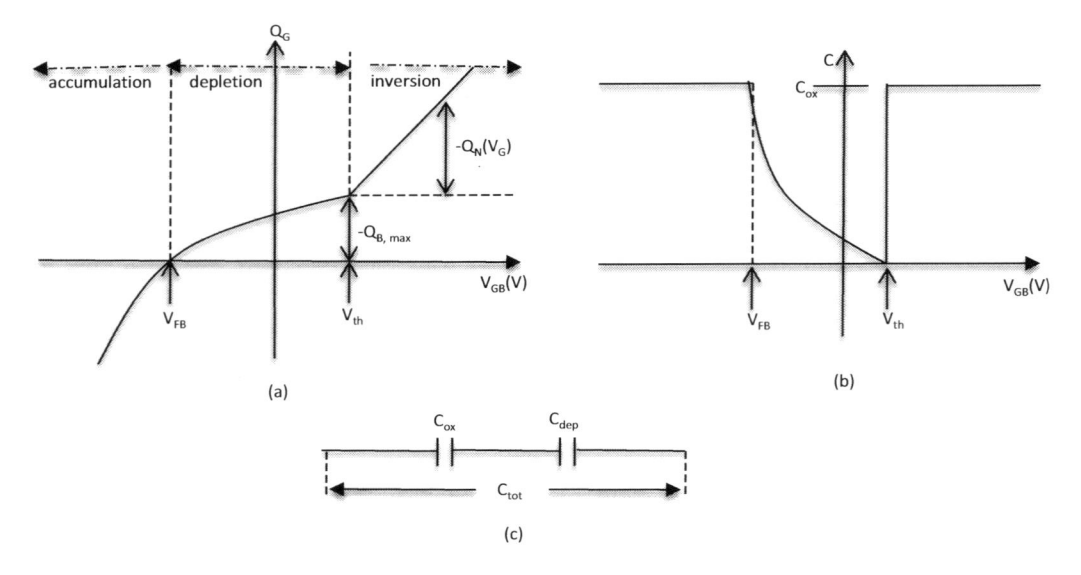

FIGURE 5.13 MOSFET Q–V and C–V curves and total capacitance: (a) MOSFET Charge–Voltage (Q–V) curve, (b) MOSFET Capacitance–Voltage (C–V) curve, and (c) total capacitance.

depletion region. In the inversion mode the capacitance C_{inv} ($= C_{ox}$) is due to the incremental charge that comes from the inversion layer, where understandably the depletion region stops growing. On the basis that the surface state capacitance is zero, the capacitance of the MOS structure when it is in the surface depletion regime is given by

$$C_{MOS,depletion} = C_{depletion} = \frac{C_{dep}C_{ox}}{C_{dep} + C_{ox}} \text{ (F)}$$ (5.39)

where

$$C_{dep} = A_\sigma \text{ (F)}$$ (5.40)

A_σ, ε_s, and x_s are substrate's area, permittivity, and effective depletion layer width, respectively. Alternatively, we can write the total capacitance Equation (5.39) as

$$C_{tot} = \frac{C_{dep}C_{ox}}{C_{dep} + C_{ox}} = \frac{C_{ox}}{1 + \dfrac{C_{ox}}{C_{dep}}} = \frac{\left(\dfrac{\varepsilon_{ox}}{t_{ox}}\right)}{1 + \left(\dfrac{\varepsilon_{ox}}{\varepsilon_s}\right)\left(\dfrac{x_s}{t_{ox}}\right)}$$ (5.41)

It is reasonable to assume this expression as suitable for determining the minimum high-frequency capacitance. The basic quadratic IV-relationship, depicted by Figure 5.7b and c, provides a reasonable model for understanding the "*direct current*, dc" behavior of MOSFET circuits. In many analytic circuits, however, the capacitance associated with the devices must also be considered so as to predict the "*alternating current*, ac" behavior. Neglecting leakage currents, the small-signal currents (i_s, i_d, i_g, and i_b) resulting from ac voltages at the terminals can, from the conservation of charge, be expressed as follows [1]:

$$i_g = \frac{dQ_G}{dt},$$

$$i_b = \frac{dQ_B}{dt}$$ (5.42)

$$i_s - i_d = \frac{d}{dt}(Q_G + Q_B)$$ (5.43)

Suppose we consider the small-signal ac terminal voltages as v_{gs}, v_{bs}, and v_{ds}; gate-source, base-source, and drain-source, respectively, then Equations (5.42) and (5.43) can be expressed as

$$i_g = C_{gs}\frac{dv_{gs}}{dt} + \left(C_{gb}\frac{dv_{gs}}{dt} - C_{bg}\frac{dv_{bs}}{dt}\right) + C_{gd}\frac{d(v_{gs} - v_{ds})}{dt}$$ (5.44)

$$i_b = C_{bs}\frac{dv_{bs}}{dt} - \left(C_{gb}\frac{dv_{gs}}{dt} - C_{bg}\frac{dv_{bs}}{dt}\right) + C_{bd}\frac{d(v_{bs} - v_{ds})}{dt}$$ (5.45)

$$i_s - i_d = C_{gs}\frac{dv_{gs}}{dt} + C_{bs}\frac{dv_{bs}}{dt} + C_{gd}\frac{d(v_{gs} - v_{ds})}{dt} + C_{bd}\frac{d(v_{bs} - v_{ds})}{dt}$$ (5.46)

where the gate-to-bulk capacitances are defined as follows:

$$C_{gs} = \left.\frac{\partial Q_G}{\partial V_{gs}}\right|_{\substack{V_{gd} \\ V_{gb}}} = \frac{2}{3} WLC_{\mathrm{ox}} \left(1 - \frac{\left[V_{gd} - V_{Th} \right]^2}{\left[V_{gs} + V_{gd} - 2V_{Th} \right]^2} \right) \tag{5.47}$$

$$C_{gd} = \left.\frac{\partial Q_G}{\partial V_{gd}}\right|_{\substack{V_{gs} \\ V_{gb}}} = \frac{2}{3} WLC_{\mathrm{ox}} \left(1 - \frac{\left[V_{gd} - V_{Th} \right]^2}{\left[V_{gs} + V_{gd} - 2V_{Th} \right]^2} \right) \tag{5.48}$$

$$C_{gb} = \left.\frac{\partial Q_G}{\partial V_{gb}}\right|_{\substack{V_{gs} \\ V_{gd}}} = \left(\frac{n-1}{n} \right)\left(1 - C_{gs} - C_{gd} \right) \tag{5.49}$$

$$C_{bs} = \left.\frac{\partial Q_B}{\partial V_{BS}}\right|_{\substack{V_{gs} \\ V_{bs}}} = (n-1)C_{gs} \tag{5.50}$$

$$C_{bd} = -\left.\frac{\partial Q_B}{\partial V_{DS}}\right|_{\substack{V_{GS} \\ V_{BS}}} = (n-1)C_{gd} \tag{5.51}$$

$$C_{bg} = -\left.\frac{\partial Q_G}{\partial V_{BS}}\right|_{\substack{V_{gs} \\ V_{ds}}} = 0 \tag{5.52}$$

where n is the slope factor defined as

$$n = 1 + \frac{C_{\mathrm{dep}}}{C_{\mathrm{ox}}} = 1 + \left(\frac{\varepsilon_s t_{\mathrm{ox}}}{\varepsilon_{\mathrm{ox}} x_s} \right) \tag{5.53}$$

Note that (i) the terminal capacitances are derived by considering the change in charge associated with each terminal with respect to a change in voltage at another terminal, under the condition that the voltage at all other terminals is constant and (ii) the signs in Equation (5.47)–(5.53) are chosen ensuring positive capacitances for an n-channel device. The considerations in these equations are for MOSFET with four nodes (i.e. the *source S, drain D, gate G*, and *bulk B*) and show that, at least, a capacitance exists between every two of the four terminals of a MOSFET, as depicted in Figure 5.14.

An additional equation is required to complete the small-signal circuit model. This is obtained by noting that the ac drain current is the sum of conductance and displacement currents given by

$$i_d = g_m v_{gs} + g_{mb} v_{bs} + g_o v_{ds} - C_{gd} \frac{d\left(v_{gs} - v_{ds} \right)}{dt} - C_{bd} \frac{d\left(v_{bs} - v_{ds} \right)}{dt} \tag{5.54}$$

where conductances g_m, g_{mb}, and g_o are as previously defined.

Intrinsic and *extrinsic* capacitances: MOSFET parasitic capacitances can be classified into two distinct groups: intrinsic capacitances and extrinsic capacitances. Intrinsic capacitances refer to the sum of capacitive effects inside the channel region: i.e. the capacitive effects under the gate oxide between the drain and the source contacts. These capacitive elements are highly dependent on the voltages at the terminals and the distribution of the electric field in the device. Unlike the linear elements' analysis, intrinsic capacitance analysis does not obey lumped analysis concept. Extrinsic capacitances,

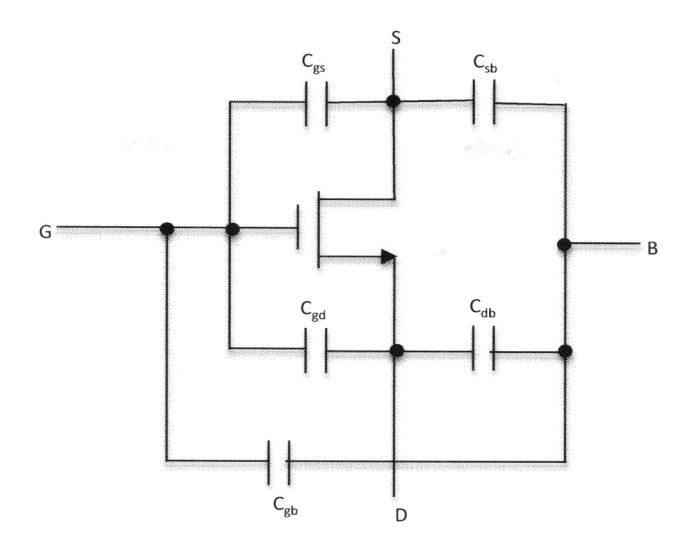

FIGURE 5.14 Intrinsic MOSFET capacitances.

on the other hand, refer to the sum of parasitic capacitive effects outside the channel region: they comprised a number of effects, mostly overlap and parasitic junction capacitances.

We are now in a position to draw the small-signal intrinsic circuit models, and following [1], we draw Figure 5.15. For completeness, we consider three possible cases:

a. When small-signals are applied simultaneously to all electrodes (nodes)—Figure 5.15a;
b. When either the substrate or gate is ac shorted to the source and signals are applied to the other electrodes (nodes)—Figure 5.15b. In this situation, the conductance $C_{gg} = C_{gb}$ for $v_{bs} = 0.0\,\text{V}$, or for $v_{gs} = 0.0\,\text{V}$.
c. A three-terminal model; i.e. when the substrate is either shorted to either the source gate or drain—Figure 5.15c. Frequently the substrate is shorted (or connected to the source. In that case, shorting when substrate B is shorted to source S, Figure 5.15b reduces to Figure 5.15c. So, $V_{SB} = 0.0\,\text{V}$, and the conductance $C_{gs*} = C_{gs} + C_{gb}$ and $C_{gd} = C_{d}$.

The intrinsic transition frequency f_{it} can be expressed as

$$f_{it} = \frac{g_m}{2\pi\left(C_{gs*}\right)} = \frac{g_m}{2\pi\left(C_{gs} + C_{gb}\right)}(\text{Hz})$$

$$\approx \frac{g_m}{2\pi WLC_o}(\text{Hz}) \tag{5.55}$$

Often f_{it} is considered as the device cutoff frequency f_c. However, due to the device's extrinsic capacitances, the switching speed of actual devices is much less than that given by Equation (5.55).

Of course, considerable simplification of the capacitance definitions is possible for saturation conditions, since, no longer is the drain-source voltage an independent variable. Provided that the channel shortening effects are negligible, both the gate-drain and substrate-drain capacitances are zero in the saturation region.

Extrinsic model: The extrinsic capacitances—also called *parasitic* or *stray* capacitances—comprise primarily of overlap (as depicted in Figure 5.16a) and junction/diffusion capacitances (as depicted in Figure 5.16b). Technically, any two conductors separated by an insulator (i.e. a dielectric) form a parallel-plate capacitor.

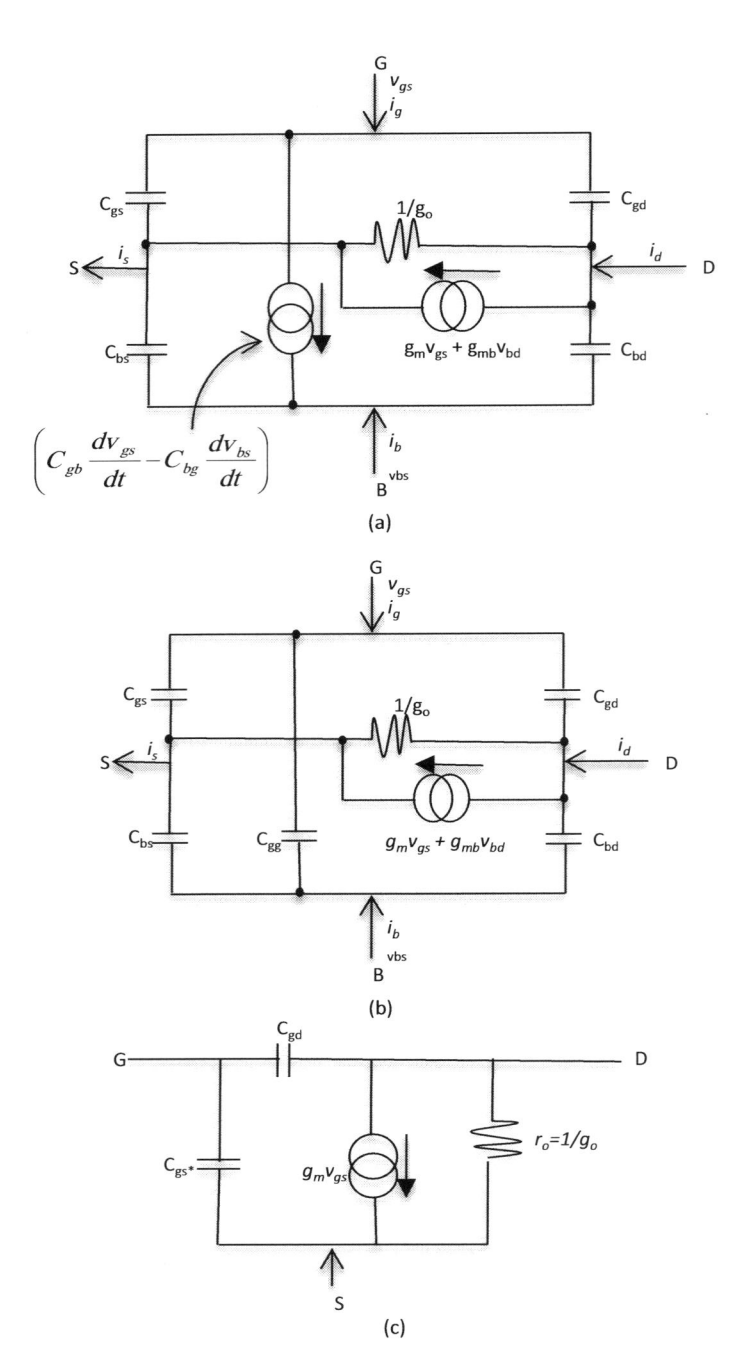

FIGURE 5.15 Small-signal intrinsic circuit models for cases: (a) general form, (b) and (c) depict reduction in substrates. {Graphs are similar to [1].}

The MOSFET overlap capacitance can be approximated by the parallel combination of (1) direct overlap capacitance C_1 between the gate and the source/drain, (2) fringing capacitance C_2 on the outer side between the gate and source/drain, (3) fringing capacitance C_3 on the channel side (inner side) between the gate and sidewalls of the source/drain junction, and direct overlap capacitance C_4 between the gate and the bulk, as depicted in Figure 5.16b.

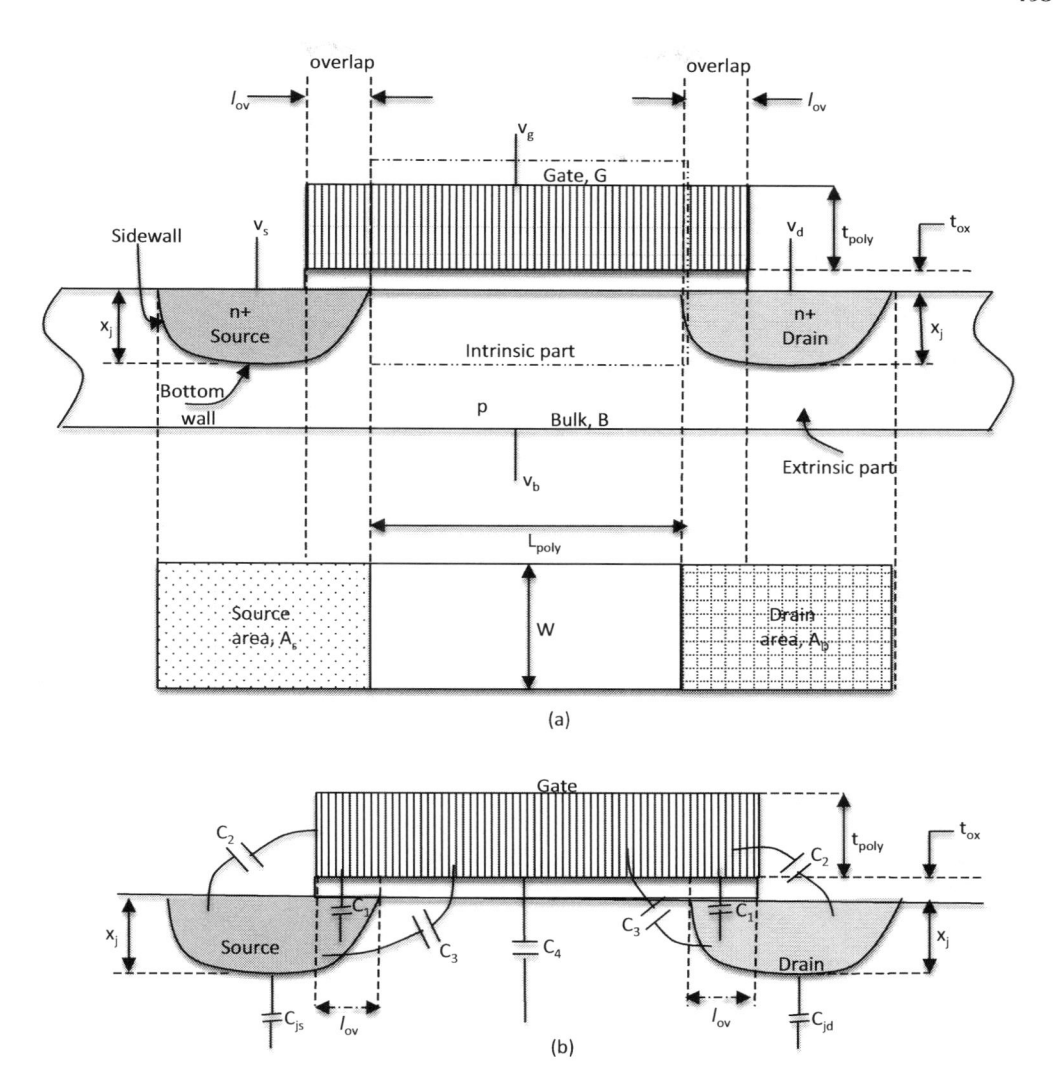

FIGURE 5.16 Outline of overlap and junction capacitances of MOSFET: (a) geometric outline of MOSFET and (b) cross-sectional view of capacitances: three-overlap capacitors—(parallel-plate C_1, outer fringing C_2, and inner fringing C_3), and junction/diffusion capacitors $C_{js,d}$.

The gate-source overlap capacitance C_{gso} and the gate-drain overlap capacitance C_{gdo} can be approximated by [3]:

$$C_{gso} = C_{gdo} = \underbrace{C_{ox}\left(l_{ov} + \Delta\right)}_{C_1} + \underbrace{\frac{\varepsilon_{ox}}{\alpha_1}\ln\left(1 + \frac{t_{poly}}{t_{ox}}\right)}_{C_2} + \underbrace{\frac{2\varepsilon_{si}}{\pi}\ln\left[1 + \frac{x_j}{t_{ox}}\sin\alpha_1\right]}_{C_3} \tag{5.56}$$

where x_j is source-drain junction depth and Δ is the correction factor that accounts for error due to variation in dielectric thickness across the wafer of slope angle α_1. Δ is a correction factor of higher order given by

$$\Delta = \frac{t_{ox}}{2}\left[\frac{1 - \cos\alpha_1}{\sin\alpha_1} + \frac{1 - \cos\alpha_2}{\sin\alpha_2}\right] \tag{5.57}$$

where $\alpha_2 = \dfrac{\pi\varepsilon_{ox}}{2\varepsilon_{si}}$, and term $(l_{ov}+\Delta)$ is the effective length of the parallel-plate component.

It is interesting to note that the fringing component C_3 (channel side) is much greater than C_2 (outer side) because ε_{si} is roughly $3\varepsilon_{ox}$, and quite often $\alpha_1 \geq \pi/2$. From Equation (5.56), it is clear that even if the overlap distance l_{ov} is reduced to zero, there will still be an "overlap" capacitance present due to fringing components. Although in Equation (5.56) the channel side fringing capacitance C_3 is assumed to be bias independent, in reality it is gate and drain voltage dependent. C_3 in Equation (5.56) gives the maximum value of the inner fringing capacitance. Note that when the device is in inversion mode C_3 is zero. So,

$$C_{GSO} = WC_{gso} = C_{GDO} \tag{5.58}$$

The gate-bulk overlap capacitance C_4 (i.e. C_{gbo}) occurs due to the overhang of the transistor gate required at one end, or by imperfect processing of MOSFET. This capacitance is a function of the defined gate length L_{poly} after etches. Normally, C_{gbo} is much smaller than C_{gso} or C_{gdo} and therefore is often neglected; i.e. $C_{gbo} \approx 0$.

Circuit design engineers do not have control over the overlap distances, and hence the overlap capacitances are the fabrication parameters that are defined by the processing steps.

Source/drain junctions or diffusion capacitances

This component is caused by the reverse-biased source-bulk and drain-bulk *pn-junctions*. These capacitances are *nonlinear* and *decrease* as the reverse bias is *increased*. The diffusion capacitance can be approximated by three contributing components: (a) *bottom-wall,* (b) *sidewalls, and* (c) gate-edge periphery. Figure 5.17 shows these components. The source N_D and bulk regions N_A form the bottom-wall junction capacitance. The sidewall junction capacitance is formed by source N_D and p+ channel-stop implant with doping N_A+. Doping concentration is higher for channel stop implant, hence the capacitance per unit area is also higher. The gate edge is primarily from the channel implant.

The total junction capacitance (source, S and drain, D) can be calculated with

$$C_{jS,D} = \underbrace{\frac{C_{j0}A_{jw}}{\left(1-\dfrac{|V_d|}{\phi_o}\right)^{m_j}}}_{\text{bottom-wall}} + \underbrace{\frac{x_j C_{jsw}P_{sw}}{\left(1-\dfrac{|V_d|}{\phi_{osw}}\right)^{m_{jsw}}}}_{\text{sidewall}} + \underbrace{\frac{C_{jgate}W}{\left(1-\dfrac{|V_d|}{\phi_{ogate}}\right)^{m_{jswg}}}}_{\text{gate-edge}} \tag{5.59}$$

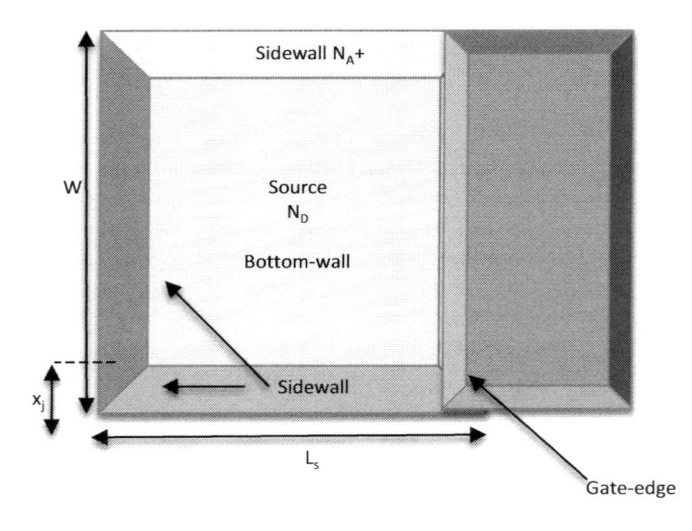

FIGURE 5.17   MOSFET junction/diffusion capacitance.

where

V_d is the junction voltage (which is reverse bias, hence $|V_d|$),

A_{jw} is the bottom-wall area, which is $A_{jw} = WL_s$

P_{sw} is the sidewall perimeter of the source/drain opening, which is the sum of three sides of the drain diffusion area, given by $P_{sw} = W + 2L_s$. All sidewalls regions $p+$ have approximately the same depth x_j.

C_{jo} is the junction capacitance (also called *transition capacitance*) at zero bias. This capacitance is highly process dependent, given by

$$C_{jo} = \sqrt{\frac{\varepsilon_s q}{2\phi_o}\left(\frac{N_A N_D}{N_A + N_D}\right)} \tag{5.60}$$

where ϕ_o, the built-in junction potential, is given by

$$\phi_o = V_T \ln\left(\frac{N_A N_D}{n_i^2}\right) \tag{5.61}$$

Suppose the sidewalls' doping density in $p+$ is $N_{A(sw)}$, then

$$C_{jsw} = \sqrt{\frac{\varepsilon_s q}{2\phi_{osw}}\left(\frac{N_{A(sw)} N_D}{N_{A(sw)} + N_D}\right)} \tag{5.62}$$

and the sidewalls' build-in potential ϕ_{osw} with drain or source regions is

$$\phi_{osw} = V_T \ln\left(\frac{N_{A(sw)} N_D}{n_i^2}\right) \tag{5.63}$$

The gate-edge junction contribution is

$$C_{jgate} = \sqrt{\frac{\varepsilon_s q N_{A(gate)}}{2\phi_{ogate}}} \tag{5.64}$$

The term m_j is called the grading coefficient, whose value depends on its doping profile of the diffusion junction. Typically, for real devices, the grading coefficient is between 0.2 and 0.6 since the doping profile is neither abrupt (step) nor linear. Typically, grading coefficients $m_{jsw} \approx m_{jswg} \approx 1.1 m_j$.

As observed in Equations (5.60)–(5.64), there are different annotated dopants for the three capacitances. However, with the rapid evolution of semiconductor electronics technology, doping concentration variation in the three junction components may be restricted or of no significance. As such, we can assume that $N_{A(sw)} = N_{A(gate)} = N_A$ and $\phi_{osw} = \phi_{ogate} = \phi_o$. Hence the total junction/diffusion capacitance (with source, S and drain, D regions) can be generalized as

$$C_{jS,D} = \underbrace{\frac{C_{j0} A_{jw}}{\left(1 - \frac{|V_d|}{\phi_o}\right)^{m_j}}}_{\text{bottom-wall}} + \underbrace{\frac{x_j C_{jsw} P_{sw}}{\left(1 - \frac{|V_d|}{\phi_o}\right)^{m_{jsw}}}}_{\text{sidewall}} + \underbrace{\frac{C_{jgate} W}{\left(1 - \frac{|V_d|}{\phi_o}\right)^{m_{jswg}}}}_{\text{gate-edge}} \tag{5.65}$$

Advanced processes use SiO_2 to isolate devices (trench isolation) instead of N_A+. For completeness, if the MOSFET is in a well, as shown in Figure 5.18, a well-to-bulk junction capacitance, C_{jbw}, which must be added to Equation (5.65). The well-bulk junction capacitance is calculated similarly

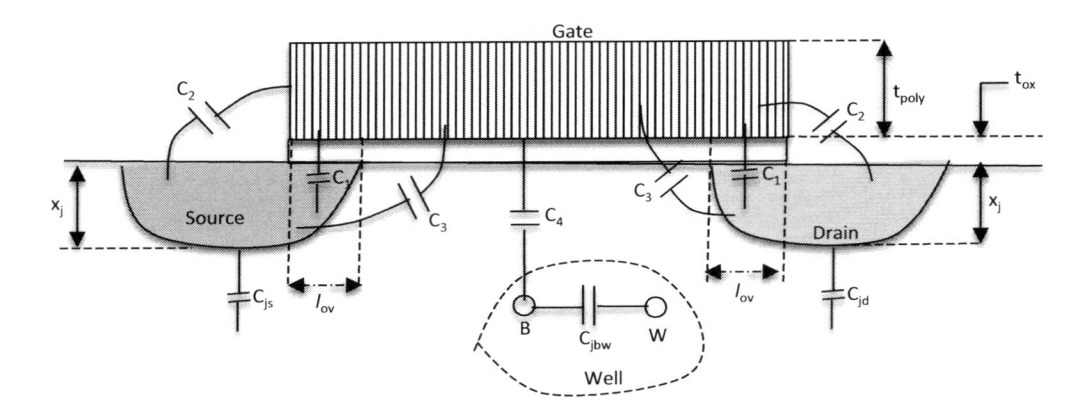

FIGURE 5.18   Cross-sectional view of capacitances with bulk-well junction capacitor C_{jbw}.

to the source and drain junction capacitances, by dividing the total well-bulk junction capacitance into sidewall and bottom-wall components [4]. If more than one transistor is placed in a well, the well-bulk junction capacitance should only be included once in the total model.

The dynamic performance of digital circuits is directly proportional to these capacitances.

We are now in a position to draw the complete small-signal model for MOS where the capacitances are "patched" onto the small-signal circuit schematic containing g_m, g_{mb}, and r_o, as defined in Equations (5.9), (5.28), and (5.32), respectively. The models for n-channel MOS (NMOS) and p-channel MOS (PMOS) are shown in Figures 5.19 and 5.20, respectively, for completeness. Note that in the PMOS, the source is the highest potential (i.e. plus, +) and is located at the top of the schematic.

In conclusion, the small-signal model lends itself well to small-signal design and circuit analysis, which can be applied to obtain results such as (a) input-to-output transfer function that shows how the signal propagates through the circuit; (b) input or output impedance that allows prediction of interactions among various parts of a larger circuit, input sensors, and output loads. In essence, small-signal models are a linear representation of the nonlinear transistor electrical behavior, as well as a linear approximation to the large signal behavior.

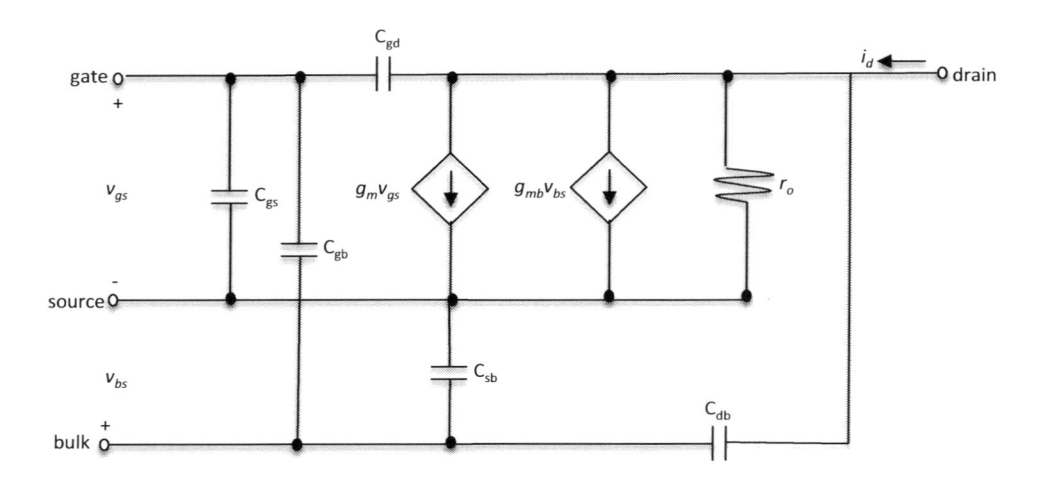

FIGURE 5.19   Complete small-signal model of *n*-type MOSFET (NMOS).

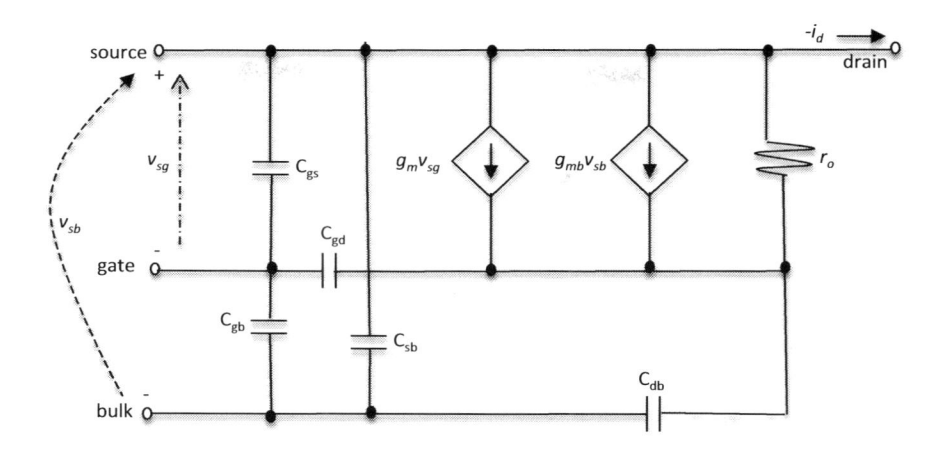

FIGURE 5.20 Complete small-signal model of p-type MOSFET (PMOS).

5.3.3 LARGE-SIGNAL MODEL OF MOSFET

Large-signal models attempt to recreate the behavior of real devices over large voltage swings, which may not be linear, and may not be terribly accurate in the details. Like for the small-signal model, the device's capacitive effects also influence its dynamic operations in the large-signal models. Figure 5.21 shows a simplified equivalent large-signal model of a MOSFET.

There are limitations to using large-signal models such as

- Large-signal models must often be greatly simplified to handle intuitively;
- Large-signal models are often nonlinear, so it is difficult to analyze circuits with more than a few elements directly; and
- Elements, such as variable stored charge, are difficult to model leading to use of a fixed capacitance, which may have a compromise value.

5.3.4 MOSFET LOGIC CIRCUITS

MOSFET devices have wide applications in digital circuits design. Unlike the bipolar transistors, MOSFET transistors can be fabricated in less area. When configured appropriately, additionally,

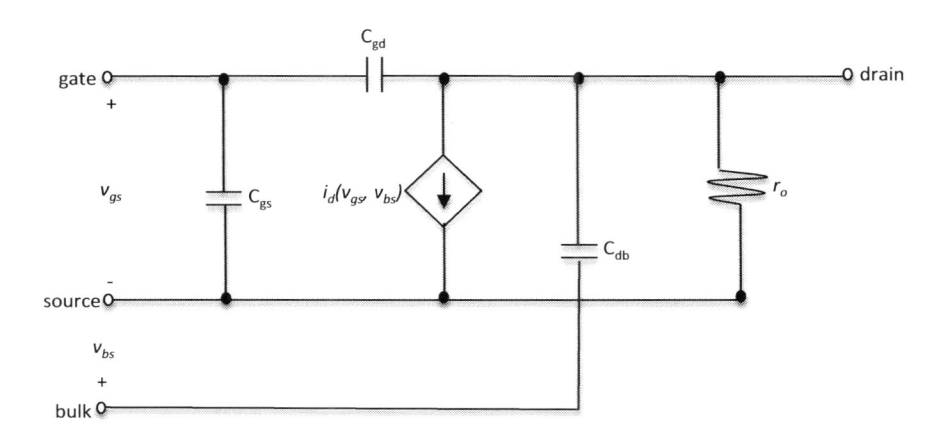

FIGURE 5.21 Complete large-signal model of n-type MOSFET (NMOS).

MOSFET devices can perform dual functions: as a load resistor and as a transistor. Figure 5.22 describes three basic logic circuits: inverter (Figure 5.22a), NOR gate (Figure 5.22b), and NAND gate (Figure 5.22c) using MOSFET devices.

Note that a *logic gate* is a physical device that implements a Boolean function that performs a logic operation on one or more logical inputs (or terminals, e.g. A, B), and produces a single logical output (e.g. Y). For any given moment, every terminal is in one of the two binary conditions *low* "0" or *high* "1." These binary conditions represent different voltage levels; i.e. the low state approximates to less than device threshold voltage V_{th}, or approximately, ground (GND) or zero volt (0 V), and the high state is approximated to $+V_{DD}$ (V).

Inverter: As seen in Figure 5.22a, two MOSFET devices $M1$ and $M2$ are used. $M1$ acts as a load resistor and $M2$ acts as the active device. The load-resistor $M1$ has its gate connected to V_{DD} ensuring that it stays in the conduction state. When the input voltage A is low; i.e. less than the threshold voltage V_{th}, device $M2$ turns off. Since $M1$ is always on, the output voltage is about V_{DD}. However, when the input voltage is high; i.e. greater than the threshold voltage V_{th}, $M2$ turns on. Current flows from V_{DD} through the load-resistor $M1$ into $M2$. It is important to note that the geometry of the $M1$ and $M2$ must be such that the resistance of $M2$, when conducting, is much less than the resistance of $M1$ to maintain the output Y at a voltage below the threshold. The basic logic expression and the associated *truth table* (Table 5.1) are

$$Y = A'$$

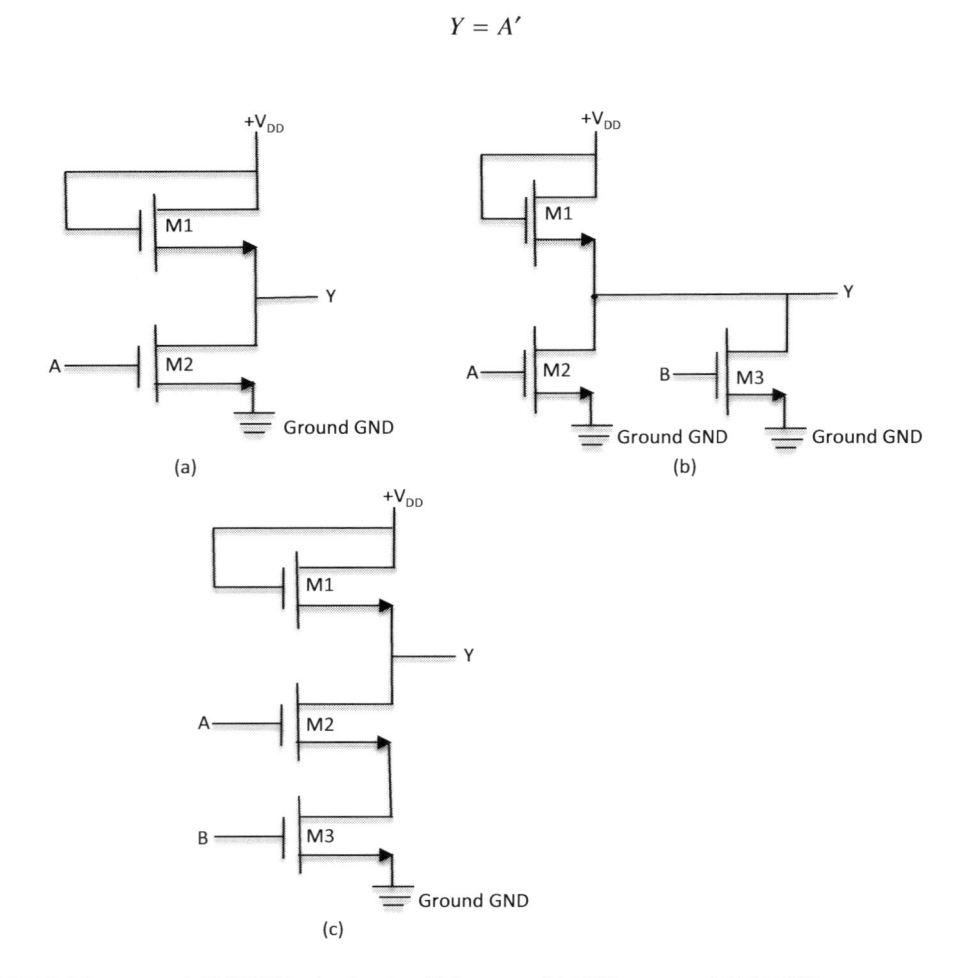

FIGURE 5.22 *n*-type MOSFET logic circuits: (a) inverter, (b) NOR gate, and (c) NAND gate.

TABLE 5.1
Truth Table of an Inverter

Y	A'
0	1
1	0

TABLE 5.2
Truth Table of NOR Gate

A	B	Y
0	0	1
0	1	0
1	0	0
1	1	0

TABLE 5.3
Truth Table of NAND Gate

A	B	Y
0	0	1
0	1	1
1	0	1
1	1	0

NOR gate: As seen in Figure 5.22b, the NOR gate uses $M2$ and $M3$ MOSFET devices connected in parallel. If either input A or B is high, the corresponding transistor conducts and the output is low. If all inputs (A, B) are low, all active resistors are off and the output Y is high. The logic function expression and truth table (Table 5.2) are

$$Y = (A + B)' = \overline{A + B}$$

NAND gate: As seen in Figure 5.22c, the NAND gate uses $M2$ and $M3$ MOSFET devices connected in series. Inputs A and B of $M2$ and $M3$, respectively, must be high for all transistors to conduct and cause the output to go low. If either input is low, the corresponding transistor is turned off and the output Y is high. As noted earlier, the series resistance formed by $M2$ and $M3$ must be less than the resistance of $M1$ (the load-resistor device). The logic expression and the associated *truth table* (Table 5.3) are

$$Y = (A \cdot B)' = \overline{AB}$$

5.4 SUMMARY

This chapter has discussed the basic principles, structures, and applications of MOSFET; a type of transistor used for amplifying and/or switching electronic signals mostly in digital circuits. In a simplified view, NMOS transistors can be treated as simple switches, where the gate voltage controls whether the path from drain to source is an open circuit (i.e. OFF) or a resistive path (i.e. ON). A PMOS transistor acts as an inverse switch; that is, it is ON when the controlling signal is low and OFF when the controlling signal is high. A summary of small-signal (incremental) models of two major types of MOSFET devices used in analog, digital, and mixed-mode integrated circuits is also considered. A very important application of MOSFETs (MOSs) is in the arrangement known as a *CMOS system*, which we consider in the next chapter (Chapter 6).

PROBLEMS

5.1. Calculate threshold voltage for a MOSFET at room temperature with an $n+$ poly-gate, a 20-nm thick oxide and substrate with $N_A = 2 \times 10^{16} \mathrm{cm}^{-3}$.

5.2. Suppose the capacitor of an MOSFET is fabricated on a p-type silicon with a doping value of $N_a = 4.5 \times 10^{16} \mathrm{cm}^{-3}$ with an oxide thickness of 1.8 nm and an $N+$ poly-gate. Determine the following:
   a. The capacitor's flatband voltage.
   b. The device threshold voltage.
   c. The maximum deletion region.
   d. The device threshold voltage if the gate constituent has changed to $P+$ poly.

5.3. Express the gain MOSFET amplifier structure of Figure 5Q.1.

5.4. For Figure 5Q.2, find voltage Vx and drain current ID for pinch-off condition assuming $\gamma = 0$.

5.5. Consider the MOSFET amplifier structure of Figure 5Q.3 assuming the NMOS transistor has $V_{th} = 0.7\,\mathrm{V}$, $\mu C_{ox} = 100\,\mathrm{mA\ V^{-2}}$, channel length $= 1\,\mu\mathrm{m}$, channel width $= 32\,\mu\mathrm{m}$, and $\gamma = 0$. Suppose the transistor operates at $I_D = 0.4\,\mathrm{mA}$, determine the values of resistors R_D and R_E.

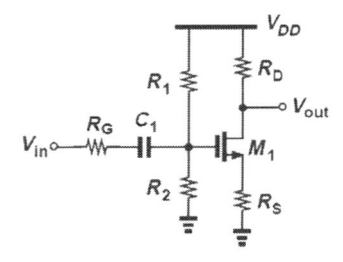

FIGURE 5.Q1

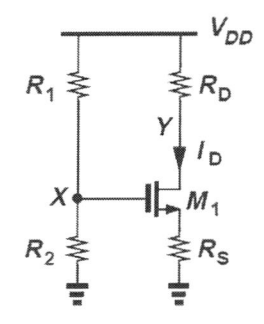

FIGURE 5.Q2

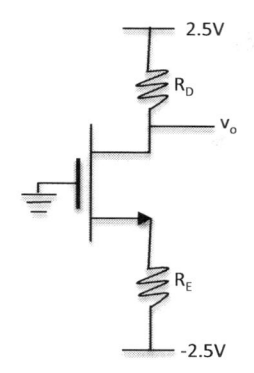

FIGURE 5.Q3

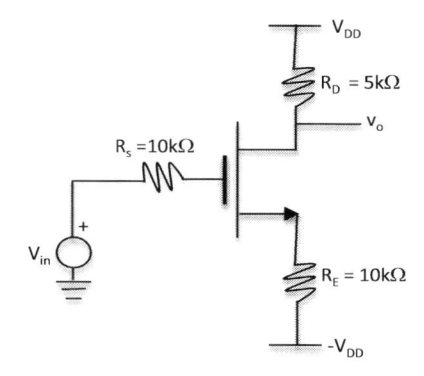

FIGURE 5.Q4

5.6. Consider the MOSFET amplifier structure of Figure 5Q.4 assuming threshold voltage is $1.0\,V$ and $\kappa = 1.0\,\text{mA V}^{-2}$. Determine
   a. the bias I_D, V_{GS}, and V_o;
   b. an analytical expression for the small-signal, low frequency, gain of the amplifier.
   c. g_m assuming the bias as determined in part (a)
   d. a numerical value for the gain of the amplifier assuming the given parameter values.

REFERENCES

1. Cobbold, R.S.C. (1970). *Theory and Applications of Field-Effect Transistors*. New York: Wiley-Interscience.
2. Cheng, Y. and Hu, C. (2002). *MOSFET Modeling & BSIM3 User's Guide*. New York: Kluwer Academic Press.
3. Schrivastava, R. and Fitzpatrick, K. (1982). A simple model for the overlap capacitance of a VLSI MOS device. *IEEE Transactions on Electron Devices*, 29(12), 1870–1875.
4. Deen, J.M. and Fjeldly, T.A. (2002). *CMOS RF Modeling, Characterization and Applications*. Singapore: World Scientific Publishing Co.

BIBLIOGRAPHY

Arora, N. (2007). *Mosfet Modeling for VLSI Simulation: Theory and Practice*. Singapore: World Scientific Publishing Co.

Cheng, Y. (2002). MOSFET modeling for RF IC design. *International Journal of High Speed Electronics and Systems*, 11(4), 1007–1084, 2001.

Enz, C.C. and Cheng, Y. (2000). MOS transistor modeling for RF IC design. *IEEE Journal of Solid-State Circuits,* 35, 186–201.

Galup-Montoro, C. and Schneider, M.C. (2007). *MOSFET Modeling for Circuit Analysis and Design.* Singapore: World Scientific Publishing Co.

Golio, M. (2002). *RF and Microwave Semiconductor Device Handbook.* Boca Raton, FL: CRC Press.

Klein, P. (1997). A compact-charge LDD-MOSFET model. *IEEE Transactions on Electron Devices,* 44, 1483–1490.

Liu, X., Jin, X., Lee, J.-H., Zhu, L., Kwon, H.-I. and Lee, J.-H. (2009). A full analytical model of fringing-field-induced parasitic capacitance for nano-scaled MOSFETs. *Semiconductor Science and Technology,* 25(1), 015008.

Mano, M.M. and Ciletti, M.D. (2007). *Digital Design,* 4th edition. Upper Saddle River, NJ: Prentice-Hall.

Tsividis, Y. (1999). *Operation and Modeling of the MOS Transistor*, 2nd edition. Boston, MA: McGraw-Hill.

6 CMOS Circuits

One very important application of metal-oxide-semiconductor field-effect transistors (MOSFETs) (or MOSs) is in the arrangement known as a *complementary metal oxide semiconductor* (CMOS). CMOS is also known as "complementary symmetry metal oxide semiconductor." The CMOS partly forms the mainstream of high-density digital system design technology, mostly used for fabricating very large-scale integrated (VLSI) or ultra-large-scale integrated (ULSI) chips because it is reliable, manufacturable, scalable, with low power and relatively low cost. CMOS is a type of MOSFET that uses complementary and symmetrical pairs of *p*-type and *n*-type MOSFETs (PMOS and NMOS transistors, respectively) to design and fabricate integrated circuits (ICs) or chips to perform logic functions. The MOSET concept has been discussed in Chapter 5.

This chapter starts by discussing the issue of noise, its sources and types, and how it affects the performance of CMOS. Afterwards, the constituents, configuration, fabrication, and design of CMOS into simple and complex logic circuits, as well as its formulation as transmission (or pass) gate and constitution into a VLSI chip are discussed. Also, the difference between static and dynamic CMOS transistors will be discussed. In the construction of static CMOS gate, two complementary networks are used, where only one of which is enabled at any time. Whilst *duality* is sufficient for static correct operation, it is not necessarily so in reality. The noise margin of a digital circuit or gate is considered: an indication of how well the circuit or gate will perform under noisy conditions. Pass-transistor logic implements a logic gate as a simple switch network. The optimal pass-transistor stages of buffered chain and its delay are derived for minimization.

6.1 NOISE

The varieties of random events that take place at microscopic scales cause fluctuations in the values of macroscopic variables such as voltage, current, and charge. These fluctuations are referred to as noise [1,2]. In some device applications, a noise source can be seen as a signal that is present in a circuit other than the preferred signal. Noise margin (or noise level) can problematically affect system performance and operations. Noise sources in devices include

1. *Inherent circuit noise*: that is, noise resulting from the discrete and random movement of charge in a wire or device. Examples of inherent circuit noise include thermal noise, shot noise, and flicker ($1/f$) noise, which we discuss in this chapter;
2. *Quantization noise*: that is, noise resulting from the finite digital word size when changing an analog signal into a digital signal, which we discuss in Chapter 7; and
3. *Coupled noise*: that is, noise resulting from the signals of adjacent circuits feeding into each other and interfering. For instance, noise generated in a digital circuit is particularly unpleasant in precision analog applications as digital noise is not purely random.

Noise sources in a CMOS transistor can be channel thermal noise, flicker ($1/f$) noise, noise in the resistive polysilicon gate, noise due to the distributed substrate resistance, and shot noise associated with the leakage current of the drain source reverse diodes. However, the noise behavior of bulk CMOS devices is dominated primarily by thermal noise and flicker ($1/f$) noise: these types of noise are always present and physically fundamental to the operation of CMOS device. Other noise mechanisms like shot noise, generation/recombination noise, and burst (or popcorn) noise—caused by the capture and emission of a channel carrier—are sometimes present in the noise spectrum and are dependent on the quality of the manufactured device, i.e. the number of defects in the bulk silicon, in the gate oxide, and in the various interfaces [3]. Flicker noise can probably appear through both quality-dependent and fundamental noise processes. The next paragraphs briefly review primary noise sources in CMOS devices.

Thermal noise is caused by the thermal agitation of charge carriers (electrons or holes) in a conductor and can be modeled as voltage or current. This noise is present in all passive resistive elements even without a current flowing in the resistive medium. Spectrum describes the *frequency content* of noise, which allows its density being used to specify noise parameters. $S(f)$ denotes the power spectral density of a noise waveform. For a series voltage source (as modeled in Figure 6.1a), the thermal noise of resistive conductor is expressed as

$$S_{tnv}(f) = \overline{V_{tn-r}^2} = \int (4kTR)\,df \tag{6.1a}$$

where k is the Boltzmann's constant (1.38×10^{-23} J K^{-1}), T is the absolute temperature in Kelvin (K), R is the resistance of the conductor in ohms (Ω), and df is the differential frequency. Equivalently, we can represent the thermal noise spectral density with a current source, as in Figure 6.1b:

$$S_{tni}(f) = \overline{i_{nt-r}^2} = \int \left(\frac{4kT}{R}\right)df \tag{6.1b}$$

The terms $4kTR$ (in Equation (6.1a)) and $4kT/R$ (in Equation (6.1b)) are voltage and current power densities having units of V^2/Hz and A^2/Hz, respectively.

MOS transistors exhibit thermal noise with the most significant source being the noise generated in the channel and can be modeled by a current source connected between the drain and source terminals, as shown in Figure 6.1c. For devices operating in saturation mode, the spectral density is given by

$$\overline{S_{tn-m}(f)} = \overline{i_{tn-m}^2} = \int (4kT\gamma g_m)\,df \tag{6.2}$$

where γ is a coefficient that varies on the channel type; that is, $\gamma = 2/3$ for long-channel transistors and higher for submicron MOS. Notionally assumed to be unity ($\gamma = 1$), g_m is the transconductance at the bias point; defined in Chapter 5, Equation (5.9).

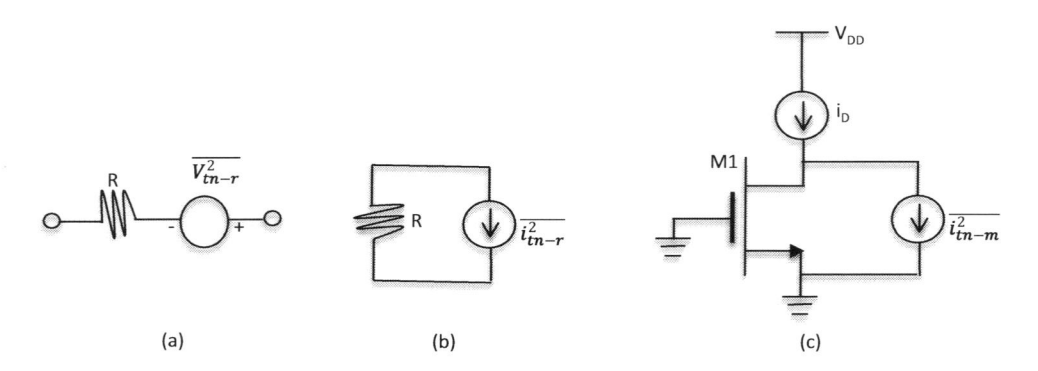

(a) (b) (c)

FIGURE 6.1 Models of thermal noise for noiseless resistor R and MOS: (a) series voltage source, (b) parallel current source, and (c) MOS thermal noise in saturation mode.

Whilst the current noise in the channel also generates noise in the gate through the gate-channel capacitance, the gate current noise is comparatively small compared with the drain noise and is thus considered negligible.

Flicker noise:

Flicker noise, also known as pink noise, and is also commonly called 1/*f* noise. Flicker noise is present in all active devices and has various origins [4]. Its spectral density is expressed by

$$\overline{v_{fn-v}^2} = \int \left(\frac{k_e^2}{f^\alpha} \right) df \tag{6.3a}$$

$$\overline{i_{fn-i}^2} = \int \left(\frac{k_i^2}{f^\alpha} \right) df \tag{6.3b}$$

where α is a constant that is very close to unity ($\alpha = 1 \pm 0.2$); k_e (in volt) and k_i (in amps) are the appropriate device constants. The terms $\left(\frac{k_e^2}{f^\alpha} \right)$ and $\left(\frac{k_i^2}{f^\alpha} \right)$ are voltage and current power densities having units of V^2/Hz and A^2/Hz, respectively.

Flicker noise is present whenever a direct current (dc) flows in a discontinuous material; for example, a material's surface or the interface between a MOSFET's gate oxide and the silicon used as the channel of the MOSFET [3,5]. Slow traps in the gate oxide can, in addition, cause flicker noise. In addition, trapping of carriers, which centers in the bulk of the device, can cause *generation/recombination* noise because these traps cause fluctuations in the number of carriers, and thus fluctuation in the resistance [6]. Flicker noise is not dependent on temperature, unlike thermal noise, but shows sensitivity to wafer orientation that correlates with interface trap density. The presence of Flicker noise has been observed in practically all electronic materials and devices including homogenous semiconductors, junction devices, metal films, liquid metals, electrolytic solutions, and even in superconducting Josephson junctions, chemical concentrated cells, mechanical, biological, geological, and even musical systems [7].

Shot noise:

Shot noise results from the discrete movement of charge across a potential barrier. Unlike thermal noise, for shot noise to be present, we must have both a potential barrier and a current i flowing. Crossing the potential barrier is a purely random event. This non-uniform flow gives rise to a broadband white noise that gets worse with an increase in average current. When in the subthreshold regime (i.e. for $0 < v_{gs} < v_T$), MOSFETs exhibit shot noise (instead of thermal noise) due to the current flowing in the channel, where v_T is the transistor threshold voltage and v_{gs} is the gate-to-source voltage. The power spectral density of shot noise is given by

$$\overline{i_{sn}^2} = \overline{(i - i_D)^2} = \int (2qi_D) df \left(A^2/Hz \right) \tag{6.4}$$

where q is the electric charge (= 1.60218×10^{-19}C) and i_D is the drain current.

Shot noise is spectrally flat or has a uniform power density (qi_D), and is independent of temperature.

6.2   CMOS BASIC STRUCTURE

CMOS technology employs MOSFET (MOS) transistors of both polarities. The basic structure of a CMOS is shown in Figure 6.2. CMOS circuits are fabricated on and in a silicon wafer—already discussed in Chapter 1. In this case, the wafer is doped with donor atoms (for example, phosphorus for an *n*-type wafer) or acceptor atoms (for example, boron for a *p*-type wafer). As seen in Figure 6.2, the NMOS transistor is formed directly in the *p*-type wafer (or *p*-well) while PMOS transistor is formed in the *p*-type wafer (*n*-well). Periodically, an epitaxial layer is grown on the wafer. In this book we will not make a distinction between this layer and the substrate. The connections made to the *p*-type body and to the *n*-well are not shown; the latter functions as the body terminal for the *p*-channel device.

In a simplified view, the MOSFET transistors can be treated as simple switches particularly with CMOS circuits where switching speeds, propagation delays, drive capability, and rise and fall times are of little concern.

Drawing from Figure 6.2, a typical static CMOS gate is built of two complementary networks, pull-up network (PUN) and pull-down network (PDN), as shown by the block diagram in Figure 6.3, where

- A PUN is composed of PMOS transistors, with sources connected to V_{DD}.
- A PDN is composed of NMOS transistors, with sources connected to V_{SS}.

Technically, the PUN and PDN are configured in a complementary topology. The function of PUN is to provide a connection between the output and V_{DD} anytime the output of the logic gate is

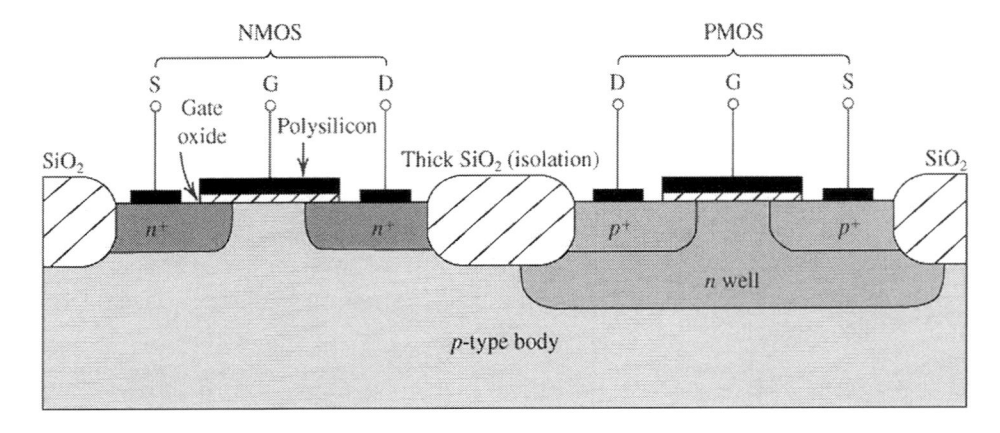

FIGURE 6.2   Cross section of a CMOS circuit where *S* (source), *G* (gate), and *D* (drain) are terminals.

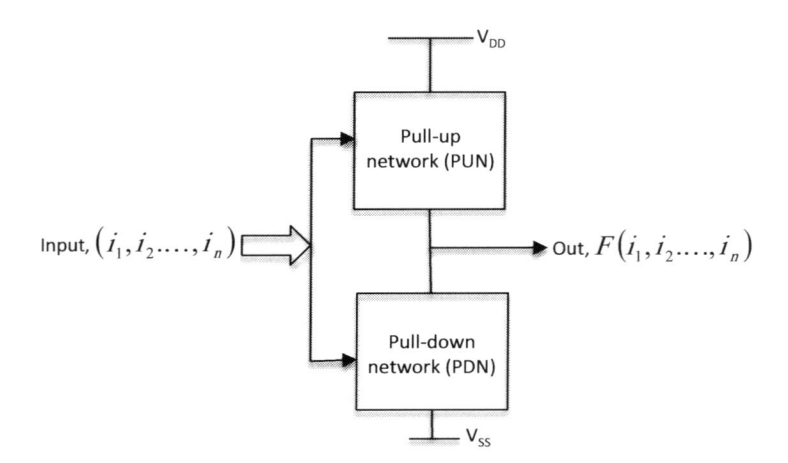

FIGURE 6.3   Generic CMOS gate.

TABLE 6.1
Static CMOS States

	Pull-up OFF	Pull-up ON
Pull-down OFF	Float	1
Pull-down ON	0	X

Note: "X" denotes "don't care" which is not allowed.

meant to be 1 $\left(\text{i.e. } F(i_1, i_2, \ldots, i_n) = 1\right)$. Similarly, the function of the PDN is to connect the output to V_{SS} when the output of the logic gate is meant to be 0 $\left(\text{i.e. } F(i_1, i_2, \ldots, i_n) = 0\right)$.

The PUN and PDN are always *duals*. For example, to construct the dual of a network, the following transformations are performed:

a. Exchange the complementary *n*-channel transistor networks (NMOSs) for complementary *p*-channel transistor networks (PMOSs) and vice versa. Note that only one of these networks may be ON at a time. In this situation, the output will be electrically connected either to V_{SS} or V_{DD}, and not to both. If both networks are ON simultaneously, the electrical path from V_{DD} to V_{SS} will cause excessive current drawn and may damage the circuit: this state is tagged *float*—see Table 6.1.

   We can represent these physical systems (V_{DD} and V_{SS}) Boolean (binary) information: voltage signal with "0" corresponding to V_{SS} and "1" corresponding to V_{DD}. Note that complementary CMOS gates always produce 0 or 1. The design of some logic gates using this exchange method is discussed in Section 6.2.1.

b. Exchange *series* connections for *parallel* connections and vice versa. In the series connection, both transistors (NMOS and PMOS) must be ON. In the *parallel* connection either transistor can be ON. Before we consider connection cases, it is appropriate to explain the *strength* of a signal by its closeness to an ideal voltage source. V_{DD} and V_{SS} rails are strongest "1" (i.e. high) and "0" (i.e. low), respectively. As a consequence, an *n*-type MOSFET (NMOS) is ON for *high* gate voltage—i.e. when NMOS is ON, it passes a "0" level cleanly—and is OFF otherwise—i.e. it degrades or weakens a "1" level. Conversely, a *p*-type MOSFET (PMOS) is ON for *low* gate voltage—i.e. it passes a "1" level cleanly—and is OFF otherwise, (i.e. it degrades or weakens a *low* "0" level). This means that "pass" transistors produce degraded outputs.

6.2.1 IMPLEMENTING CMOS LOGIC CIRCUITS

To construct static CMOS circuits, it could then be argued that we need to implement dual *N*- and *P* networks, meaning PUN or PDN, respectively. In reality, *duality* is sufficient for correct operation but not necessarily.

CMOS gates are built on the same chip such that both have the same threshold voltage V_{th}; that is, $Vthn = V_{th}$, and $Vthp = -Vth$, for NMOS and PMOS, respectively, and same process parameter κ (one needs different channel length, *L*, and width, *W*, to get the same κ for PMOS and NMOS). Chapter 5, Section 5.3.4 has described three basic logic circuits: inverter, NOR gate, and NAND gate using MOSFET devices. We can use same MOSFETs (MOSs) to configure basic and combinational CMOS digital circuits, as follows.

Inverter: The simplest design is the CMOS inverter with one NMOS and one PMOS, which is shown in Figure 6.4. Remember that PMOS (*M*2) in the pull-up is activated by false inputs ($V_i = 0$), whereas NMOS (*M*1) in the pull-down is activated by true inputs ($V_i = 1$).

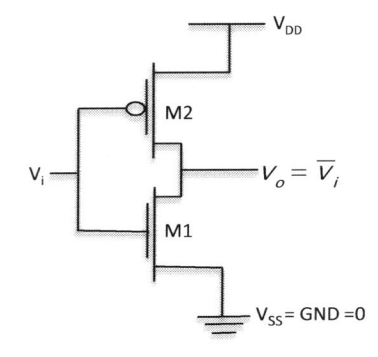

FIGURE 6.4   CMOS inverter.

Other inverter configurations: Aside Figure 6.4, other inverter configurations can be arranged. Some are shown in Figure 6.5. Figure 6.5a is an NMOS-only inverter, no PMOS load. It is useful in avoiding latch-up. Figure 6.5b and c use a PMOS load, which, in general, is most useful in logic gates with a large number of inputs.

A logic "1" input signal results in a dc current flowing in all Figure 6.5 configurations. An important design concern is ensuring that this dc current is contained (that is, not too large). The output logic low V_{OL} will never reach $0\,V$ in these inverters, which makes their noise margins poorer than the basic CMOS inverter of Figure 6.4. The output high level V_{OH} of the inverter of Figure 6.5c will reach V_{DD}, while the other inverters' output high level will be less than V_{DD}, i.e. $(V_{DD} - V_{th})$. At high operating frequencies, the basic CMOS inverter (Figure 6.4) dissipates the most power. While at dc or low frequencies, the inverter configurations outlined in Figure 6.5 dissipate more power.

Inverter voltage transfer characteristics and noise margins

Figure 6.6 can be used to explain, unlike Figure 3.18 of Chapter 3, the basic inverter voltage transfer characteristics and the definition of the noise margins.

Three stages of operation are when: $M2$ is on and $M1$ is off; $M1$ and $M2$ are on; and when $M1$ is on and $M2$ is off. The location "X" corresponds to the point when the input voltage V_i is equal to the output voltage V_o. This point "X" is called the *inverter switching point* voltage V_{sp}, where both $M1$ and $M2$ are in the saturation region. Since the drain currents in $M1$ and $M2$ is equal, then

$$\xi_{(M1)}\left(V_{sp} - V_{th(M1)}\right)^2 = \xi_{(M2)}\left(V_{DD} - V_{sp} - V_{th(M2)}\right)^2 \tag{6.5}$$

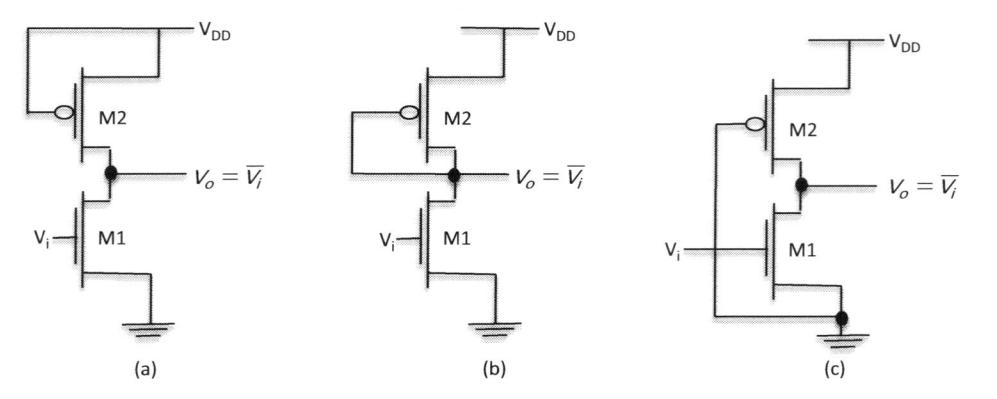

        (a)   (b)   (c)

FIGURE 6.5   Other CMOS inverter circuits' arrangements: (a) inverter circuit with no PMOS load, (b) inverter circuit with PMOS load, and (c) another inverter circuit arrangement with PMOS load.

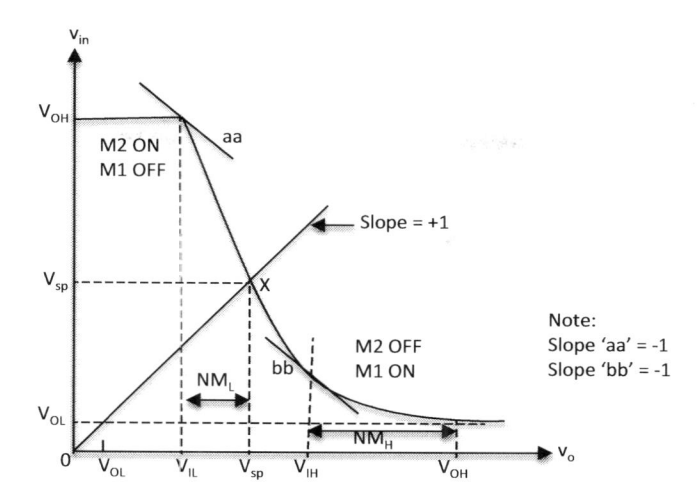

FIGURE 6.6   Voltage transfer characteristics of the CMOS inverter.

Solving for V_{sp} to give

$$V_{sp} = \frac{\left(V_{th(M1)}\sqrt{\dfrac{\xi_{(M1)}}{\xi_{(M2)}}}\right) + \left(V_{DD} - V_{th(M2)}\right)}{1 + \sqrt{\dfrac{\xi_{(M1)}}{\xi_{(M2)}}}} \tag{6.6}$$

where ξ is the inverter *transconductance parameter* defined in Chapter 5, Equation (5.7a).

The noise margin of a digital circuit or gate indicates how well the circuit or gate will perform under noisy conditions. As noted in Chapter 3, Section 3.4.1, and depicted in Figure 3.20, two types of *noise margins* occur, namely: *noise margin high* (NM_H) and *noise margin low* (NM_L). Figure 3.20 is similar to Figure 6.7. So, using Figure 6.7 (and/or Figure 6.6) as a guide, the noise margin associated with a high input level is expressed as

$$NM_H = V_{OH} - V_{IH} \tag{6.7}$$

and the noise margin associated with a low input level is expressed as

$$NM_L = V_{IL} - V_{OL} \tag{6.8}$$

where V_{OH}, V_{OL}, V_{IH}, and V_{IL} are the output high level, output low level, the minimum value of input interpreted by the inverter as a logic 1, and the minimum value of input interpreted by the inverter as a logic 0, respectively. For an ideal logic,

$$NM_H = NM_L = V_{DD}/2 \tag{6.9}$$

Typically, $V_{OL} \approx V_{min} = 0\,\text{V}$, and $V_{OH} \approx V_{max} = V_{DD}$. As a result, Equation (6.7) becomes

$$NM_H = V_{DD} - V_{IH} \tag{6.10}$$

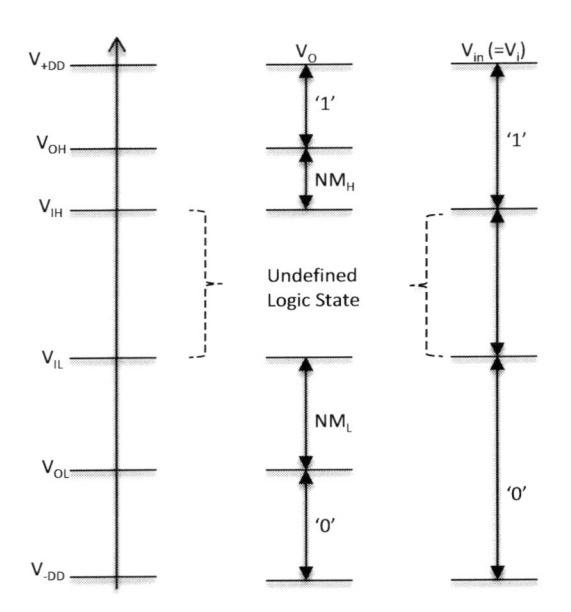

FIGURE 6.7 Noise margins and corresponding logics for input V_{in} and output V_o gate voltages.

while Equation (6.8) becomes

$$NM_L = V_{IL} - 0 = V_{IL} \tag{6.11}$$

So, the only terms to solve are V_{IL} and V_{IH}.

Derivation for V_{IL}: As seen in Figure 6.6, at the point V_{IL}, the NMOS device ($M1$) is in saturated mode (i.e. OFF) while PMOS ($M2$) is non-saturated (i.e. ON). To calculate for V_{IL}, we equate currents for saturated NMOS ($M1$)—drawing from Chapter 5, Equation (5.30) and neglecting the *body-effect* factor (i.e. $\gamma \approx 0$)—and the non-saturated PMOS ($M2$) expression as

$$\xi_{(M1)}\left(V_{sp} - V_{th(M1)}\right)^2 = \xi_{(M2)}\left[2\left(V_{DD} - V_{sp} - \left|V_{th(M2)}\right|\right)\left(V_{DD} - V_0\right) - \left(V_{DD} - V_0\right)^2\right] \tag{6.12}$$

A derivation condition $\left\{dV_o/dV_i = -1\right\}$ needs to be applied in order to evaluate $I_{D(M2)}\left(V_i\right) = I_{D(M1)}\left(V_i, V_o\right)$; that is,

$$\frac{dV_o}{dV_i} = \frac{\left(\partial I_{D(M2)}/\partial V_i\right) - \left(\partial I_{D(M1)}/\partial V_i\right)}{\left(\partial I_{D(M1)}/\partial V_o\right)} = -1 \tag{6.13}$$

to give

$$V_{IL} = \frac{2V_o - V_{DD} - \left|V_{th(M2)}\right| + \dfrac{\xi_{(M1)}}{\xi_{(M2)}}V_{th(M1)}}{1 + \dfrac{\xi_{(M1)}}{\xi_{(M2)}}} \tag{6.14}$$

Derivation for V_{IH}: As seen in Figure 6.6, at the point V_{IH}, the NMOS device ($M1$) is non-saturated (i.e. ON) while PMOS ($M2$) is saturated (i.e. OFF). Equating corresponding "state" condition we write

$$\xi_{(M1)}\left[2\left(V_{IH}-V_{\text{th}(M1)}\right)V_0-V_o^2\right]=\xi_{(M2)}\left(\left(V_{DD}-V_{IH}-\left|V_{\text{th}(M2)}\right|\right)\right)^2 \tag{6.15}$$

and following a similar derivation condition that we applied to V_{IL}, we solve

$$V_{IH}=\frac{2V_o+V_{\text{th}(M1)}+\dfrac{\xi_{(M2)}}{\xi_{(M1)}}\left(V_{DD}-\left|V_{\text{th}(M2)}\right|\right)}{1+\dfrac{\xi_{(M2)}}{\xi_{(M1)}}} \tag{6.16}$$

The voltage threshold V_{th} for the inverter is different from the individual CMOS threshold. We can derive the inverter threshold by assuming that both transistors (NMOS and PMOS) are saturated, again neglecting the *body-effect* factor (i.e. $\gamma \approx 0$):

$$\xi_{(M1)}\left(\left(V_{\text{th}}-V_{\text{th}(M1)}\right)\right)^2=\xi_{(M2)}\left(\left(V_{DD}-V_{\text{th}}-\left|V_{\text{th}(M2)}\right|\right)\right)^2 \tag{6.17}$$

and solving for V_{th}:

$$V_{\text{th}}=\frac{V_{\text{th}(M1)}+\left[\left(V_{DD}-\left|V_{\text{th}(M2)}\right|\right)\sqrt{\dfrac{\xi_{(M2)}}{\xi_{(M1)}}}\right]}{1+\sqrt{\dfrac{\xi_{(M2)}}{\xi_{(M1)}}}} \tag{6.18}$$

Remember that *transconductance parameter* ζ, defined in Chapter 5, Equation (5.7a), is dependent on each MOSFET's channel width W and channel length L. As a result, the aspect ratio (W/L) of each device has influence on the CMOS inverter logic threshold voltage V_{th}. Usually both MOSFETs ($M1$ and $M2$) have a similar channel length set by the minimum value of the technology. Thus, in CMOS design, the width ratio $\left(W_{(M2)}/W_{(M1)}\right)$ is used to set the level of threshold voltage V_{th}:

$$V_{\text{th}}=\frac{V_{\text{th}(M1)}+\left[\left(V_{DD}-\left|V_{\text{th}(M2)}\right|\right)\sqrt{\dfrac{W_{(M2)}}{W_{(M1)}}}\right]}{1+\sqrt{\dfrac{W_{(M2)}}{W_{(M1)}}}} \tag{6.19}$$

In designing a CMOS inverter, the ratio required to establish a given inverter threshold voltage is

$$\sqrt{\frac{\xi_{(M2)}}{\xi_{(M1)}}}=\frac{V_{DD}-V_{\text{th}}-\left|V_{\text{th}(M2)}\right|}{V_{\text{th}}-\left|V_{\text{th}(M1)}\right|} \tag{6.20}$$

Example 6.1

Estimate $\zeta_{(M1)}$ and $\zeta_{(M2)}$ so that the switching point voltage Vsg of a CMOS inverter designed in the long-channel CMOS process is $0.5VDD$ for $VDD = 5$ V and threshold voltage 0.6 V.

Solution:

Using Equation (6.6) and noting $V_{\text{th}(M1)}=V_{\text{th}}$, and $V_{\text{th}(M2)}=-V_{\text{th}}$, then plugging in numbers $V_{\text{th}}=0.6$ V and $VDD=5$ V, we solve

$$V_{th} = \frac{\left(V_{th(M1)}\sqrt{\dfrac{\xi_{(M1)}}{\xi_{(M2)}}}\right) + \left(V_{DD} - V_{th(M2)}\right)}{1 + \sqrt{\dfrac{\xi_{(M1)}}{\xi_{(M2)}}}} \tag{6.21}$$

$$2.5 = \frac{\left(0.6\sqrt{\dfrac{\xi_{(M1)}}{\xi_{(M2)}}}\right) + \left(5.0 - (-0.6)\right)}{1 + \sqrt{\dfrac{\xi_{(M1)}}{\xi_{(M2)}}}}$$

So,

$$\frac{\xi_{(M1)}}{\xi_{(M2)}} = \left(\frac{3.1}{1.9}\right)^2$$

Resulting $\xi_{(M1)} \approx 3\xi_{(M2)}$. Note that $\xi = \kappa\left(\dfrac{W}{L}\right)$ from Chapter 5, Equation (5.7a), where κ is a process parameter. Assuming same process parameters and equal length, i.e. $L_{(M1)} = L_{(M2)}$, then the width $W_{(M1)} \approx 3W_{(M2)}$. This is an important practical result. It shows how the electron and hole mobilities are related to the switching point voltage, as well as to effective switching resistances (R_{M1}, R_{M2}).

6.2.1.1 Dynamic Behavior of CMOS Inverter

Figure 6.8 represents a simple CMOS logic ac model. The source current i_c represents the MOSFET transistor in saturation since the bias state dominates the transition time and the load capacitance C_L. Neglecting the *body-effect* factor (i.e. $\gamma \approx 0$) in Chapter 5, Equation (5.30), we write the expression for source current i_c, in the saturation/active mode, as

$$i_c = i_D = \kappa\left(\frac{W}{L}\right)\left(V_{GS} - V_{th}\right)^2 = \xi\left(V_{GS} - V_{th}\right)^2 \tag{6.22}$$

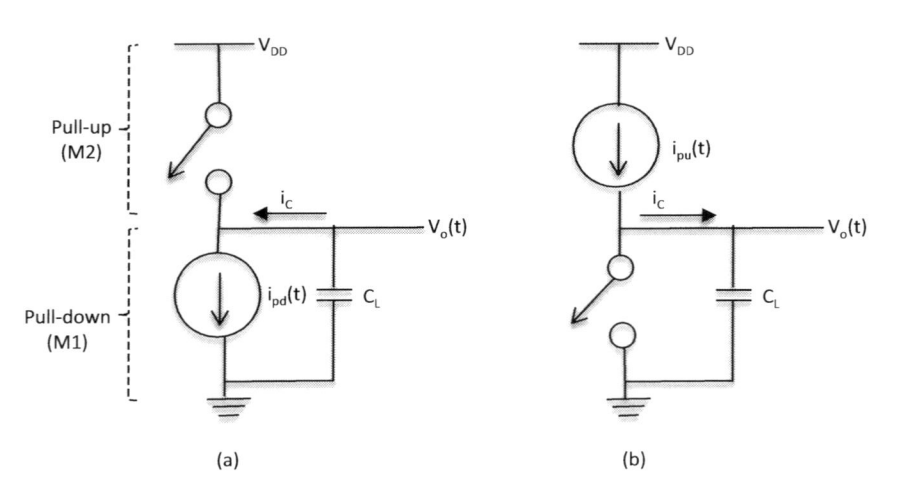

(a)   (b)

FIGURE 6.8 Simple transient response model for CMOS logic gate: (a) high-to-low model and (b) low-to-high model.

From basic circuit theory, the current through capacitor can be expressed as

$$i(t) = C_L \frac{dV(t)}{dt} \tag{6.23}$$

Equating $i(t)$ to $i_c(t)$ and approximating $\frac{dV(t)}{dt} = \frac{\Delta V(t)}{\Delta t}$, then

$$\Delta V(t) = \frac{i_c}{C_L} \Delta t \tag{6.24}$$

where Δt is the propagation delay (i.e. time delay τ_D between input and output signals) for the signal to rise to $\Delta V(t)$, which approximates to V_{DD}. Rearranging Equation (6.24) to have time delay

$$\tau_D = \frac{C_L}{i_c} V_{DD} \tag{6.25}$$

Substituting Equation (6.22) in Equation (6.25) with $V_{SG} = V_{DD}$ to give

$$\tau_D = \frac{2LC_L}{W\mu_{(M1)}C_{ox}} \left(\frac{1}{(V_{DD} - V_{th})^2} \right)$$

$$= \frac{C_L}{\xi(V_{DD} - V_{th})^2} \tag{6.26}$$

where ζ is the inverter *transconductance parameter* defined in Chapter 5, Equation (5.7a).

The load capacitance C_L comprises of parasitic interconnect capacitance and the gate and drain capacitances of each MOSFET, which can be expressed as

$$C_L = \left(C_{d(M1)} + C_{d(M2)} \right) + C_W \tag{6.27}$$

where C_d corresponds to source-drain capacitance for each MOSFET, and C_W is the interconnect wire's capacitance. C_W is a function of shape of the wire, environment, distance substrate, and surrounding wires. Simplistically, C_W is defined as

$$C_W = C_{ox}LW = \frac{\varepsilon_{ox}}{t_{ox}} LW \tag{6.28}$$

where C_{ox}, defined by Equation (5.5), is the thin-oxide field capacitance under the gate region of length L and width W. Equation (6.28) is valid for parallel-plate capacitance. However, when the capacitance between sidewall of the wires and the substrate (fringing capacitance) is considered, C_W becomes

$$C_W = C_{pp} + C_{fringe}$$

$$\approx \left[\frac{(W - x_j/2)\varepsilon_{ox}}{t_{ox}} + \frac{2\pi\varepsilon_{ox}}{\log(t_{ox}/x_j)} \right] L \tag{6.29}$$

where x_j is source-drain junction depth (as depicted in Chapter 5, Figure 5.16). For large (W/x_j) ratio, the fringing capacitance C_{fringe} is less than parallel-plate capacitance, C_{pp}. As such, the wire capacitance C_W approximates to C_{pp} as in Equation (6.28). However, when the (W/x_j) ratio is less than 1.5, C_{fringe} is greater than the parallel-plate capacitance, C_{pp}. Advanced processes have enabled a reduced (W/x_j) ratio of less than unity (i.e. $W/x_j < 1$). For large digital systems, the parasitic wiring capacitance can dominate the load capacitance. For design performance, we have the option of keeping the capacitance small, thereby reducing transistor sizes. Smaller devices yield faster designs at the expense of symmetry and noise margin—to be discussed in Section 6.3. It should be noted that intrinsic device delay τ_D is determined by the technology and components layout.

As noted in Chapter 1 by Equation (1.11), wire (or line) resistance R_W can be expressed as

$$R_W = \left(\frac{\rho}{x_j}\right)\frac{L}{W} = R_s \frac{L}{W} \qquad (6.30)$$

where R_s is the sheet resistance and ρ is the resistivity of the process material (in ohm m).

From the design perspective, we deal with wire resistance R_W by selective technology scaling, and/or use better interconnect materials such as silicides with better conductivity properties than polysilicon, e.g. Tungsten silicide (WSi_2), Titanium silicide ($TiSi_2$), Tantalum silicide ($TaSi_2$), Platinum silicide ($PtSi_2$), Nickel silicide ($NiSi$), and Cobalt silicide ($CoSi$). These silicides result from metal deposition on silicon Si and formation by thermal heating, laser irradiation, or ion beam mixing with the hope of having low resistance but at high thermal budget. Other properties of silicides include good process compatibility with Silicon (Si); e.g. ability to withstand high temperatures, oxidizing ambience, various chemical cleans used during processing, little or no electromigration, easy to dry etch, and good contacts to other materials. Also critical is having sufficiently low diffusivity in silicon in order to prevent the metals from diffusing into the silicon p-n junction depletion regions and/or to the gate oxides. Two types of silicide processes are currently used: *polycide*—a method of patterning the silicide on the polysilicon gate electrode, and *silicide*—a method of self-aligning a silicide layer to all exposed silicon regions.

Silicides have become widely used in VLSI processing in order to avoid the increase in resistance R_W otherwise associated with reduced polysilicon linewidths and film thicknesses.

When a transistor is fabricated, its effective channel length L_{eff} is different from the drawn length L_{drawn}, that is,

$$L_{eff} = L_{drawn} - 2\Delta L_{diff} - 2\Delta L_{poly} \qquad (6.31)$$

where

Δ is the correction factor that accounts for error due to variation in dielectric thickness across the wafer, defined in Chapter 5, by Equation (5.57).

ΔL_{diff} is a parameter that accounts for the channel length offset due to mask effect, which approximates to $\Delta L_{diff} \approx 0.7x_j$.

L_{poly} is the defined polysilicon gate length after etches.

L_{drawn} is the drawn length. This length is what fabrication engineers sometimes refer to in a transistor. Since photolithography steps often alter the length of the channel, the actual channel length may not be the drawn length. The drawn length is usually the parameter of interest to the digital designer because it is used to get a transistor of the desired size [8].

The effective channel width W_{eff} can be expressed by accounting for the offset due to etch effect.

$$W_{eff} = \frac{W_{drawn}}{n_f} - 2\Delta W_{diff} - 2\Delta W_{poly} \qquad (6.32)$$

where n_f is the number of device fingers.

In the transistors-parameters' calculation, the wire length L is implied L_{eff}, and channel width W is implied W_{eff}.

The load capacitance C_L expression (Equation (6.27)) changes if a fanout circuit (as shown in Figure 6.9) follows the inverter of Figure 6.4. In such an arrangement, the output of stage 1 must charge the Source/Drain capacitances of the first stage (i.e. $C_{d(M1) \text{ and } Cd(M2)}$) and the gate capacitances of the second stage (i.e. $C_{g(M3)}$ and $C_{g(M4)}$). So, the load capacitance C_L expression becomes

$$C_L = \left(C_{d(M1)} + C_{d(M2)}\right) + \left(C_{g(M3)} + C_{g(M4)}\right) + C_W \tag{6.33}$$

Power dissipation

There are two kinds of power dissipation in digital electronics, namely:

i. Static power dissipation—that is, when logic gate output is stable—as a result there is nearly no static power dissipation present with CMOS device and
ii. Dynamic power dissipation—that is, during switching of logic gate—when power dissipation is due to charge and discharge of capacitances.

Following basic circuit theory, the total energy supplied by V_{DD} is also equal to the current in load capacitor C_L; so, the total energy delivered E_{Tdy} by the source is expressed by

$$E_{Tdy} = \int_0^\infty V_{DD} i(t) dt = V_{DD} \int_{V_c(0)}^{V_c(\infty)} C_L \frac{dv_c}{dt} dt \tag{6.34}$$

where the limits $V_c(0)$ and $V_c(\infty)$ are the lower and upper voltages across the capacitor C_L, which leads to $0\,V$ and V_{DD}, respectively, when integrating from $t = 0$ to $t = \infty$. Solving Equation (6.34) to have total energy dissipated, E_{Tdy}, in the process of first charging and then discharging the capacitor:

$$E_{Tdy} = C_L V_{DD}^2 = \underbrace{\left(\frac{1}{2} C_L V_{DD}^2\right)}_{\text{charging}} + \underbrace{\left(\frac{1}{2} C_L V_{DD}^2\right)}_{\text{discharging}} \tag{6.35}$$

Note that half of this energy is stored in the load capacitor—as demonstrated in Chapter 1, Equation (1.15)—while the other half is dissipated as heat in the equivalent resistance of the PMOS transistor

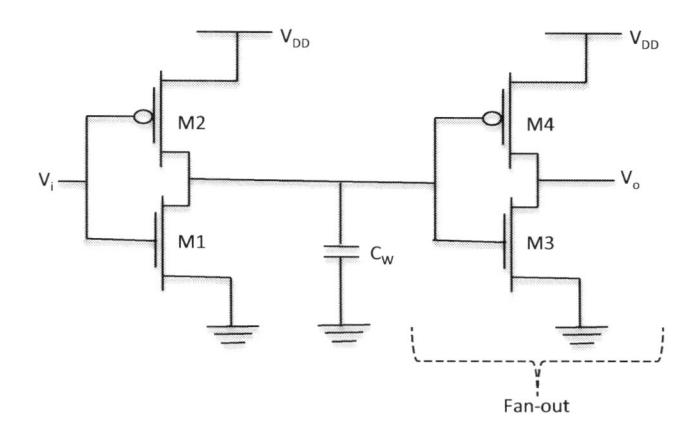

FIGURE 6.9 CMOS inverter with fanout.

while charging the load capacitance. The same happens when the load capacitance is discharging; in this case, it is the NMOS transistor that dissipates half of that energy. Thus, every time a logic gate goes through a complete switching cycle, the MOSFET transistors within the gate dissipate energy are equal to E_{Tdy}.

Logic gates normally switch states at relatively high frequency f, and the dynamic power P_{dy} dissipated by the logic gate may equal to

$$P_{dy} = E_{Tdy}f = fC_L V_{DD}^2 \qquad (6.36)$$

In the case of other conventional CMOS gates, the situation is the same, but considering the PMOS-based PUN and the dual NMOS-based PDN. More precisely, since most signals in a circuit do not commute at the cock frequency but at a lower rate, an activity factor α is usually included in the dynamic power expression [9]:

$$P_{dy} = \alpha fC_L V_{DD}^2 \qquad (6.37)$$

The activity factor is the probability that the circuit node transitions from 0 to 1, because that is the only time the circuit consumes power. The value assigned to α varies with activities. For example, a clock has $\alpha = 1$ because it rises and falls every cycle. Static CMOS logic has been empirically determined to have activity factors closer to 0.1 because some gates maintain one output state more often than another and because real data inputs to some portions of a system often remain constant from one cycle to the next [10]. Most data have a maximum activity factor of 0.5 because it transitions only once in each cycle.

In essence, PDN, PUN, and the load capacitance C_L determine the ac-performance of CMOS logic circuits.

The CMOS inverter has many advantages compared with the NMOS inverter discussed in Chapter 5, Section 5.3.4. Its transfer function is much closer to an ideal inverter transfer function. More complex combinational gates can be designed but will require more transistors; e.g. NAND and NOR.

6.2.1.2 NOR
The CMOS implementation of a NOR gate is shown in Figure 6.10.

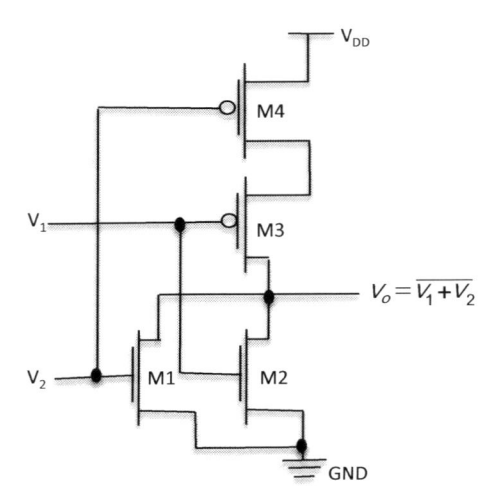

FIGURE 6.10   CMOS 2-input NOR gate.

Arbitrary shapes/layouts are not allowed in some nanoscale design rules. The output V_o of this network is high, if and only if both inputs V_1 and V_2 are low (i.e. $V_1 = V_2 = 0$). The worst-case pull-down transition happens when only one of the NMOS devices is high or turns on (i.e. if either V_1 or V_2 is "1"). There is always a path from either 1 or 0 to the output for all input combinations, but no direct path from 1 to 0 (low power dissipation). No ratioing is necessary. More is said about ratioed logic later.

6.2.1.3 NAND

Complementary CMOS gates inherit all the nice properties of the basic CMOS inverter, discussed earlier. They exhibit rail-to-rail swing with $V_{O(H)} = V_{DD}$ and $V_{O(L)} = $ Ground (GND). Since the CMOS NAND gates circuits are designed such that the PDN and PUN are mutually exclusive, they have no static power dissipation. The CMOS implementation of a NAND gate is shown in Figure 6.11.

The NAND arrangement in Figure 6.11 has three possible input combinations switch to obtain the output V_o of the gate from "1" to "0" (i.e. "high" to "low"): (a) when $V_1 = V_2 = 0$, (b) $V_1 = 0$, $V_2 = 1$, and (c) $V_1 = 1$, $V_2 = 0$. With the exception of (a) where both transistors in the PUN are ON simultaneously, only one of the pull-up devices is ON. From the state of the internal node (between $M1$ and $M2$), both gate-to-source V_{GS} voltages of transistors $M1$ and $M2$ must be above their threshold voltages V_{th}, making that of $M2$ higher than $M1$ due to the body effect. It is possible to share transistors between the two PDNs, but the sizing of the PMOS devices relative to the pull-down devices is critical to its functionality as well as performance.

In essence, we can implement any logic function using inverters, NORs, and NANDs. However, the timing and area of the standard sum-of-products (SOP) approach can be improved further by creating the entire logic function in one circuit. We can design a complex function using the duality approach; i.e. PUN to be the *complement* of the PDN, in series and parallel formats, as shown in the next few examples.

6.2.1.4 CMOS SR Flip-Flops

The simplest well-known *set–reset* (SR) flip-flop is shown in Figure 2.17 (Chapter 2, Section 2.2.1), which is a cross-coupled inverter that provides an approach to storing a binary variable in a stable way. However, extra circuitry must be added to enable control of the memory states—as discussed in Section 2.2.1 and represented by Figure 2.18.

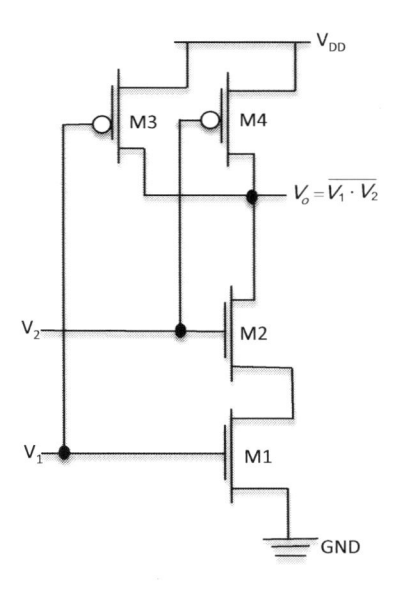

FIGURE 6.11   CMOS 2-input NAND gate.

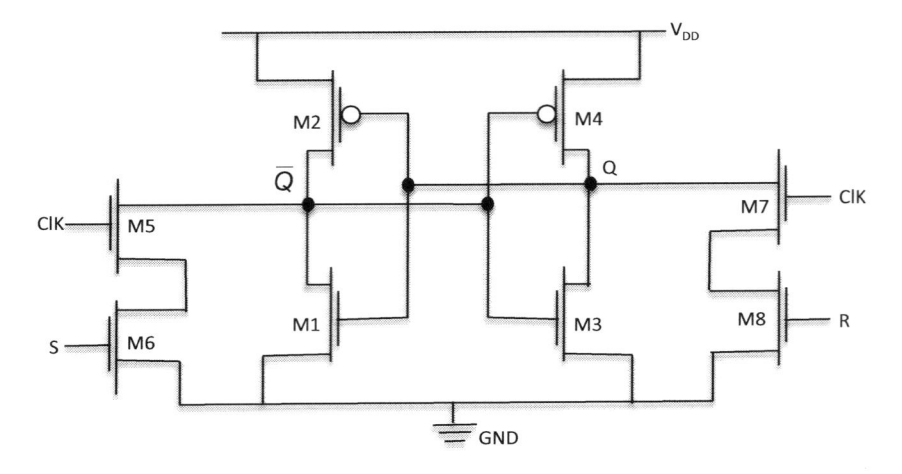

FIGURE 6.12 CMOS clocked SR flip-flop.

Since most systems operate in a synchronous fashion with transition events referenced to a clock (*Clk*), one possible realization of a level-sensitive positive latch—clocked SR flip-flop—is a fully complimentary CMOS implementation, shown in Figure 6.12, which consists of a cross-coupled inverter pair, plus four extra transistors to drive the flip-flop from one state to another and to provide clocked operation.

Observe that the number of transistors in this CMOS implementation is comparable to that used to implement Figure 2.18, but the circuit has the added feature of being clocked. The drawback of saving some transistors over a fully complimentary CMOS implementation is that transistor sizing becomes critical in ensuring proper functionality. Consider the case where Q is high and an R pulse is applied. The combination of transistors $M4$, $M7$, and $M8$ forms a ratioed inverter. In order to make the latch switch, we must succeed in bringing Q below the switching threshold of the inverter $M1/M2$. Once this is achieved, the positive feedback causes the flip-flop to invert states (i.e. $0 \rightarrow 1$, or $1 \rightarrow 0$). This requirement forces us to increase the sizes of transistors $M5$, $M6$, $M7$, and $M8$. In a steady state, one inverter resides in the high state, while the other one resides in the low state. There are no static paths between V_{DD} and GND that can exist except during switching.

Example 6.2

Describe the function of the circuit in Figure 6.13.

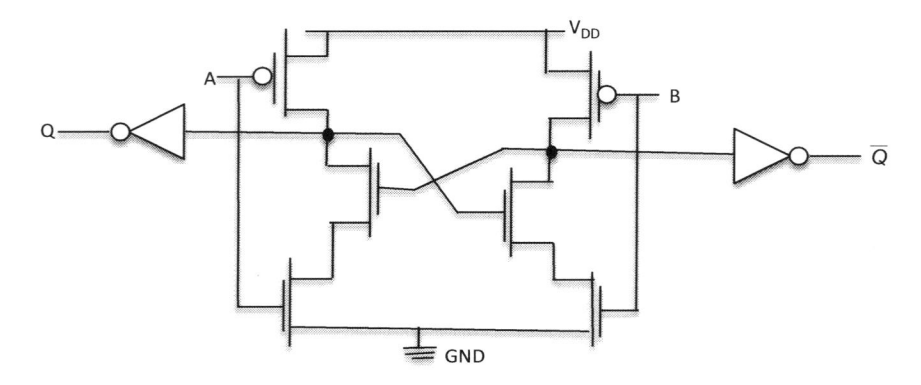

FIGURE 6.13 A CMOS circuit.

Solution:

The circuit in Figure 6.15 implements an SR latch. Input A is *Set* and Input B is *Reset*. The invalid state is when both A and B is 0 (i.e. $A = B = 0$).

6.2.1.5 Ratioed Logic

The CMOS logic design described previously is robust and scalable but requires $2n$ transistors to implement an n-input logic gate. Ratioed logic is an attempt to reduce the number of (NMOS and PMOS) transistors required to implement a given logic function, at the cost of reduced power dissipation, as well as reduced robustness. As earlier noted, the purpose of the PUN in complementary CMOS is to provide a conditional path between V_{DD} and the output when the PDN is turned off. Schmitt trigger is a simple example of ratioed logic.

Schmitt Trigger

Chapter 4, Section 4.2.5, discussed a form of a comparator: Schmitt trigger, a regenerative circuit. It has the useful property of showing *hysteresis* in its dc characteristics; that is, it has variable switching threshold V_m that depends upon the direction of the transition from low-to-high (V_{M+}) or high-to-low (V_{M-}). This peculiar feature can come in handy in noisy environments. An implementation of Schmitt trigger using CMOS is shown in Figure 6.14.

As noted in Chapter 5, Section 5.2, each transistor (NMOS or PMOS) has different *aspect ratios* $\left(W/_L\right)$. Aspect ratio serves as a scale factor for the MOSFET drain current. In order to distinguish between the aspect ratio of NMOS and PMOS, we represent the ratio of NMOS as $k_n = \left.\dfrac{W}{L}\right|_{\text{NMOS}}$ and PMOS as $k_p = \left.\dfrac{W}{L}\right|_{\text{PMOS}}$.

The switching threshold V_m of the CMOS inverter, of Figure 6.14, is determined by the aspect ratio $\left(k_n/_{k_p}\right)$ between the NMOS and PMOS transistors. Increasing the ratio results in a reduction of the threshold, while decreasing it results in an increase in V_M. Adaptation of the ratio results in a shift in the switching threshold and a hysteresis effect. This adaptation is achieved with the aid of feedback.

Suppose initially that $V_i = 0$, so that $V_o = 0$. The feedback loop biases $M3$ in the conductive mode while $M4$ is off. The input signal effectively connects to an inverter ($M1$ and $M3$) as a PUN, and with $M2$ in the pull-down chain, thus modifying the effective transistor ratio of the inverter to $k_{M2}/k_{M1} + k_{M3}$, and moving the switching threshold upwards.

Once the inverter switches, the feedback loop turns off $M3$ while $M4$ is activated. This extra pull-down device speeds up the transition from low-to-high and produces a clean output signal with steep slopes.

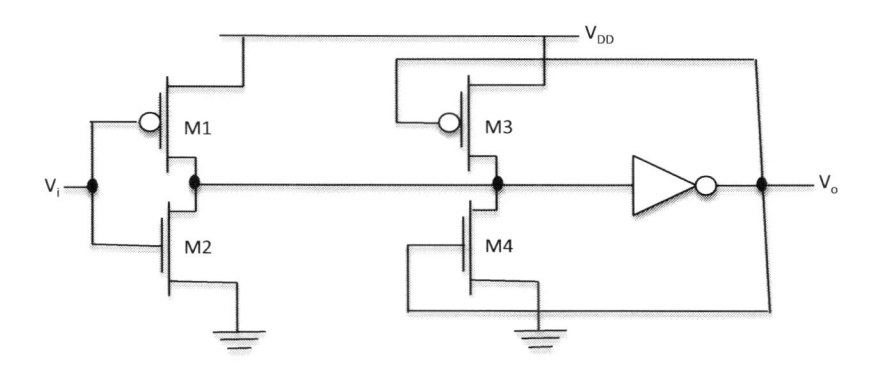

FIGURE 6.14 CMOS implementation of a Schmitt trigger.

Similarly, for the high-to-low transition, the PDN originally consists of $M2$ and $M4$ in parallel, while the PUN is formed by $M1$. This reduces the value of the switching threshold to V_{M-}.

6.2.1.6 Complex CMOS Logic Circuits

Just like the logic design approach discussed in Chapter 2, Section 2.3, we can design a complex CMOS functions by creating the NMOS PDN and creating PMOS PUN (i.e. *complement* of the PDN). We can build circuits of any complexity out of logic functions using NANDs, NORs, and inverters. We can also create equivalent aspect ratios to enable understanding of the circuit performance.

Suppose we have a function given by

$$V_o = \overline{A \cdot (D + E) + B \cdot C} \tag{6.38}$$

then we can design the function's circuit as follows.

First, create the PDN, as shown in Figure 6.15a.

Note that an equivalent aspect ratio (or gain factor) of network for N-series-connected transistors with the same sizes can be summed up as

$$\left. \frac{W}{L} \right|_{\text{equv,series}} = \left[\frac{1}{(W/L)_1} + \frac{1}{(W/L)_2} + \frac{1}{(W/L)_3} + \cdots \right]^{-1}$$

$$= \left[\sum_N \frac{1}{(W/L)_N} \right]^{-1} = \frac{1}{\sum_N \frac{1}{(W/L)_N}} \tag{6.39}$$

Similarly, the equivalent aspect ratio of network for N-parallel-connected transistors with the same size is summed up as

$$\left. \frac{W}{L} \right|_{\text{equv,||}} = (W/L)_1 + (W/L)_2 + (W/L)_3 + \cdots$$

$$= \sum_N (W/L)_N \tag{6.40}$$

Now consider the equivalent for PDN *aspect ratio* of Figure 6.15a:

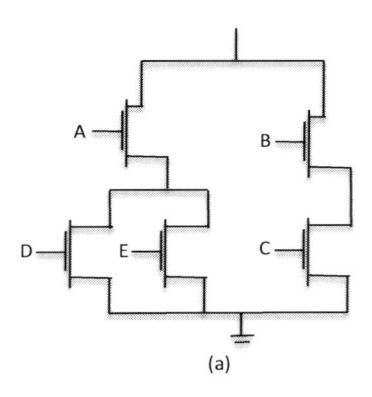

(a)

FIGURE 6.15 (a) Pull-down network, (b) pull-up network, and (c) Combining *pull-up and pull-down cir-cuits* to form CMOS circuit for function V_o.

(Continued)

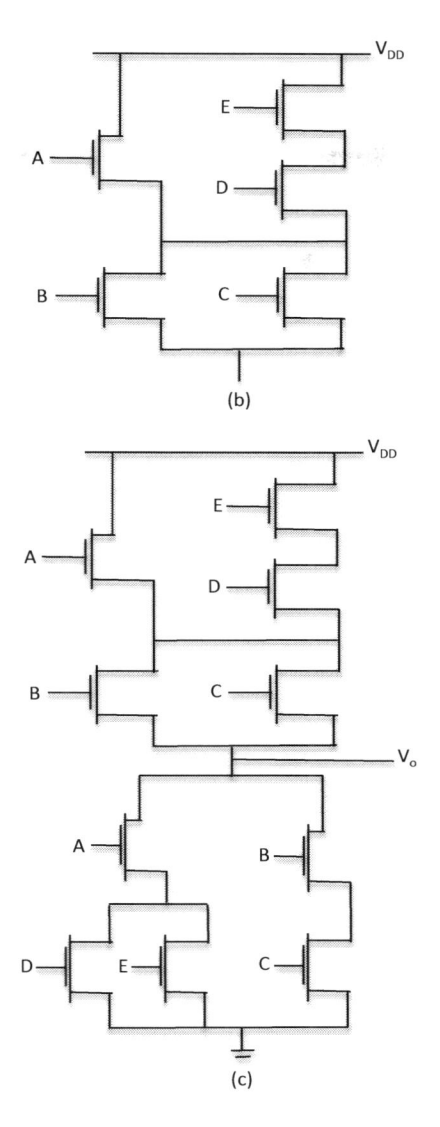

FIGURE 6.15 (CONTINUED)   (a) Pull-down network, (b) pull-up network, and (c) Combining *pull-up and pull-down circuits* to form CMOS circuit for function V_o.

$$\left.\frac{W}{L}\right|_{\text{equv-PDN}} = \underbrace{\left[\left(W/L\right)_A + \left(W/L\right)_{DE}\right]}_{ADE} \| \underbrace{\left[W/L\right]}_{BC} \tag{6.41}$$

where

$$\left(W/L\right)_{ADE} = \frac{1}{1/\left(W/L\right)_A} + \frac{1}{\left(W/L\right)_D + \left(W/L\right)_E} \tag{6.42a}$$

$$\left(W/L\right)_{BC} = \frac{1}{1/\left(W/L\right)_B} + \frac{1}{1/\left(W/L\right)_C} \tag{6.42b}$$

So, PDN *aspect ratio is*

$$\left.\frac{W}{L}\right|_{\text{equv-PDN}} = \left(W/L\right)_{ADE} + \left(W/L\right)_{BC} \tag{6.43}$$

Second, create an equivalent aspect ratio PUN using Figure 6.15b:

From the parallel and series connections of CMOSs, of Figure 6.15b, we derive the PUN *aspect ratio*, thus

$$\left.\frac{W}{L}\right|_{\text{equv-PDN}} = \underbrace{\left[\left(W/L\right)\right]}_{\widetilde{ADE}} + \underbrace{\left[W/L\right]}_{BC} \tag{6.44}$$

where

$$\left(W/L\right)_{ADE} = \left[\frac{1}{\left(W/L\right)_A} + \frac{1}{1/\left(W/L\right)_D + 1/\left(W/L\right)_E}\right] \tag{6.45a}$$

$$\left(W/L\right)_{BC} = \left(W/L\right)_B + \left(W/L\right)_C \tag{6.45b}$$

So, PUN *aspect ratio is*

$$\left.\frac{W}{L}\right|_{\text{equv-PUN}} = \left[\frac{1}{\left(W/L\right)_{ADE}} + \frac{1}{\left(W/L\right)_{BC}}\right]^{-1} \tag{6.46}$$

Finally, create the *PUN circuit* on top of the *PDN circuit* to have the CMOS circuit of Figure 6.15c. Following Equations (6.43) and (6.46), the equivalent circuit ratio becomes

$$k_{\text{equiv}} = \frac{k_{\text{equv-PND}}}{k_{\text{equv-PUN}}} = \frac{\left.\dfrac{W}{L}\right|_{\text{equv-PND}}}{\left.\dfrac{W}{L}\right|_{\text{equv-PUN}}} \tag{6.47}$$

With Equation (6.47), the circuit's *transconductance parameter* can be found (following Equation (5.9)).

6.2.1.7 Layout of CMOS Logics

Advances in fabrication technology have enabled the use of near sub-0.1μm technology node of CMOS for VLSI, and even ULSI and beyond, ICs. More research studies are continuing in miniaturization of transistors to nanoscale. Miniaturization requires efficient layout of complex logic circuits; that is, drawing to scale, which is time-consuming. Before committing to a full layout, designers use stick diagrams (like "rough sketches") to plan cells and estimate area. CMOS is used in most VLSI or ULSI circuit chips.

Figure 6.16 shows a general layout by *stick diagram*. Typically, the general layout configuration is drawn to figure out how the diffusion layer contacts are placed, and how the inputs and outputs are connected to the diffusion regions. As seen in Figure 6.16, a stick diagram has (i) diffusion regions (ii) metal traces or interconnects, (iii) contacts, and (iv) columns of polysilicon inputs (A_1, A_2). By extension, for CMOS circuits with multiple inputs (A_i, where $i = 3, 4, …, n$) can be visualized and drawn to scale—conforming to constraints imposed by the manufacturing process, design flow, and performance requirements—prior to going into fabrication process.

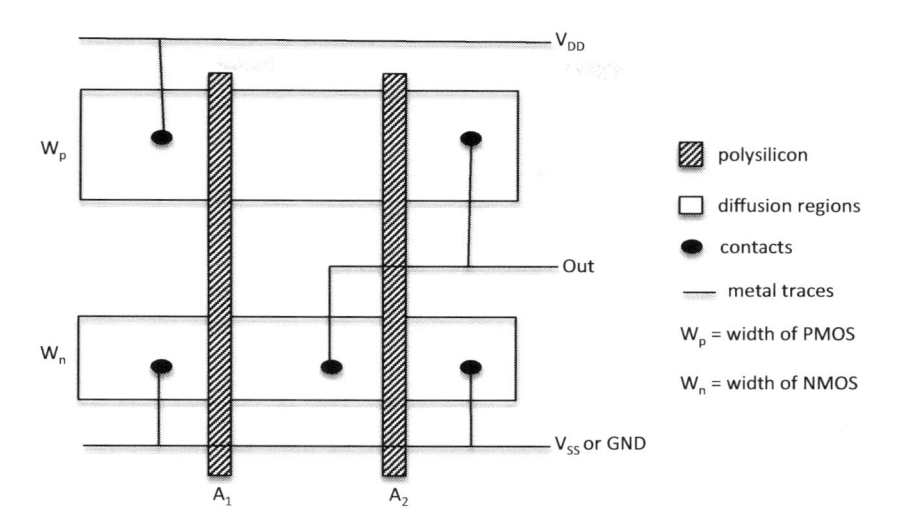

FIGURE 6.16 Stick diagram to mapping complex CMOS circuit into a layout diagram.

Interconnects are the metal wires which pass current between the transistors and are oriented both vertically and horizontally, see Figure 6.16. Metals are used for many purposes including interconnecting terminals of CMOS and other devices, interconnecting circuit blocks, distributing clock signals across chip, providing a power supply grid, and connecting to chip inputs/outputs. Metal wire resistances and capacitances are dependent on the physical properties of the metal and dielectric layer composites, layer dimensions, and near neighbor interactions. Using carbon nanotubes (CNTs) as interconnects in future ICs is receiving greater attention, including their growth techniques and methods of characterization. Except graphene, CNTs are the only material that can carry a much higher current density before failure [11,12]. Their (CNTs) use as interconnects is not trivial because of quantum effects. The overall resistance of a nanotube interconnect is given by [13]

$$R_W = \frac{1}{2n}\left(R_q + rL\right) \tag{6.48}$$

where R_q is the quantum resistance (approx. $R_q = 12.6$ kΩ/channel), r is the incremental resistivity/unit length of the nanotube, L is its length, and n is the number of walls in the interconnect. More is said about nanotubes in Chapter 8.

6.3 DYNAMIC CMOS LOGIC

The preceding CMOS logics can be described as static because they actively restore the logic output values. Figure 6.17 shows graphical differences between static and dynamic profile subject to noise with time. Both static and dynamic CMOS logics have advantages. For instance, static logics have high functional reliability, easy circuit design, and unlimited validity of logic outputs. While dynamic CMOS logics can result in improving performance significantly compared to static circuits, but they can be very sensitive to noise providing the noise is less than the noise margin. Whilst glitches cause extra delay and extra power from false transitions, static CMOS logic will eventually settle to correct output even if disturbed by large noise spikes. In addition, dynamic logics have high switching speed, small area consumption, and low power dissipation.

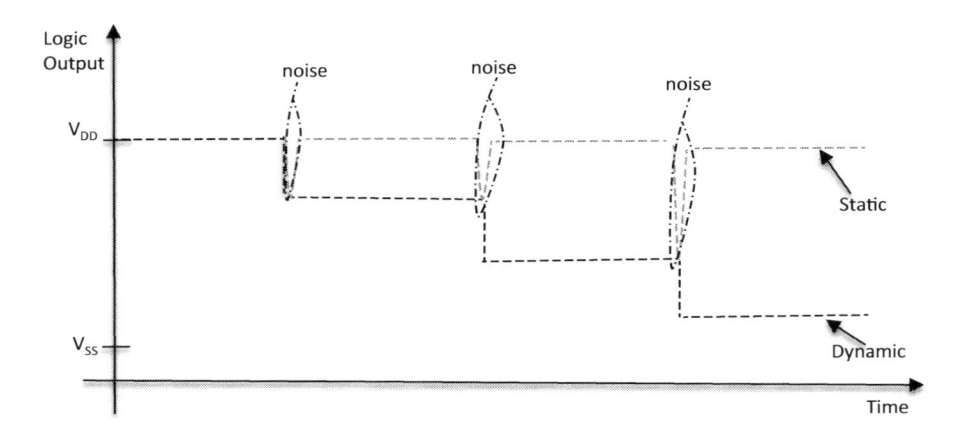

FIGURE 6.17   Representation of static and dynamic CMOS logics.

Basic Principles of dynamic CMOS gates

The basic construction of a *N*-type dynamic logic gate is shown in Figure 6.18. The construction of the PDN proceeds just as it does for static CMOS earlier discussed. However, with the addition of a *clock (Clk)* input, the dynamic logic circuit uses a sequence of *precharge* and *conditional evaluation* phases to realize complex logic functions. This process is discussed as follows.

Precharge Operation When *Clk* = 0, the output node Out is precharged to V_{DD} by the PMOS transistor *M2*. During that time, *M1*—the evaluate NMOS transistor—is off. This prevents the pull-down path fighting the pull-up path. In essence, the evaluation transistors (if more than *M1*) attempt to eliminate any static power that would be consumed during the precharge period; that is, not allowing the pull-down path and the precharge device turning on simultaneously.

Evaluation Operation When *Clk* = 1, the precharge transistor *M2* is off, and the evaluation transistor *M1* is turned on. The output is conditionally discharged based on the input values and the pull-down topology. If the inputs are such that the PDN conducts, then a low resistance path exists between Out and GND and the output is discharged to GND. If the PDN is turned off, the precharged value remains stored on the output load capacitance C_L. During the evaluation phase, the

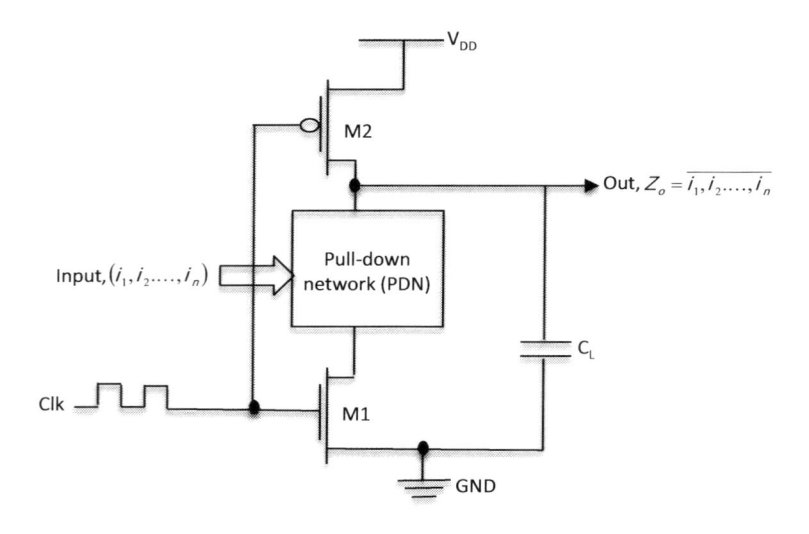

FIGURE 6.18   Basic concept of a dynamic gate using *N*-type network.

only possible path between the output node and a supply rail (V_{DD}) is to GND. Consequently, once Out is discharged, it cannot be charged again till the next precharge operation. The inputs to the gate can therefore make at most one transition during evaluation. Notice that the output can be in the high impedance state during the evaluation period if the PDN is turned off, and this behavior is fundamentally different to the static counterpart that always has a low resistance path between the output and the power rail.

An example of a dynamic logic circuit is shown in Figure 6.19: a fan-in equals five. When $Clk = 0$, the precharge output is 1 and when $Clk = 1$, the PDN evaluates.

Ideally, no static current path ever exists between V_{DD} and GND. The overall power dissipation, however, can be significantly higher compared to a static logic gate. Given that there is reduced load capacitance attributed to the number of transistors per gate and dynamic gate does not have short circuit current since all the current provided by the pull-down devices go into discharging the load capacitance, then the dynamic logic gates would have faster switching speeds.

As observed in Chapter 2 that combinational and sequential logics can be cascaded by connecting the output of one logic circuit to the input of another. Whilst cascading can be down with dynamic logic circuits, some problems arise in cascading process because the output (and hence the input to the next stage) is precharged to 1. Setting the inputs to 0 during precharge could solve this problem. In doing so, all logic transistors of the next function block are turned off after precharge, and so ensuring that the inputs can only make a single $0 \rightarrow 1$ transition during the evaluation period and preventing inadvertent discharging of the storage capacitors occurring during evaluation. A number of design styles complying with the above rule have been developed, but the two most important design styles, namely: (a) CMOS domino logic, introduces a buffer at the output of proceeding logics in the pipeline, thereby restoring the elements, as shown in Figure 6.20. The buffer furthermore reduces the capacitance of the dynamic output node by separating internal and load capacitances. CMOS domino logic uses only PDNs; and (b) NP Domino Logic alternates the cascaded precharge/evaluate-logic with PUN/PDNs, as shown in Figure 6.21. Both styles have advantages and disadvantages. For instance, (a) CMOS domino logic achieves a good balance of switching speed, area/power consumption, and design reliability, whereas NP domino logic achieves low power consumption and high speed, but difficult to design.

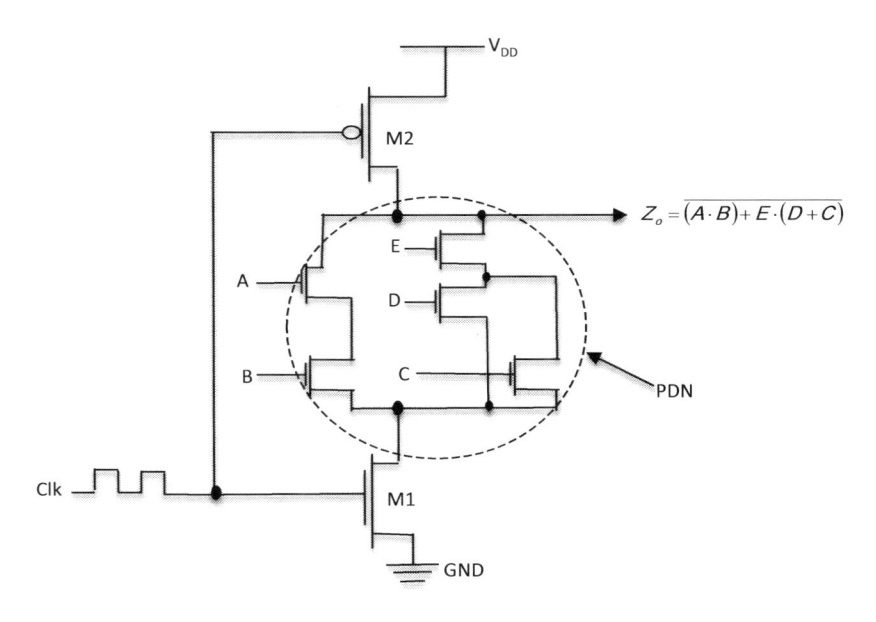

FIGURE 6.19 Example of dynamic gate with fan-in equals 5.

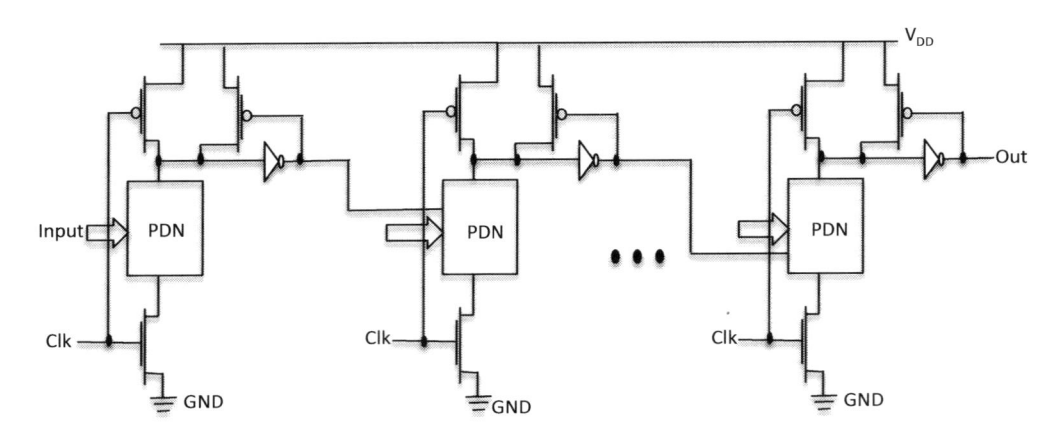

FIGURE 6.20   CMOS domino logic circuit.

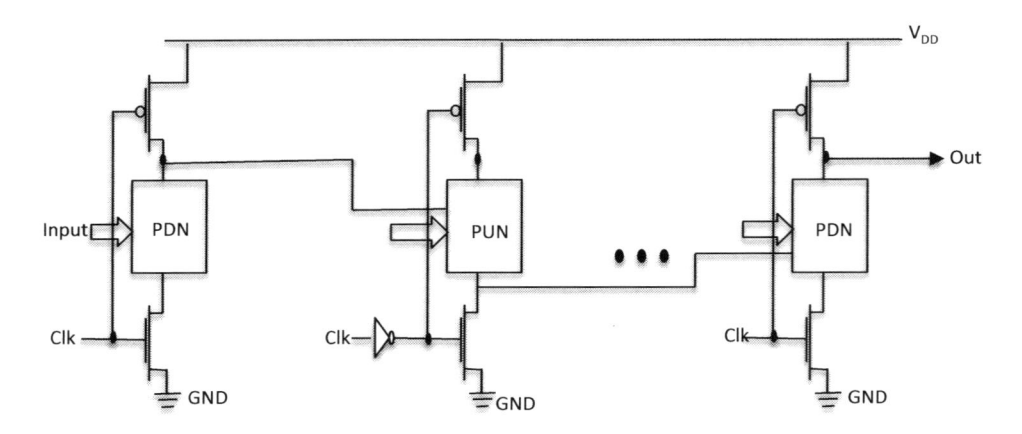

FIGURE 6.21   NP domino logic circuit.

Dynamic logic clearly can result in high performance solutions compared to static circuits. However, there are several important considerations that must be considered to make dynamic circuits function properly. These include charge leakage, charge sharing, backgate (and in general capacitive) coupling, and clock feed through.

Optimization considerations have brought to fore different variations to the generic Domino logics. In some cases, large dynamic stacks are replaced using parallel small fan-in structures and complex CMOS gates.

6.3.1   ISSUES WITH CMOS AS ELECTRONIC SWITCH

Transistors can be used as switches; however, they could produce degraded outputs in part due to parasitics. In fact, static and dynamic performance depend on a switch with low parasitic resistance and capacitance. Suppose an NMOS, in Figure 6.22, is used as a pass transistor that switches data from input A to output B. This figure can be used to explain the issue of parasitic resistance and capacitance (illustrated by dotted lines in Figure 6.22) in a simple pass transistor.

When the control voltage V_g at terminal C is high, the voltage V_{in} at the input terminal A is delivered to the output terminal B, no matter whether the voltage V_{in} is high or low. However, if the input voltage V_{in} (at the input A) is high, current flows through the NMOS to charge up the parasitic

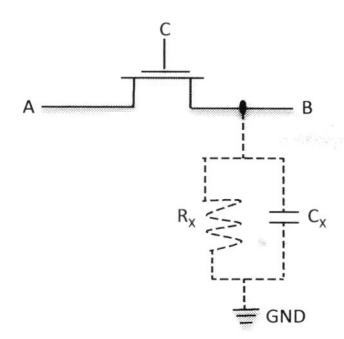

FIGURE 6.22   Pass transistor as a switch with parasitic elements.

capacitor C_x and stops when the difference between the output voltage V_o (at B) and the gate voltage V_g (at c) reaches the threshold voltage V_{th} (i.e. when $V_o - V_g \approx V_{th}$), thereby making the voltage V_o (at B) somewhat lower than the voltage V_{in} (at A). When V_g becomes low, the pass transistor becomes non-conductive and the electric charge stored in the parasitic capacitor C_x gradually leaks to the GND through the parasitic resistor R_x. If V_{in} is low when the control voltage V_g is high, the electric charge stored on the parasitic capacitance C_x flows to the ground through the pass transistor, as well as through the input terminal A if it has not yet completely leaked to the ground. This complex electronic behavior of the pass transistor makes a circuit with pass transistor unreliable. The intermediate value of the voltage V_o (at B), which is lower by the threshold voltage than the input voltage V_{in} or partially leaked voltage, causes unpredictable operations of the logic network when the voltage V_o (at B) is fed to ordinary CMOS logic gates in the next stage. Moreover, it degrades the switching speed of the CMOS logic gates. In the worst case, the circuit loses noise margin or it does not operate properly.

Three techniques can be used to avoid this drawback, namely:

i. Using a pair of NMOS and PMOS pass transistors;
ii. Using a PMOS feedback circuit at the output of the NMOS pass transistor; and
iii. Raising the gate voltage swing above the normal voltage plus threshold voltage.

which are discussed under appropriate headings in the next few pages.

i. *Using a pair of NMOS and PMOS pass transistors*: Combine an NMOS pass transistor and a PMOS pass transistor in parallel, as shown in Figure 6.23, so that when the pass transistor is conductive, the output voltage at B reaches exactly the same value as the input voltage at A, no matter whether the input voltage is high or low. This parallel pair of NMOS and PMOS pass transistors is called a *transmission gate* (TG). Although this has better stability over the pass-transistor circuit of Figure 6.22, it consumes roughly twice as large an area.

Generally, a transmission gate (T-gate or TG or pass gate) is a bidirectional switch. Pull-ups, pull-downs, and dual networks are not necessary. The configurations, shown in Figure 6.23, are often used to denote TGs.

As seen in Figure 6.24, the T-gate switches between inputs A and B when controlled by control signal C connected to the gate of the NMOS and its complimentary $\bar{C}$ connected to the PMOS gate. When the control signal is *high* (V_{DD}), both transistors are turned on. The converse is true: that is, when the control signal is *low* (0 v). This type of operation is commonly used in bus situations where only one gate can drive the bus line at the same time. T-gates are put on the output of each gate on the bus. The circuit that drives will use a T-gate to connect to the bus with a low impedance path. All other circuits that are not driving will switch their T-gates to be at high impedance. Complex logic using T-gates has

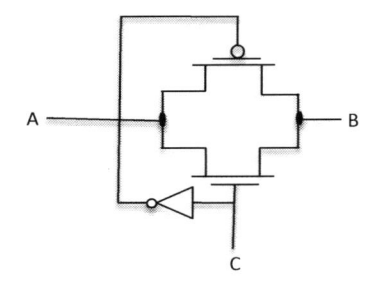

FIGURE 6.23   TG as a switch.

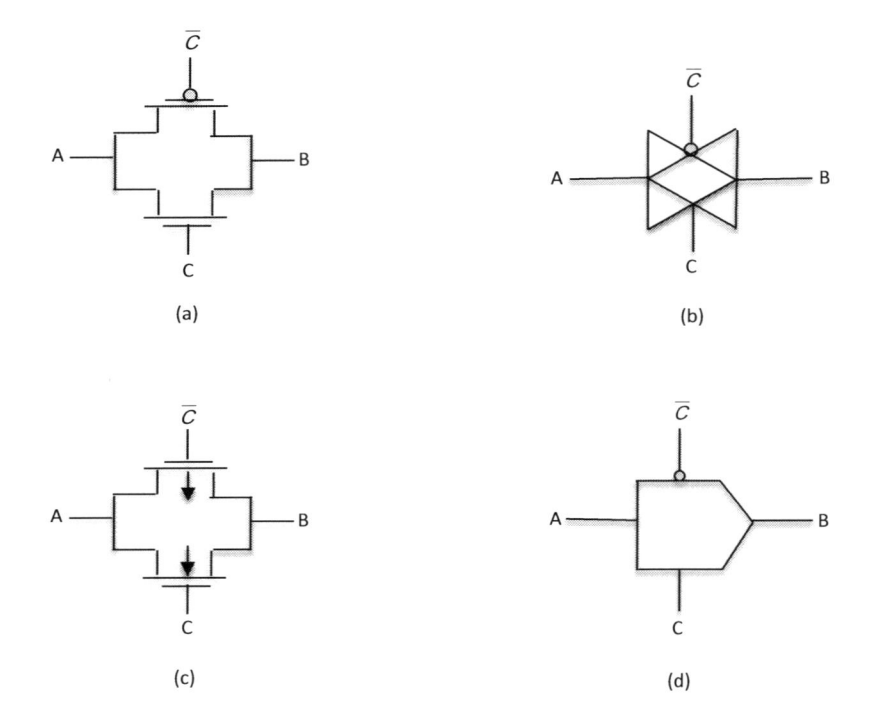

FIGURE 6.24   TGs for 3- and 4-terminal CMOS: (a) TG arrangement for 3-terminal CMOS, (b) 3-terminal TG symbol, (c) TG arrangement for 4-terminal CMOS, and (d) 4-terminal TG symbol.

an advantage when it comes to layout because of the symmetry of the PMOS and NMOS. One n-well can be used for all PMOSs.

The effective resistance R_{eq} of the TG can be estimated by individually obtaining the PMOS resistance (R_p) and NMOS resistance (R_n) in parallel. From simple circuit theory:

$$R_{eq} = R_p \parallel R_n \tag{6.49}$$

where, neglecting the *body-effect* factor (also called *channel length modulation factor*),

$$R_n \approx \frac{1}{\zeta_n \left(V_{DD} - \left| V_{\text{th}(n)} \right| \right)} \tag{6.50a}$$

$$R_p \approx \frac{1}{\zeta_p \left(V_{DD} - |V_{\text{th}(p)}|\right)} \tag{6.50b}$$

where ζ_n and $V_{\text{th}(n)}$ are, respectively, the NMOS *transconductance parameter* and *threshold voltage*, while ζ_p and $V_{\text{th}(p)}$ are, respectively, the PMOS *transconductance parameter* and *threshold voltage*. So,

$$R_{\text{eq}} = R_n \parallel R_p = \frac{R_n R_p}{R_n + R_p} \approx \frac{1}{\zeta_n \left(V_{DD} - |V_{\text{th}(n)}|\right) + \zeta_p \left(V_{DD} - |V_{\text{th}(p)}|\right)} \tag{6.51}$$

Thus, the simplifying assumption that the switch has a relatively constant resistive value is acceptable when analyzing transmission-gate networks.

ii. Using PMOS feedback circuit at the output of the NMOS pass transistor

In this approach, a PMOS is used as a feedback circuit at the output of the NMOS pass transistor, as in Figure 6.25. The gate of PMOS is driven by the CMOS inverter, which works as an amplifier. When the CMOS inverter discharges the electric charge at the output, it turns on the feedback PMOS to raise the pass transistor output to the supply voltage V_{DD}, eliminating the unreliable operation. One limitation of this approach is that it does not solve the degradation of switching speed due to low voltage because the speed is determined by the initial voltage swing before the PMOS turns on. Fabrication-area increase with this approach is smaller than the TG (in Figure 6.23).

iii. Raising the gate voltage swing above the normal voltage

In this approach, a gate signal voltage V_g that is higher than the power supply voltage is applied to the gate C of the MOSFET. This requires a boost driver every time the gate signal is generated. As a result, the gate signal would require raising its voltage V_g swing up to the normal voltage plus threshold voltage (i.e. $V_g \geq V_{DD} + V_{\text{th}}$). The circuit for this process is shown in Figure 6.26.

Whilst this "boosting" has application in dynamic random access memory (DRAM) and referred to as "word boost," it is difficult to use in logic functions in general. In addition, if a voltage higher than the power supply voltage is applied to the MOSFET gate, special care against breakdown and reliability problems would be required. This calls on chip designers to increase the thickness of gate insulation.

The preceding examples show that pass-transistor logic implements a logic gate as a simple switch network. This results in very efficient implementations for some logic functions. Their design is modular because all gates use a similar topology; only inputs are permuted. This facilitates the design of a library of simple and complex gates, such as

a. OR (as shown in Figure 6.27);
b. Tri-state inverter (as shown in Figure 6.28, where S is "select line." Note that a high on the select input S allows the circuit to operate normally as an inverter, whereas a low on S input forces the output into *high-impedance state* (also called Hi-Z). This circuit

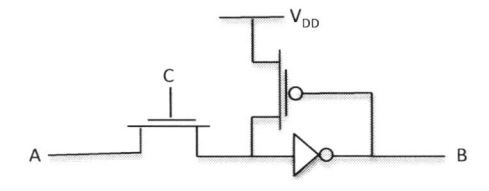

FIGURE 6.25 Modified pass transistor with PMOS and inverter as a switch.

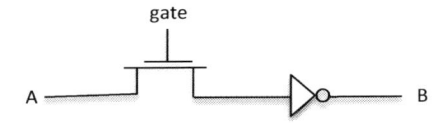

FIGURE 6.26   Modified pass transistor with inverter as a switch.

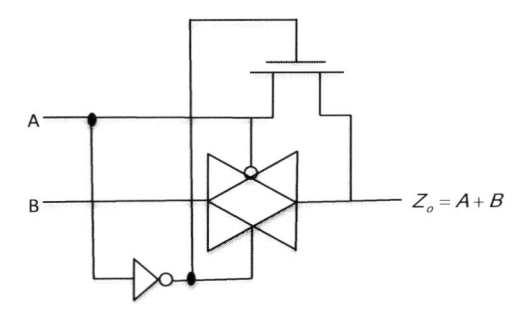

FIGURE 6.27   An OR operation using TG.

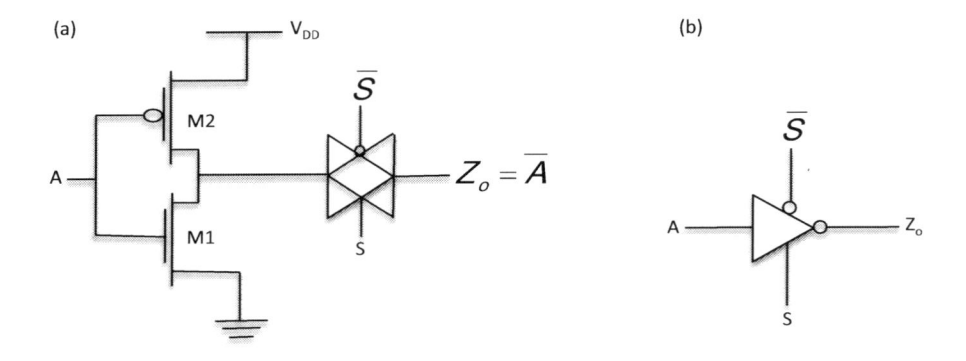

FIGURE 6.28   Tri-state inverter: (a) tri-state inverter circuit and (b) symbol of tri-state inverter.

is useful when data is shared on a communication bus. Its disadvantage is that it dissipates power even when the select signal S is low. Moreover, it can be difficult to switch enable (select line) signals at exactly the same time when they are distributed across a large chip. As a result, delay between different enabled switching can cause contention. Given these problems, multiplexers (MUXs) are now preferred over tri-state buses [10];

   c. *XORs* (as shown in Figure 6.29);

   d. *MUXs* or two-input path selectors (as shown in Figure 6.30, where S is the "select line");

   e. *Adders,* to name a few. Drawing from the basic adder logic expressions; i.e. Equations (2.22)–(2.24) in Chapter 2, CMOS full-adder circuit using TGs is implemented as shown in Figure 6.31, where x_i, y_i are inputs, C_i and C_{i+1} are *carry in* and *carry-out*, respectively, and sum S_i. Efficiency in this case is the number of transistors used in comparison to the standard CMOS logic gates. TGs enable rail-to-rail swing.

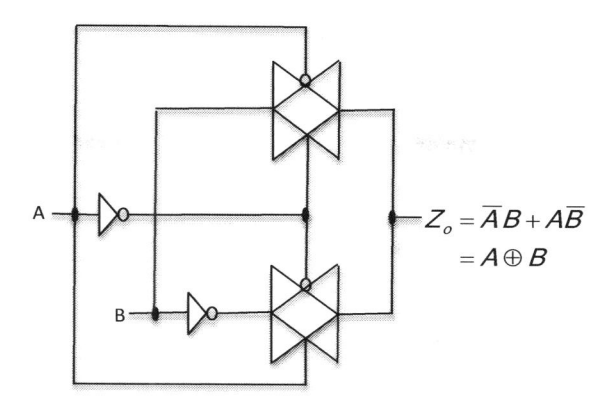

FIGURE 6.29 XOR operation using TGs.

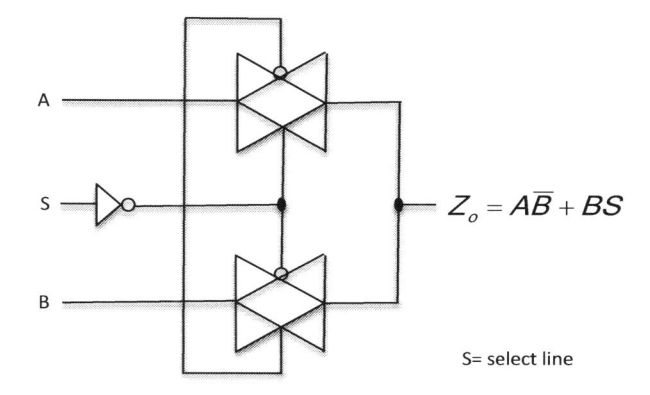

S= select line

FIGURE 6.30 2-Input MUX operation using TGs.

6.3.2 Propagation Delay of a Chain of Transmission (Pass) Gates

In multiprocessor interconnection networks, particularly in switch-based networks, processing elements are connected to switches that are organized in some topology. In switch-based networks, for instance, processing elements communicate with each other over a connection that includes one or more switches [14]. In some cases, a chain of pass transistors is used to implement the carry chain. In doing so, an important consideration is the delay associated with a chain of TGs.

Figure 6.32a shows a chain of n TGs. Such a configuration often occurs in circuits such as adders or deep MUXs. Assume that all TGs are turned on and a step is applied at the input and in order to analyze the propagation delay of this network, the TGs are replaced by each stage-switch equivalent resistance R_{eq}, which was noted earlier to have been assumed to have a constant resistive value, leading to a resistor–capacitor (RC) network of Figure 6.32b, where C_L represents each gate's output load capacitance. The exact analysis of delay is not simple, but as discussed earlier, we can estimate the dominant time constant at the output of a chain of n TGs as follows.

Suppose we replace the pass gates with their effective resistances R_{eq}. Time delay can be found by solving a set of differential equations of the form:

$$\frac{\partial V_i}{\partial t} = \frac{1}{R_{eq}C_L}\left(V_{i+1} + V_{i-1} - 2V_i\right) \tag{6.52}$$

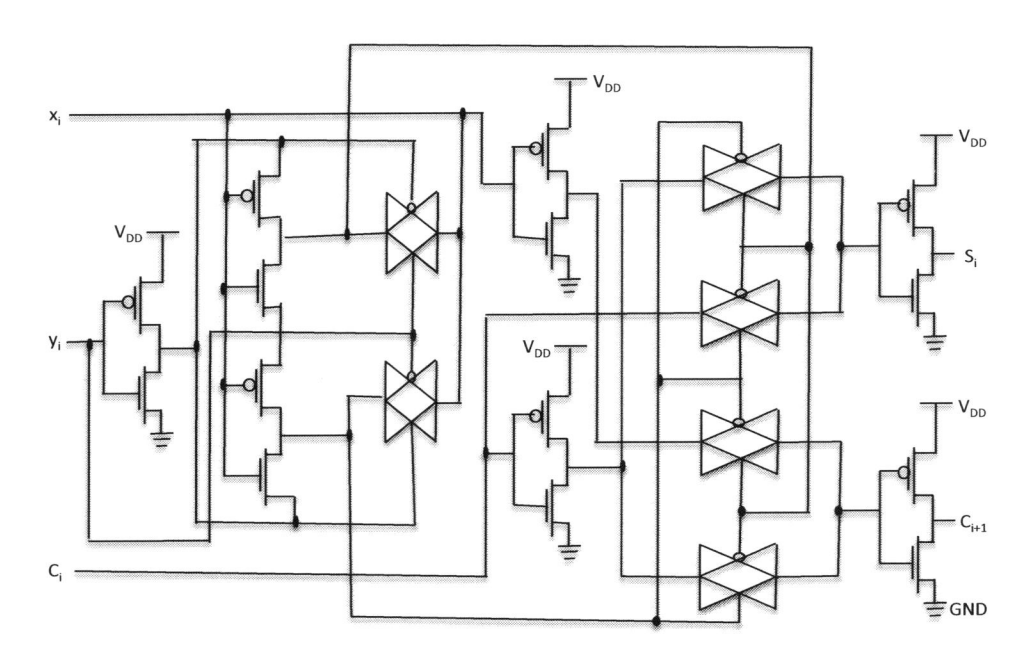

FIGURE 6.31 Using CMOS TGs to implement binary full adder.

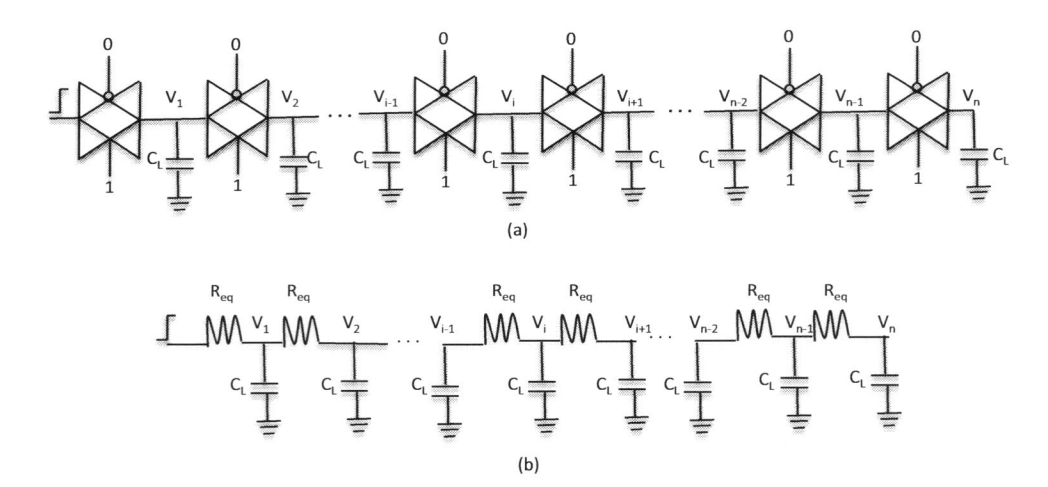

FIGURE 6.32 Cascaded pass gates and their response analysis equivalents: (a) cascaded pass gates and (b) cascaded pass-gate replacement with their equivalent resistances, R_{eq}.

At the output of n pass gates, an estimate of the dominant time constant as

$$\tau_D(V_n) = \sum_{k=0}^{n} kC_L R_{eq} = C_L R_{eq} \frac{n(n+1)}{2} \tag{6.53}$$

From Equation (6.53), we can infer that the delay of this circuit is proportional to the square of the number of pass transistors, i.e. n^2. This means that it is not beneficial to increase the number of pass transistors to many. As a consequence, for large n, it is better to break the chain every m switches and insert buffers, as shown in Figure 6.33.

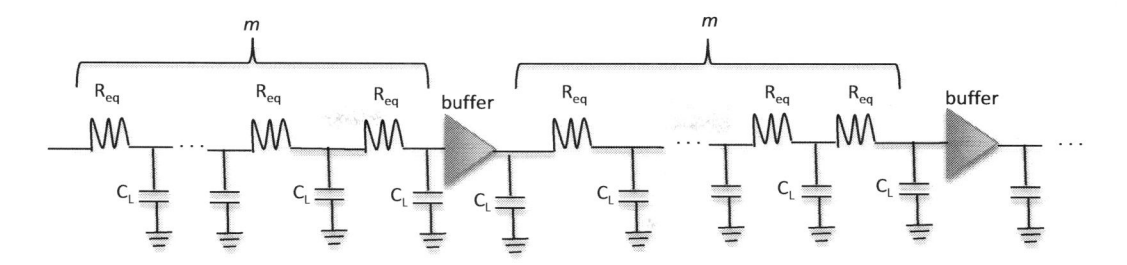

FIGURE 6.33   Cascaded pass gates with buffers.

A buffer is a circuit designed to drive large loads with a minimum delay time, which can also be used to improve the signal strength. Suppose the buffer delay is t_{buf}, the total delay τ_D through the distributed RC circuit can be estimated using

$$\tau_D = 0.69\left[\frac{n}{m}C_L R_{\text{eq}}\frac{m(m+1)}{2}\right] + \left(\frac{n}{m}-1\right)t_{\text{buf}}$$

$$= 0.69\left[C_L R_{\text{eq}}\frac{n(m+1)}{2}\right] + \left(\frac{n}{m}-1\right)t_{\text{buf}} \qquad (6.54)$$

Equation (6.54) shows a linear dependence on n instead of n^2, which allows the optimal number of switches, m_{opt}, between buffers to be found by differentiating Equation (6.54) with respect to m and equating it to zero; i.e. $\left(\partial\tau_D/\partial m = 0\right)$ to yield:

$$m_{\text{opt}} = 1.7\sqrt{\frac{t_{\text{buf}}}{C_L R_{\text{eq}}}} \qquad (6.55)$$

Empirically, the optimal pass-transistor stages for delay minimization is about two or three.

Example 6.3

Suppose each pass gate has an effective resistance of 15 kΩ, a load capacitance of 10 fF, and a buffer delay of 500 ps, find the number of TGs required between buffers.

Solution:

Given:
$R_{\text{eq}} = 15 \times 10^3\ \Omega$,
$C = 10 \times 10^{-15}\ \text{F}$,
$t_{\text{buf}} = 500 \times 10^{-12}\ \text{s}$,
By substituting these terms' values in Equation (6.55) we have

$$m_{\text{opt}} = 1.7\sqrt{\frac{5\times10^{-10}}{1.5\times10^4\times10^{-14}}} = 1.7\sqrt{5/1.5} = 3.1 \approx 3 \qquad (6.56)$$

From this analysis, a buffer is required for every 3 TGs.
    In design practice, the number of pass transistors cannot be arbitrarily chosen because designers want to have a certain number of fanout connections and a buffer cannot support too many fanouts and pass transistors. Also, the structure of a logic network cannot be arbitrarily chosen because of area size and power consumption.

6.4   LARGE-SCALE INTEGRATION AND FABRICATION OF CMOS TRANSISTORS

As discussed in Chapter 1, under *Classification of ICs* in Section 1.1, *large-scale integration* is described as the process of creating ICs, or "chips" as is more commonly called, by combining thousands of transistors (or circuit elements) into a single chip. These elements are either physically embedded in or layered on top of a crystalline silicon surface. By classification, examples of large-scale integration devices are VLSI and ULSI chips. They are complex structures, which can be grouped into two categories: memory chips and logic chips. In a memory chip, transistors are packed into a rectangular array, where individual storage elements (transistors) are arranged like a matrix. In contrast, logic chips have an individual design where mathematical methods are essential. Chapter 1 has discussed the processes involved in producing memory and logic chips from the design and device formulation phase to the device fabrication stage. VLSI design is probably the most fascinating application area of combinatorial optimization. CMOS transistor is used in most VLSI or ULSI circuit chips. VLSI and ULSI depend on the smallest-size feature permitted by the current technology. The size of the transistors has to be as small as possible. Although the internal operating physics of the downscaled MOSFET (MOS) transistor changes, the characteristics of the scaled MOSFET device are similar to that of the original one. As such, designers need to understand the physical implementation of circuits because it has a significant effect on chips' performance, power, and cost. Any chip should, with high probability, operate reliably for its intended lifetime.

Fabrication of such large numbers of very large and ultra large logic circuits into chips is governed by both theoretic and practical limits. High-resolution lithography will be required towards fabricating quantum chips. Avoidance of drain-source junction punch-through may determine the theoretical limit via the minimization of channel length. The entirety of practical limits is described by three parameters: feature size, die processing and area, and packing efficiency. Generalized fabrication processing techniques have also been adequately discussed in Chapter 1. The level of complexity of the fabrication process is increasing, and the device's subtle mechanisms that govern their properties need to be appreciated as to empower the circuit designers with abilities to fully utilize the potential of existing and future technologies.

Of course, fabrication of VLSI chips has advanced considerably with advances in technology. The fabrication-learning process derived from working with a few smaller representative elements is extremely useful in developing a methodology for addressing large-scale multifaceted programs.

As noted in Chapter 1, chips fabrication projection into the ULSI and beyond would depend on practical and analogical limits in terms of material properties and configuration. There is a correspondence between structural materials and technological advances in different fields; thus, allowing for directional change on material limits from silicon base to other materials with conductive properties amongst other transitional elements of the *Periodic Table* (shown in Chapter 1, Figure 1.2).

As the degree of integration increases, more components are integrated on a chip and more wires are necessary to interconnect them. As a result, more power would be consumed on on-chip interconnections, which could impair the performance of the chip. Efforts are being made by innovative technology to lower power consumption, reduce latency, and improve performance on-chip network. Some examples of this innovative technology include "quantum logic gate" in silicon [15], the use of silicon photonics—the application of photonic systems with silicon as the optical medium [14], graphene [16], and CNTs [13]. A quantum logic gate (or simply quantum gate) is a basic quantum circuit operating on a small number of qubits. Traditionally, bits are either 1 or 0, while in a quantum gate; qubits can be both numbers at the same time. Quantum logic gates are reversible, unlike many classical (traditional) logic gates. (More is said about quantum gates and electronics in Chapter 9.)

6.5 SUMMARY

This chapter has discussed the issue of noise, its sources and afterwards, and how it affects the performance of CMOS. CMOS is a type of MOSFET (or MOS) that uses complementary and symmetrical pairs of p-type and n-type MOSFETs (PMOS and NMOS transistors, respectively). The constituents, configuration, and design of CMOS into simple and complex logic circuits, as well as its formulation as transmission (or pass) gate and constitution into a VLSI chip were discussed. The difference between static and dynamic CMOS transistors is discussed. In the construction of static CMOS gate, two complementary networks are used where only one of which is enabled at any time. Whilst *duality* is sufficient for static correct operation, it is not necessarily so in reality. The noise margin of a digital circuit or gate is considered: an indication of how well the circuit or gate will perform under noisy conditions. Pass-transistor logic implements a logic gate as a simple switch network. The optimal pass-transistor stages of buffered chain and its delay are derived for minimization.

PROBLEMS

6.1 Draw a transistor-level schematic for a compound CMOS logic gate for each of the following functions:

    f. $F = \overline{ABC + D}$.

    g. $F = \overline{(AB + C) \cdot D}$.

    h. $F = \overline{AB + C + (A + B)}$.

6.2 Design a precharge-evaluate logic gate that will implement the logical functions:

    a. $F = \overline{ABCD + E}$.

    b. $F = \overline{A + BC + D + E}$.

6.3 Suppose an inverter has the transfer characteristic depicted by Figure 6Q.1. Estimate the values for V_{IL}, V_{IH}, V_{OL}, and V_{OH} that give best noise margins. What are the high and low noise margins?

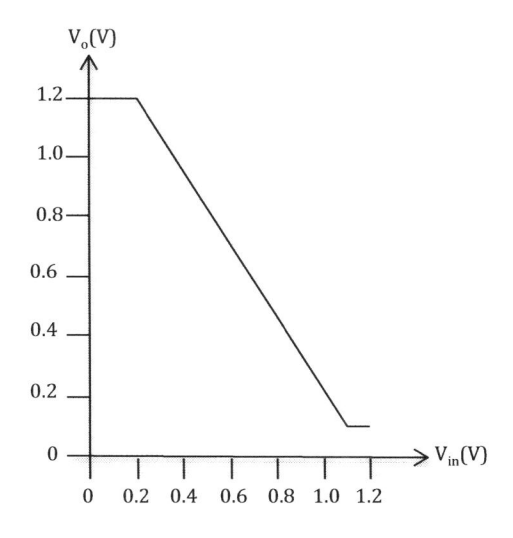

FIGURE 6Q.1 A transfer characteristics.

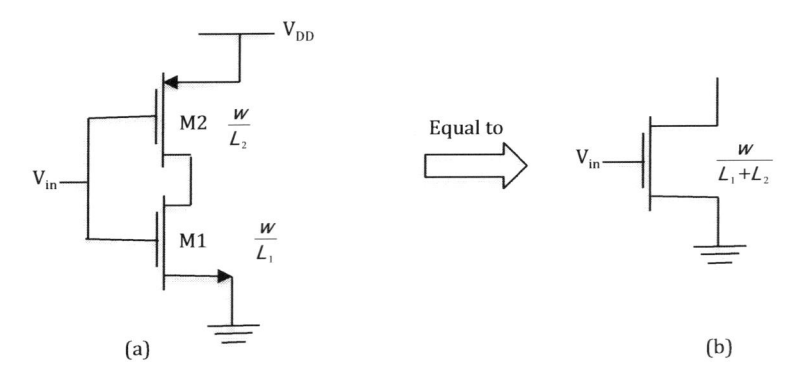

(a)   **(b)**

FIGURE 6Q.2   (a) MOSFET circuit operating in series and (b) approximated equivalent of MOSFET circuit.

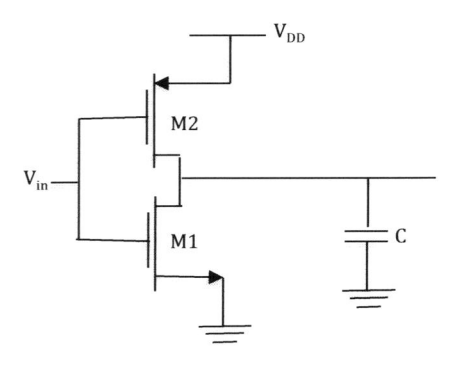

FIGURE 6Q.3   CMOS inverter.

6.4   Show that (i) Figure 6Q.2a can be approximated as Figure 6Q.2b, and (ii) neglecting the body effect, show that the MOSFET $M1$ in Figure 6Q.2a when in a series connection with $M2$ cannot operate in the saturation region. *Hint: Show that M1 is always in cutoff mode, i.e.* $V_{GS(M1)} < v_{th(M1)}.$

6.5   Consider the inverter shown in Figure 6Q.3. Suppose we model resistance of the turn-on device as

$$R_{M1} = \frac{12.5}{\left(W/L\right)_{M1}} \, (\text{k}\Omega) \quad R_{M2} = \frac{29.5}{\left(W/L\right)_{M2}} \, (\text{k}\Omega)$$

The CMOS widths and lengths are specified as: $W_{M2} = W_{M1} = 3\,\text{nm}$, $L_{M1} = 1.5\,\text{nm}$, and $L_{M2} = 1.5 L_{M1}$, and the capacitor $C = 5\,\text{pF}$. Calculate the inverter's propagation delay.

6.6   Design a TG suitable switch-based network. Find the number of pass gates required between buffers if each pass gate has an effective resistance of 12.5 kΩ, a load capacitance of 12 fF, and a buffer delay of 450 ps.

REFERENCES

1. van der Ziel, A. (1954). *Noise*. New York: Prentice-Hall.
2. Sarpeshkar, R., DelbrGck, T. and Mead, C.A. (1993). White noise in MOS transistors and resistors. *IEEE* Circuits and Devices Magazine, 9(6), 23–29.
3. Baker, R.J. (2010). *CMOS Circuit Design, Layout, and Simulation*, 3rd Edition. NJ: Wiley.
4. Johnson, J.B. (1925). The Schottky effect in low frequency circuits. *Physical Review*, 26(1), 71–85.

5. Gross, B.J. and Sodini, C.G. (1992). 1/f noise in MOSFETs with ultrathin gate dielectrics. In *Proceedings of the International Technical Digest on Electron Devices Meeting*, San Francisco, CA, USA, 881–884.

6. Hooge, F.N. (1994). 1/f noise sources. *IEEE Transactions on Electron Devices*, 41(11), 1926–1935.

7. Schroder, D.K. (2006). *Semiconductor Material and Device Characterization*. NJ: Wiley.

8. Wolf, W. (2008). *Modern VLSI Design: IP-Based Design*. NJ: Prentice Hall.

9. Iniewski, K. (2012). *Advanced Circuits for Emerging Technologies*. Hoboken, NJ: John Wiley.

10. Weste, N.H.E and Harris, D.M. (2011). *CMOS VSLI Design: A Circuits and Systems perspective*. Boston, MA: Addison-Wesley.

11. Kong, J., Yenilmez, E., Tombler, T.W., Kim, W., Dai, H., Laughlin, R.B., Liu, L., Jayanthi, C.S. and Wu, S.Y. (2001). Quantum interference and ballistic transmission in nanotube electron waveguides. *Physical Review Letters,* 87, 106801.

12. Yao, Z., Kane, C.L. and Dekker, C. (2000). High-field electrical transport in single-wall carbon nanotubes. *Physical Review Letters,* 84(2000), 2941–2944.

13. Robertson, J., Zhong, G., Hofmann, S., Bayer, B.C., Esconjauregui, C.S., Telg, H. and Thomsen, C. (2009). Use of carbon nanotubes for VLSI interconnects. *Diamond & Related Materials,* 18, 957–962.14. Serpanos, D. and Wolf, T. (2011). *Architecture of Network Systems*. New York: Elsevier.

15. UNSW. (2015). http://newsroom.unsw.edu.au/news/science-tech/crucial-hurdle-overcome-quantum-computing. Accessed 11 February 2016.

16. Caine, T.H. (2012). *Graphene and Carbon Nanotube Field Effect Transistors*. New York: Nova Science Publishers.

BIBLIOGRAPHY

Bhushan, M. and Ketchen, M.B. (2014): *CMOS Test and Evaluation: A Physical Perspective*. New York: Springer.

Chen, L.J. (ed.) (2004). *Silicide Technology for Integrated Circuits (Processing)*. London, UK: The Institution of Electrical Engineers.

Chen, W.-K (ed.) (2000). *The VLSI Handbook*. Boca Raton, FL: CRC Press.

Krambeck, R.H., Lee, C.M. and Law, H.-F.S. (1982). High-speed compact circuits with CMOS. *IEEE Journal of Solid State Circuits*, SC-17(3), 614–619.

Li, Z., Gordon, R.G., Li, H., Shenai, D.V. and Lavoie, C. (2010). Formation of nickel silicide from direct-liquid-injection chemical-vapor-deposited nickel nitride films. *Journal of the Electrochemical Society,* 157(6), H679–H683.

Segura, J. and Hawkins, C.F. (2004). *CMOS Electronics: How It Works, How It Fails*. Hoboken, NJ: John Wiley-Interscience.

7 Data Conversion Process

There is a strong motivation behind exploring a novel frontier of data processing that could benefit from cutting-edge miniature and power-efficient nanostructured silicon photonic devices. Recent example is photonic accelerator (PAXEL)—a processor that can process time-serial data either in an analog or digital fashion on a real-time basis [1]. Data processing is a way of converting data into a machine-readable form using a predefined sequence of operations. Communications signals can be analog or digital, and information can be transmitted using analog or digital signals. Analog signals are continuously changing in time (or frequency), while digital signals are discrete in time and amplitude. Interchangeability of information transfer allows the development of conversion processes without loss of detail. The challenge is to achieve a high sampling rate and high conversion accuracy in the presence of component mismatch, nonlinearity errors, and noise. Although the electronic circuits required to do this conversion processing can be quite complex, the basic idea is fairly simple. The basic concepts of data conversion and their inherent errors, as well as the choice of the converter types that strongly influence the architecture of the overall system, which are fundamental to the continuing revolution in information technology and communication systems, are explained in this chapter.

7.1 INTRODUCTION

Analog signals are continuously changing in time. By "continuous" it means that no matter when one looks at the signal, there is a voltage or current value that represents the instantaneous value of the signal. Music and speech are examples of analog signals that vary continuously in frequency (or time), amplitude, or both. Figure 7.1 represents an analog signal.

Suppose an agreed voltage or current level—called *threshold*—is applied to the analog signal, as seen in Figure 7.2a, where any amplitude (value) below the threshold is counted as an OFF signal, and that above the threshold is counted as an ON signal. There are no values in between. Typically, the *threshold* is a *reference* voltage V_{ref} (or current I_{ref}). If we represent ON state as *logic* "1" and OFF state as *logic* "0," such signal representations are called *discrete* or *digital* because they have a fixed number of definite states. A trace of a digital signal is shown in Figure 7.2b. Thus, a digital signal is a sequence of discrete symbols, and since these symbols are zeros and ones, they are called *bits*.

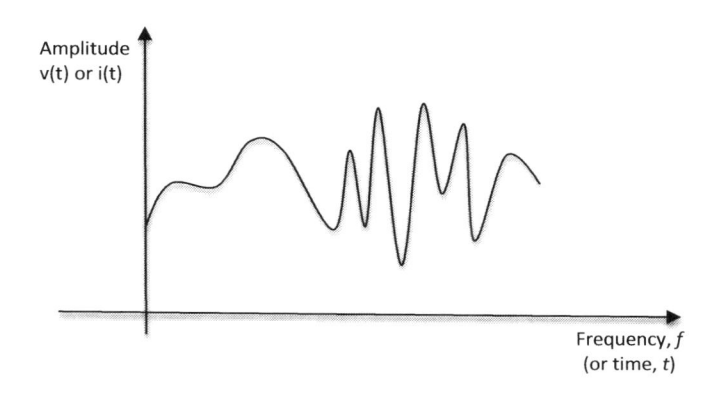

FIGURE 7.1   An analog signal.

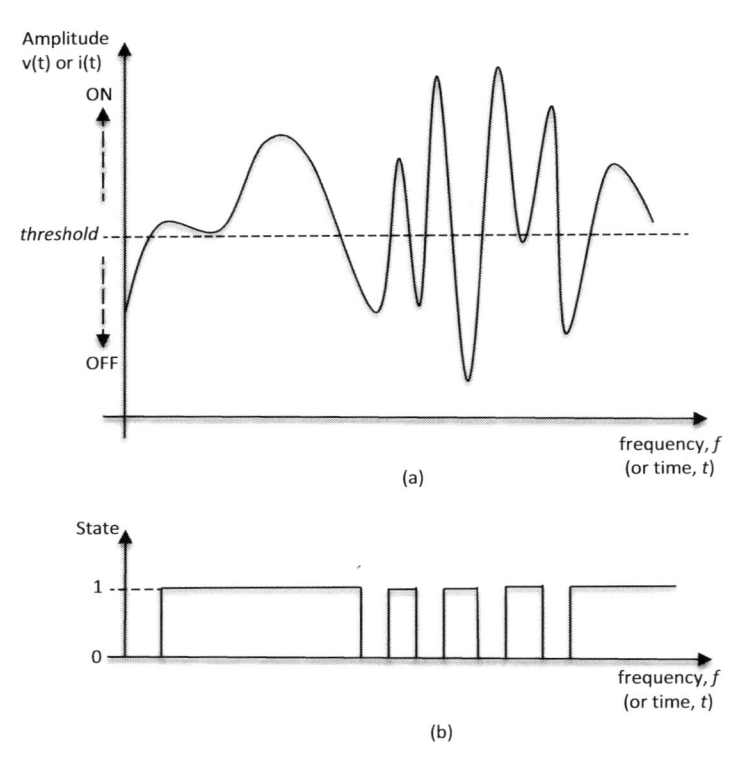

FIGURE 7.2   A trace of digital signal: (a) an analog signal with threshold applied and (b) digital signal representation of (a).

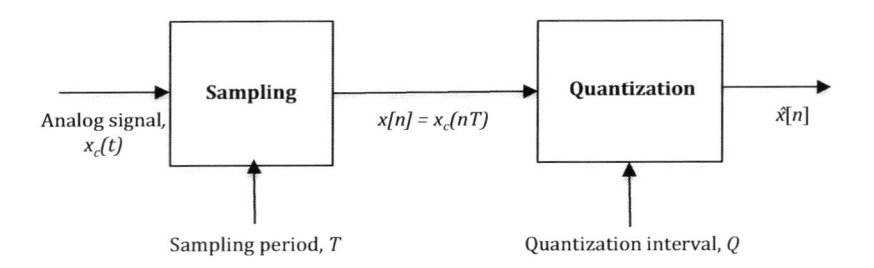

FIGURE 7.3   Concept of sampling and quantization.

The process of converting analog to digital signals is known as analog-to-digital (A/D) conversion. As a result, by using an analog-to-digital converter (ADC) two basic functions are performed: sampling and quantization of an incoming continuous-time signal, conceptually illustrated by Figure 7.3. These consist of the discretization of a signal in time and amplitude, respectively, as illustrated in Figure 7.2. The complimentary conversion back to analog is called digital-to-analog (D/A) conversion; i.e. using digital-to-analog converter (DAC). It is possible to convert information from A/D form, and back again, without losing details. The analog signal is first sampled and then quantized in levels and then each level is converted to a binary number (or digit).

Sampling transforms a continuous-time signal $x_c(t)$ into a discrete-time signal or sequence $x[n]$. Whereas, quantization maps a given amplitude $x[n]$ at time $t\,(=nT)$ into a value $\hat{x}[n]$ that belongs to a finite set of values. Both processes—sampling and quantization—are governed by sampling theory. These processes are explained further in the next subsections.

7.2 SAMPLING THEOREM

This is one of the theorems or concepts needed in all electrical engineering fields. To explain this sampling theorem, suppose a continuous-time signal $x_c(t)$ is periodic function with period T, which is being sampled at regular time intervals $x_c(t)$, as in Figure 7.4a, such that $x_c(t_{-7}), x_c(t_{-6}), \ldots, x_c(t_0),$ $x_c(t_1), \ldots, x_c(t_7)$ are discretized signal in the form $x[-7], x[-6], \ldots, x[0], x[1], \ldots, x[7]$, as shown in Figure 7.4b.

To faithfully reproduce the original signal, certain conditions need to be fulfilled:

i. Sampling interval (also called *time interval* between samples) Δt must be very small and evenly distributed; and

ii. Sampling frequency f_s should be at least twice as large as the bandwidth of the input signal. This bandwidth constraint is known as Nyquist sampling rate. As such,

$$f_s = \frac{1}{\Delta t} > 2f_c \tag{7.1}$$

where f_c is the carrier frequency. This implies that sampling a signal at f_s will cause the frequency spectrum of the input signal to be repeated at all multiples of f_s. If the sampling frequency is less than twice the maximum analog signal frequency, i.e. $f_s < 2f_c$, *aliasing* will occur. Aliasing is an effect that causes different signals to become indistinguishable when sampled. A signal that has frequency components between, say, f_k and f_c must be sampled at a rate $f_s > 2(f_k - f_c)$ in order to prevent alias components from overlapping the signal frequencies. It is plausible to sample below the Nyquist rate without aliasing occurring. Uniform sampling is simply not capable of achieving this goal with a single ADC channel. On the other hand, non-uniform sampling can be applied, if appropriate aliasing reduction (or reconstruction) algorithms are coupled with the ADC allowing the sample rate requirement of the system to be dictated by the bandwidth of the encoded data on the incoming signal rather than by its carrier frequency [2,3].

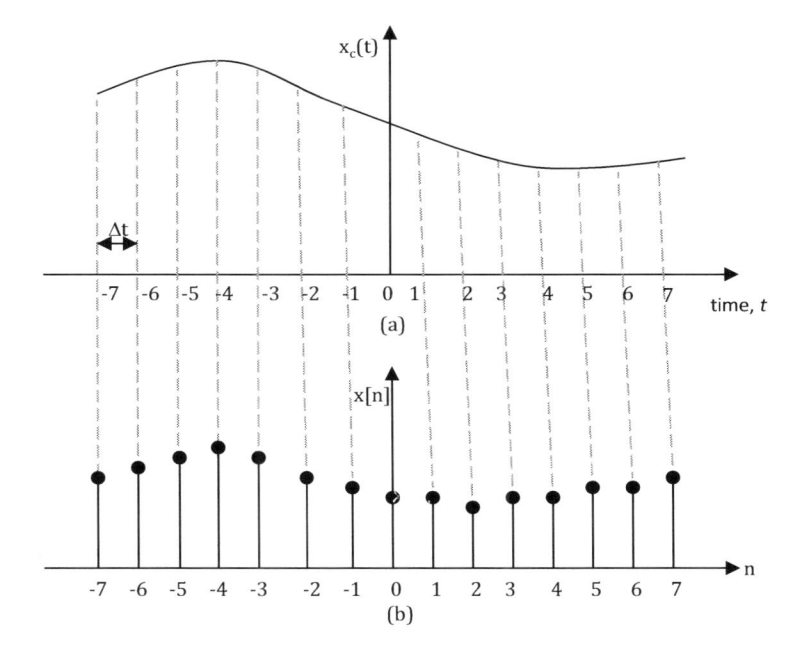

FIGURE 7.4 Sampling concept: (a) continuous signal and (b) discrete equivalent of (a).

Note that signal period T depends on the signal frequency range $[0, f_{max}]$. In order to digitize the analog data accurately, we need to sample the analog signal as fast as possible with an ADC that has a large number of bits. This implies that the sampling interval Δt becomes smaller. Of course, the actual sample value obtained will rarely be the exact value of the underlying analog signal at some time because of the effect of noise. Noise in the samples is often modeled as adding (usually small) random values to the samples. Additionally, the sampling of $x_c(t)$ introduces another error (Δx_n) due to the time involved in performing the sampling. This error is discussed further in the next sections.

7.3 QUANTIZATION PROCESS

Quantization is a process of breaking analog waveform $x_c(t)$ into prescribed number of discrete amplitude levels of length L and interval Δq. Figure 7.5a shows an analog waveform, while Figure 7.5b is the quantized version of the waveform (i.e. $x_q[n]$), which looks like an ascending and descending staircase.

Quantization rounds off the value of $x_q[n]$ to fit into the available register length. The rounding operation creates an error, called *quantization error, e[n]*, expressed by

$$e[n] = x_c[n] - x_q[n] \tag{7.2}$$

We can interpret the error (e) as having an approximately uniform distribution over an ith cell of width Δ_i when the signal is encoded into the cell.

If the quantized signals $x_q[n]$ are transmitted as pulses of varying heights, the resultant system could be a modified form of *pulse amplitude modulation* (PAM). PAM is one of modulation schemes where the message information is encoded in the amplitude of a series of signal pulses to facilitate transmission of digital information. For more details on PAM, the reader is referred to one of this author's books [4].

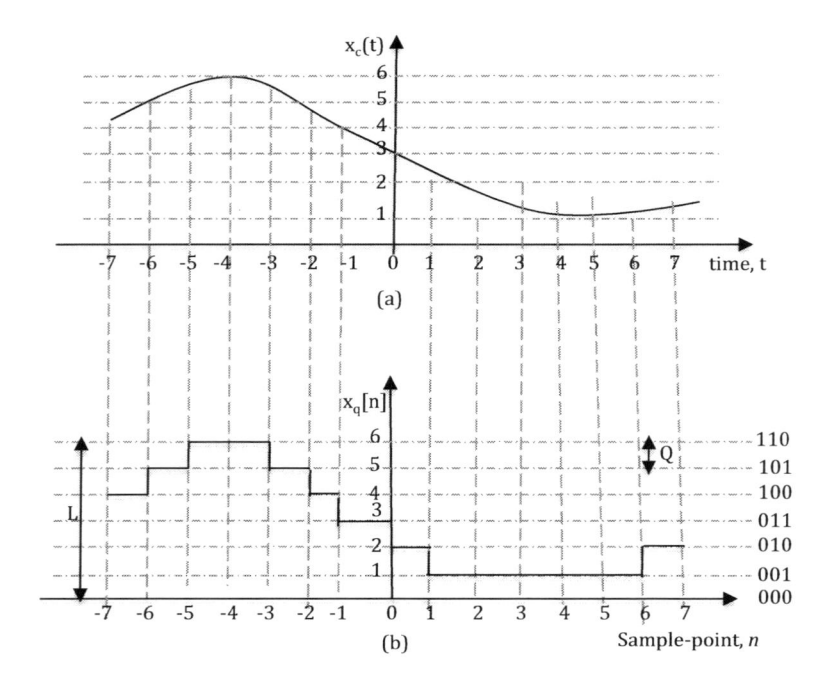

FIGURE 7.5 Quantization concept: (a) continuous signal and (b) sampled and continuous signal $x_q[n]$.

Quantization interval depends on the dynamic range of the signal amplitude and perceptual sensitivity. Quantization is generally irreversible and results in loss of information: it introduces distortion into the quantized signal, which cannot be eliminated.

7.4 ADC AND DAC

Both ADC and DAC are sampled data systems; as such, they require the use of circuit techniques that can handle discrete-time analog signals. The ability to convert analog signals to digital and vice versa is very important in signal processing as well as in communication engineering. Figure 7.6 depicts the functional block of A/D and D/A converters, where the A/D converter—also called an ADC—accepts an analog sample signal $x_c(t)$ and produces an n-bit digital word (i.e. b_0, b_1, b_2, ..., b_{n-1}), as shown in Figure 7.6a. On the other hand, the D/A converter—also called a DAC—accepts an n-bit digital word (i.e. b_0, b_1, b_2, ..., b_{n-1}) and produces a reconstructed analog signal $x_c(t)$, as shown in Figure 7.6b. The discussion that follows considers linear converters in which the digital signal is directly proportional to the amplitude $|x_c(t)|$ of the analog signal, and the reference amplitude x_{ref} of the signal acts as *high* logic level of the converters.

The data conversion process can be described by the expression

$$x_n = x_{ref} \left(b_0 2^{-1} + b_1 2^{-2} + b_2 2^{-3} + \cdots + b_{n-1} 2^n \right) \tag{7.3}$$

This expression (Equation 7.3) applies to both ADC and DAC. In the case of ADC, an analog waveform x_n is converted into an n-bit binary number, and for the DAC, the input bits b_0, b_1, b_2, ..., b_{n-1} are converted to an analog waveform or signal by the scaling factor x_{ref}. The resolution represents the quantization error inherent in the conversion of the signal to digital form. Following Equation (7.3), the *quantization noise* can be expressed by

$$\delta x_n = \frac{x_{ref}}{2^n} \tag{7.4}$$

This *quantization noise* is often called quantization error. The quantization noise is, however, strongly dependent on and, in general, correlated with the input signal, and hence cannot be precisely modeled as signal-independent additive noise [5].

As noted in Section 7.2, the sampling of $x_c(t)$ in A/D conversion introduces an error ($\pm\Delta x_n$) due to the time involved in performing the sampling, expressed by

$$\Delta x_n = \int_0^{T_s} \frac{\delta x_n}{\delta t} dt \tag{7.5}$$

where T_s (= $1/f_s$) is the sampling time of measurement.

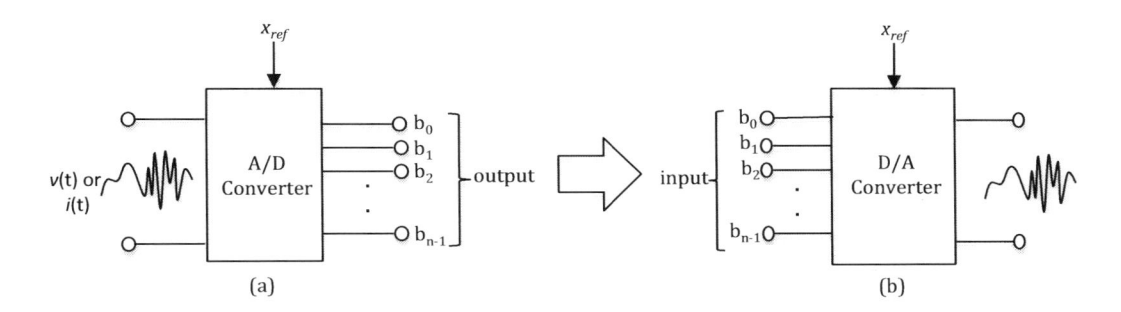

FIGURE 7.6 Functional block of ADC and DAC: (a) A/D and (b) D/A.

The parameters characterizing the performance of data conversion are resolution, accuracy, and dynamic response.

The next subsections examine some circuits developed to perform A/D and D/A conversions.

7.4.1 A/D CONVERSION CIRCUITS

The main goal of an ADC is to digitize the analog signals; that is, processing, recording, and storing the analog signals as numbers. Typical examples of an ADC usage are cell (mobile) phones, thermocouples, digital voltmeters, and digital oscilloscope.

Conversion from A/D form inherently involves comparator action where the value of the analog voltage at some point in time is compared with some standard. In Chapter 4, Sections 4.2.5 and 4.2.8 have discussed *Comparators* and *Integrators*, respectively.

A very popular high-resolution A/D conversion scheme is illustrated in Figure 7.7. As designed by switch S_1 nodes, two phases would be involved in the conversion process: Phase 1—when S_1 switches to v_{in}; and Phase 2—when S_1 switches to $+V_{ref}$. Suppose the analog input signal v_{in} is negative. Prior to the start of the conversion cycle, switch S_2 is closed, consequently discharging capacitor C and setting v_1 to zero (i.e. $v_1 = 0$).

The Phase 1 conversion cycle begins with an opening switch S_2 and connecting the integrator input through switch S_1 to the analog input signal. Since v_{in} is negative, a current I (i.e. $I = v_{in}/R$) will flow through R in the direction away from the integrator. Of course, the value of v_1 rises linearly with I having a slope Δ_1 giving by

$$\Delta_1 = \frac{I}{C} = \frac{v_{in}}{RC} \tag{7.6}$$

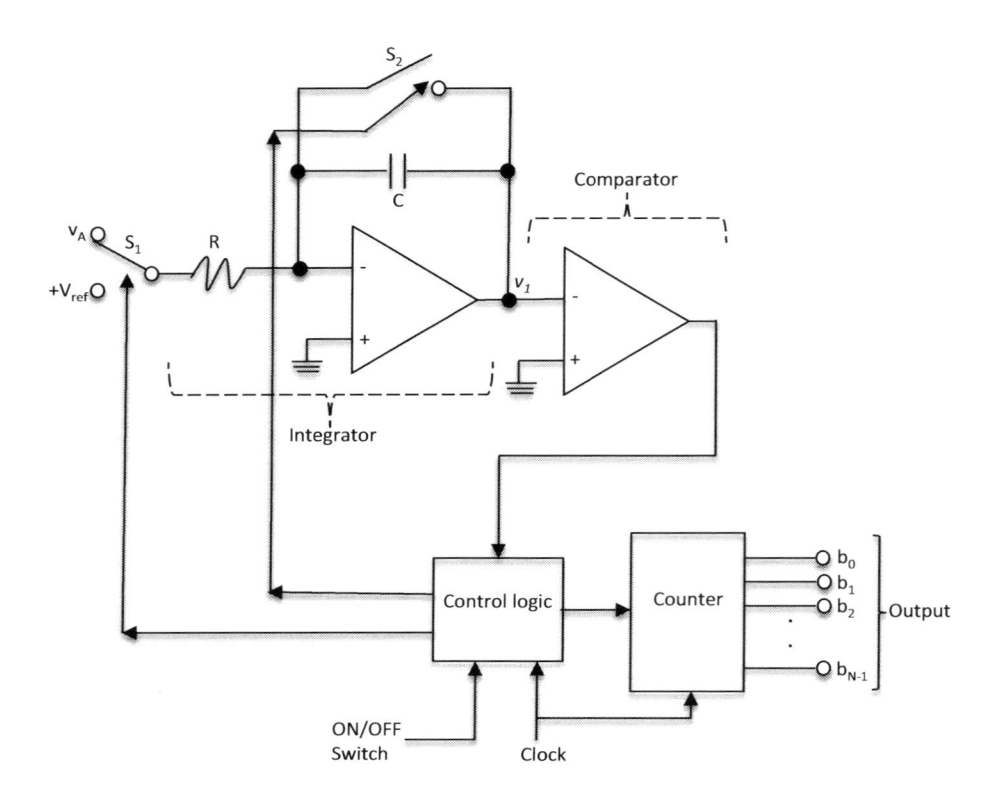

FIGURE 7.7 An A/D conversion technique.

Simultaneously, the counter is enabled and it counts the pulses from the Clock, which has a fixed frequency and duration, say, T_1. It ends when the counter has accumulated a fixed count n_1, which in this case of N-bit word, equals 2^N. Denoting the peak voltage at the output of the integrator as V_{peak}, we can write

$$V_{peak} = v_1 = \frac{v_{in}}{RC} T_1 \tag{7.7}$$

At the end of this phase, the counter is reset to zero, while Phase 2 begins at $t = T_1$.

At $t = T_1$, when Phase 2 of the conversion begins, the integrator input is connected through switch S_1 to the positive reference voltage V_{ref}. The current into the integrator reverses direction and, by Ohms law, is equal to V_{ref}/R, and v_1 decreases linearly with a slope Δ_2 given by

$$\Delta_2 = \frac{V_{ref}}{RC} \tag{7.8}$$

Simultaneously the counter is enabled and it counts the pulses from the fixed-frequency clock. When v_1 reaches zero volt, the comparator signals the control logic to stop the counter at n_2. If we denote the duration of Phase 2 by T_2, then we can write

$$T_2 = \frac{v_{in}}{V_{ref}} T_1 \tag{7.9}$$

The counter reading n_2 equals

$$n_2 = \frac{v_{in}}{V_{ref}} n_1 = 2^N \left(\frac{v_{in}}{V_{ref}} \right) \tag{7.10}$$

This basic A/D operation is sometimes called a *dual-slope A/D conversion* process because it has a different slope Δ for each phase (i.e. Δ_1, Δ_2), and features high accuracy, since its performance is independent of the exact values of R and C.

Speed and accuracy are two critical measures of ADC performance. There exists many commercial implementations of the dual-slope method and other methods, some of which utilize complementary metal oxide semiconductor (CMOS) technology culminated in high speed, low voltage CMOS digital circuits with low gate delays at relatively low voltage supply. The most popular methods are flash (also called *parallel, pipelined,* or *simultaneous*), and charge-redistribution (also called *successive approximation-register*) converters; each method having different architecture and offering certain advantages with respect to conversion speed, accuracy, and other parameters. These methods—flash and charge redistribution—are discussed in the next subsections.

7.4.2 FLASH OR PARALLEL ADC

Conceptually, flash (parallel, or simultaneous) A/D conversion is simple. The basic principle of operation is to use the comparator principle to determine whether or not to turn on a particular bit of the binary number output. Chapter 4, Section 4.2.5, has discussed elements of a *comparator amplifier*. Its threshold levels are usually generated by resistively dividing one or more reference voltages into a series of equally spaced voltages, which are applied to one input of each comparator. Figure 7.8 illustrates a flash (parallel, or simultaneous) A/D conversion process.

As seen in Figure 7.8, a flash converter utilizes $2^N - 1$ analog comparators, $2^N - 1$ reference voltages, and a digital encoder (encoding-logic block). Each reference voltage $v_1, v_2, v_3, \ldots, v_{2^N-1}$—each

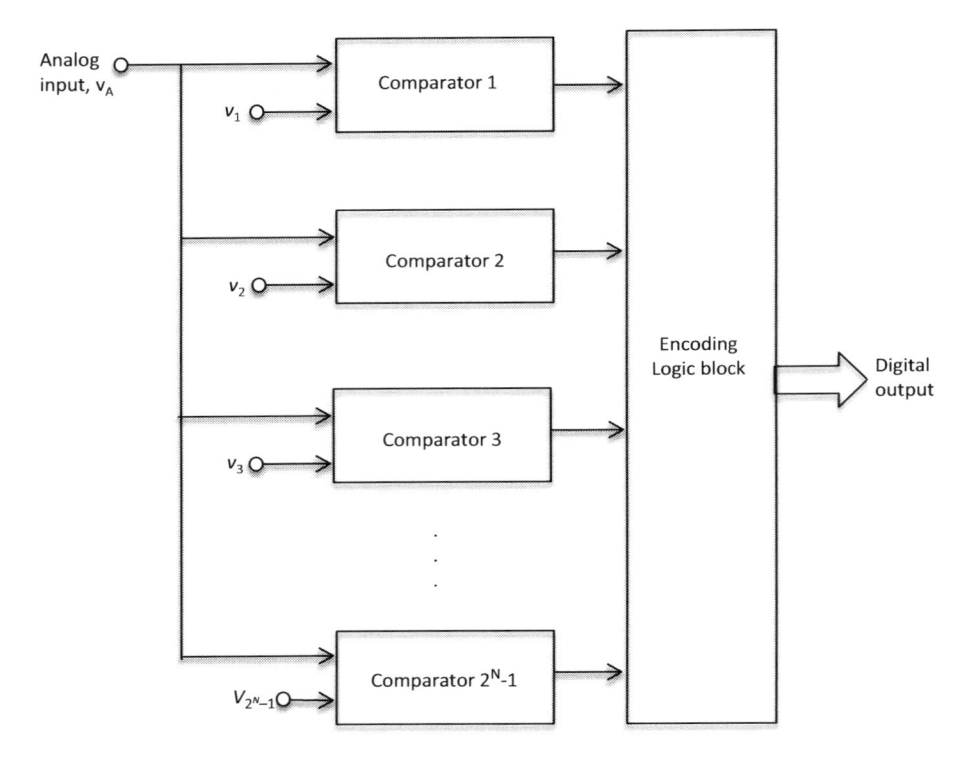

FIGURE 7.8 Flash, parallel, or simultaneous A/D conversion process.

is a unique fraction of the reference voltage—is applied to the negative input of a comparator and the analog signal input voltage v_{in} is applied simultaneously to all the comparators. For $2^N - 1$ analog comparators, reference voltages v_i (where $i = 1, 2, 3, ..., 2^N - 1$) are expressed thus

$$v_1 = \frac{2^N - 1}{2^N} V_{ref}$$

$$v_2 = \frac{2^N - 2}{2^N} V_{ref}$$

$$v_3 = \frac{2^N - 3}{2^N} V_{ref} \qquad (7.11)$$

$$\vdots$$

$$v_{2^N - 1} = \frac{1}{2^N} V_{ref}$$

The digital encoder processes the outputs of the comparators to provide n bits digital output. Thus, the comparators with reference voltages less than the analog input (v_{in}); i.e. $v_i < v_{in}$, will output a digital "1," and when the comparators with reference voltages greater than the analog input; i.e. $v_i > v_{in}$, will output a digital "0." When all the digital outputs are read together, the digital outputs present a "thermometer code," which the *encoding-logic* converts to a standard binary code (like 10101........0000) as its output. Note that each comparator must be precisely matched to achieve acceptable performance at a given resolution.

Naturally, the flash resolution is determined by the number of comparators used, which in this case, has n-bit resolution. Note that a complete conversion can be obtained within one clock cycle, making *flash* a fast converter. The downside of flash conversion techniques is a rather complex circuit implementation—which grows exponentially as the resolution bit increases—as well as high power dissipation—due to increase in chip area—and possibly comparator and reference mismatch—which may limit the flash's useful resolution. Generally, the physical limits of monolithic integration allow only up to 8–10 bits of useful resolution per A/D chip [6].

Variations on the basic technique have been successfully employed in the design of integrated circuit (IC) converters.

7.4.3 THE CHARGE-REDISTRIBUTION CONVERTER

The charge-redistribution (also called *successive approximation-register*) A/D conversion technique is particularly suited for CMOS implementation. The implementation of this converter is in three phases: (a) sample phase, (b) hold phase, and (c) charge-redistribution phase. To understand the three phases involved, it is important to understand the basic concept of "sample and hold" (S/H).

7.4.3.1 Basic Concept of S/H

A sample-and-hold (S/H) circuit samples (or captures) its input voltage v_{in} and holds that voltage at its output until the next sample is taken. S/H circuits are also referred to as track-and-hold circuits. Commonly, an S/H circuit precedes ADCs in systems where an analog signal, such as speech, is to be converted to digital form for long-distance transmission or signal processing [7]. In principle, the operation of this circuit is very simple. Figure 7.9 shows a simple S/H circuit with its timing diagram.

To sample the input signal, the referencing switch, CMOS, is closed. The capacitor C_h charges up to the value of the input voltage v_{in} as long as the switch remains closed. Typically, the samples are taken at the clock time intervals. To hold the sampled value, the switch is opened (i.e. the CMOS is switched off), and node "X" is held at high impedance with C_h still charged, thereby isolating the capacitor C_h from the input and allowing the capacitor to hold the sampled voltage. Ideally, the voltage at node "X" would stay constant having a value equal to v_{in} at the instance of clock going low. The unity-gain buffer provides output current to the input of the A/D converter without drawing current from the capacitor. In an ideal S/H circuit, the capacitor would charge instantaneously when the switch is closed and would hold the stored voltage precisely for an indefinite time when the switch is open. In a real S/H circuit, the time taken to charge the capacitor (also called the acquisition time) is not zero and the charge stored on the capacitor leaks away when the switch is open. As a consequence, the output voltage V_o is said to *droop* during the hold state. The amount of drift exhibited by the output when in hold mode is called the *droop rate* (measured in mV s^{-1}).

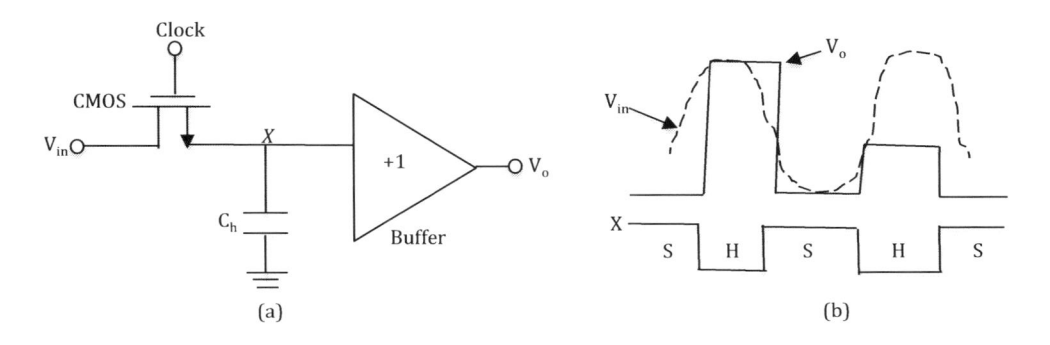

FIGURE 7.9 A simple S/H block: (a) S/H circuit and (b) sample (S) and hold (H) timing diagram.

A good operational amplifier buffer can minimize droop, even with a small capacitor C_h. But a small C_h makes the circuit vulnerable to another error that is harder to handle: the effects of charge injection. Charge injection is due to the gate overlap capacitance that happens during the S/H transition resulting in a voltage error, which can be expressed as

$$\Delta V_x = \frac{Q}{C_h} = -\frac{C_{\text{ox}} WL}{2C_h}(V_{DD} - V_{\text{in}} - V_{\text{th}})\qquad(7.12)$$

where Q, V_{DD}, V_{th}, C_{ox}, W, and L are the charge injected into the capacitor, supply voltage, CMOS threshold voltage, gate capacitance per square, channel width, and channel length, respectively. This voltage error ΔV_x is typically small, and may be considered ineffectual.

The droop rate can be estimated as

$$V_{\text{droop}} = \frac{i_{\text{leak}}}{C_h} T_{\text{conv}}\qquad(7.13)$$

where T_{conv} and i_{leap} are the total conversion time and total leakage current, respectively. The conversion time T_{conv} is set by the conversion time of the converter. The droop rate is an important performance parameter in high-resolution A/D conversion applications [7]. In a precision S/H IC, if the droop rate is not to be significantly increased by printed circuit board (PCB) leakage, the insulation resistance R_{ins} between the hold capacitor and ground on the PCB must be greater than five times the leakage resistance R_{leak} (i.e. $R_{\text{ins}} \gg 5R_{\text{leak}}$).

Example 7.1

A precision S/H IC with 10 nF hold capacitor has a stated maximum droop rate of 25 mV s⁻¹. If the maximum input voltage is 5 V, what is the leakage current from the capacitor? The hold capacitor is connected to an external pin of the IC. What order of insulation resistance is needed on the PCB if the droop rate is not to be significantly increased?

Solution:

From basic circuit theory, the capacitor voltage and the leakage current are related by

$$i_{\text{leak}} = C_h \frac{dv}{dt}\qquad(7.14)$$

Given: $C_h = 10 \times 10^{-9}$ F, and droop rate: $dv/dt = 0.025$ Vs⁻¹.

   i. Substituting these parameters in Equation (7.14), the leakage current is estimated as

$$i_{\text{leak}} = C_h \frac{dv}{dt} = 0.25 \text{ nA}$$

   ii. At a maximum hold voltage of 5 V, and from Ohms law, the leakage resistance

$$R_{\text{leak}} = \frac{V}{i_{\text{leak}}} = 20 \text{ G}\Omega.$$

If the droop rate is not to be significantly increased by PCB leakage, the insulation resistance between the hold capacitor and ground on the PCB must, in this case, be five times higher (i.e. $R_{\text{ins}} \gg 5R_{\text{leak}}$, or about 100 GΩ. This is a very high resistance value and may not be achievable.

7.4.3.2 Charge-Redistribution Successive Approximation

Having studied the basic S/H concept as well as its associated errors, we turn our attention to charge-redistribution successive approximation type ADCs where no external S/H is needed.

The classical design of early successive approximation ADC was based on D/A converters with R-2R resistor ladder, which is discussed in Section 7.4.4, operating as a string of current dividers. McCreary and Gray [8] later proposed the successive approximation ADC based on charge redistribution in the binary-weighted capacitor array. This scheme consists of a simple architecture and a lack of high-performance operational amplifier. However, due to easier fabrication of capacitors, zero quiescent current and lower mismatch and tolerance, the binary-weighted capacitor array has successfully replaced the R-2R resistor ladder in MOS technology. As a consequence, charge substitutes current as the working medium in the implementation of successive converters. The charge-distribution scheme is still successfully implemented in commercial successive approximation ADCs, and importantly, its simplicity of design allows for both high speed and high resolution while maintaining a relatively small area.

Figure 7.10 shows the basic configuration of an N-bit charge-redistribution successive approximation ADC. Basically, this configuration consists of a comparator, successive approximation

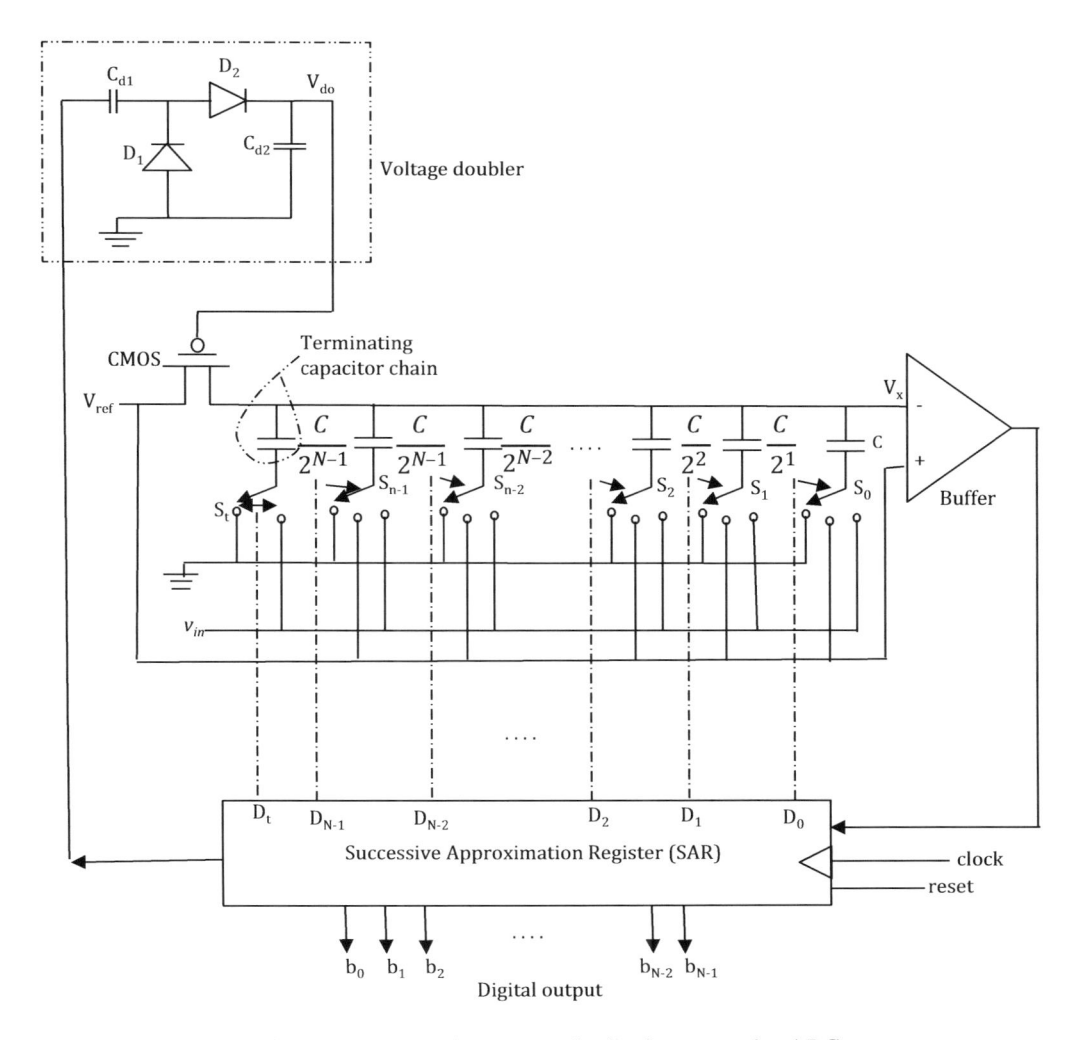

FIGURE 7.10   Basic configuration of an *n*-bit charge-redistribution successive ADC.

register (SAR), an array of binary-weighted capacitors C_i, {where $i = 0, 1, ..., N-1$} with an extra capacitor C_{n-1} as a terminating capacitor, a set of switches S_i controlled by SAR via D_i connecting the capacitor plates to certain voltages, and a voltage doubler.

SAR control logic technically encompasses a ring counter (shift register) and a code register. In fact, SAR operation can be represented using a sequential *finite state machine* (FSM). FSM is one way of describing the behavior of a circuit with state, or with a finite set of states. For more details on *sequential logic circuits,* see Chapter 2, Section 2.2.1.

A simple *voltage doubler* circuit, denoted by dotted block in Figure 7.10, produces a voltage V_{do} that is twice its voltage supply, ensuring that reference switch (CMOS) will be in deep triode when switched on. The doubler's capacitors C_{d1} and C_{d2} are typically of the same value, and diodes D_1 and D_2 are of the same type.

The conversion process begins by discharging the capacitor array via the reset switch. Although this may appear to be an insignificant action, the converter is also performing automatic offset cancelation [9]. Once the reset switch is closed, the comparator acts as a unity-gain buffer. Thus, the capacitor array charges to the offset voltage of the comparator. This requires that the comparator be designed to be stable: this would require internal compensation be switched in during the reset period. Next, the input voltage, v_{in}, is sampled onto the capacitor array. The reset switch is still closed, for the top plate of the capacitor array needs to be connected to virtual ground.

The conversion cycle is performed in three phases, namely: the sampling phase, the charge holding phase, and the charge-redistribution phase where the actual conversion is performed.

Sampling phase: In the first half cycle of the first clock cycle, the reference switch is closed; the converter samples the input signal and then performs the binary search based on the amount of charge on each of the capacitors. The binary-weighted capacitor array also samples the input voltage v_{in}; so, no external S/H is needed. As a result, the top plates of capacitors are connected to the ground, and the bottom plates of the capacitor array are connected to the input voltage v_{in} via the switching network. Note that the voltage drop across the reference switch must be sufficiently small such that the voltage of the top plates of the capacitor array is V_{ref} [10].

Charge holding phase: In the second half cycle of the clock cycle, the reference switch is open and the bottom plates of the capacitors are connected to the ground. Since the charge on the top plates is conserved, the potential of the capacitor top plate goes to $-v_{in}$.

Charge-redistribution phase: The actual conversion is performed in this phase. In the first step of N-bit conversion cycle, the bottom plate of the largest capacitor C_{N-1} is connected via the switch S_{n-1} to the reference voltage V_{ref}, which corresponds to the full-scale range of the converter. The capacitor C_{N-1} forms the one-to-one capacitance divider with the remaining capacitors connected to the ground. The comparator input voltage becomes

$$V_x = -v_{in} + V_{ref}/2^1 \tag{7.15}$$

and the state of the bit b_{N-1} is tested by comparing v_{in} to $V_{ref}/2^1$.

If $v_{in} > V_{ref}/2^1$, then $V_x < 0$, and the comparator output goes high providing the most significant bit (MSB) b_{N-1} to "1." Furthermore, the bottom plate of the capacitor C_{N-1} is left connected to the reference voltage V_{ref}.

If $v_{in} < V_{ref}/2^1$, then $V_x > 0$, and the bit b_{N-1} is set to "0," and the bottom plate of the capacitor C_{N-1} is returned to the ground to discharge the capacitor C_{N-1}.

In the next step, the state of the bit b_{N-2} is tested by comparing v_{in} to $V_{ref}/2^2$ through the difference voltage dividers depending on the state of b_{N-1}. This is performed by setting the bottom plate of the next largest capacitor C_{N-1} to V_{ref}, and checking the polarity of the resulting value of V_x. In this case, however, the voltage division property of the array causes $V_{ref}/2^2$ to be added to V_x, thus

$$V_x = -v_{in} + b_{N-1}\frac{V_{ref}}{2^1} + \frac{V_{ref}}{2^2} \tag{7.16}$$

If

$$v_{\text{in}} > b_{N-1} \frac{V_{\text{ref}}}{2^1} + \frac{V_{\text{ref}}}{2^2} \tag{7.17}$$

then $V_x < 0$ and the bit b_{N-2} is set to "1." The bottom plate of the capacitor C_{N-2} is left connected to the reference voltage V_{ref}.

If

$$v_{\text{in}} < b_{N-1} \frac{V_{\text{ref}}}{2^1} + \frac{V_{\text{ref}}}{2^2} \tag{7.18}$$

then $V_x > 0$ and the bit b_{N-2} is evaluated to "0," and the bottom plate of the capacitor C_{N-2} is returned to the ground to discharge the capacitor C_{N-2}.

This conversion process proceeds accordingly with all capacitors in the array, in N-clock cycles, until all the bits have been determined. After the final conversion step, the comparator input voltage equals

$$\begin{aligned} V_x &= -v_{\text{in}} + b_{N-1} \frac{V_{\text{ref}}}{2^1} + b_{N-2} \frac{V_{\text{ref}}}{2^2} + b_{N-3} \frac{V_{\text{ref}}}{2^3} + \cdots + b_2 \frac{V_{\text{ref}}}{2^{N-2}} + b_1 \frac{V_{\text{ref}}}{2^{N-1}} + b_0 \frac{V_{\text{ref}}}{2^N} \\ &= -v_{\text{in}} + V_{\text{ref}} \left(\frac{b_{N-1}}{2^1} + \frac{b_{N-2}}{2^2} + \frac{b_{N-3}}{2^3} + \cdots + \frac{b_2}{2^{N-2}} + \frac{b_1}{2^{N-1}} + \frac{b_0}{2^N} \right) \end{aligned} \tag{7.19}$$

At the nth clock cycle, the output of the comparator corresponds to nth bit of the output ADC word. This bit value is saved by the SAR and after the end of the charge-redistribution mode; the digital output of the SAR (b_0, b_1, b_2, ..., b_{N-2}, b_{N-1}) is equal to the digitally approximated input voltage v_{in}.

Conversion Accuracy:

The accuracy of the SAR-ADC is determined by the accuracy of implementation of binary-scaled capacitances. The sum of the capacitances $C_{N-1} + C_{N-2} + \cdots + C_0$ defines the conversion range. The limitation of the charge-redistribution successive approximation ADC is their capacitors' matching. Mismatch causes deviation from actual step width resulting in two errors: the *integral* nonlinearity (INL) and *differential* nonlinearity (DNL). The INL error is analogous to the linearity error of an amplifier, and is defined as the maximum deviation of the actual transfer characteristic of the converter from a straight line. The DNL error is the linearity between code transitions of the converter, and is a measure of the monotonicity of the converter. A converter is said to be monotonic if an increase in input values results in an increase in output values. *Glitches* can occur during changes in the output at major transitions. Both DAC and ADCs can be non-monotonic, but a more common result of excess DNL in ADCs is missing codes. Missing codes (or non-monotonicity) in an ADC are as objectionable as non-monotonicity in a DAC [11]. DAC is non-monotonic if its transfer characteristic contains one or more localized maxima or minima [12]. In many applications that require converters with closed-loop systems—meaning systems that utilize feedback—their non-monotonicity can change a negative feedback to a positive feedback. As such, it is critically important that DACs, especially, are monotonic.

The worst-case INL and DNL errors for a number of N binary-scaled capacitors implemented with a tolerance $\Delta C/C$, that is less than 1 least significant bit (LSB) values, can be expressed [13,14] by

$$INL_{\text{max}} = \pm 2^{N-1} \frac{\Delta C}{C} \text{ LBSs} \tag{7.20}$$

$$DNL_{\text{max}} = \pm \left(2^N - 1 \right) \frac{\Delta C}{C} \text{ LBSs} \tag{7.21}$$

The relative tolerance of capacitance ratio $\Delta C/C$ in MOS technology can be as low as $\pm 0.1\%$ [14]. For a given tolerance $\Delta C/C$, DNL will rise exponentially with the number of bits. A large C is desirable from a resolution point of view.

In essence, traditional switched capacitor techniques take advantage of the excellent properties of on-chip capacitors and MOS switches and permit the realization of numerous analog sampled-data circuits [15].

7.4.4 D/A CONVERSION TECHNIQUES

Two methods are commonly used for D/A conversion: the method using weighted resistors and the one using the R-2R ladder network. The former is simple in configuration, but its accuracy may not be very good. The latter is a little complicated in configuration but is more accurate.

Weighted resistors method: A DAC can be constructed by using a *summing amplifier*, as depicted in Figure 7.11, where the resistors are scaled to represent weights for the different input bits (i.e. b_i, where $i = 0, 1, 2, \ldots, N–1$) to produce corresponding analog voltage. Chapter 4, Section 4.2.4, has discussed the elements of a *summing amplifier*.

Following Equation (4.67), we write the relationship between the digital inputs (b_0 to b_{N-1}) and the analog output V_o. For $b_i = 0$ or 1 ($i = 0, 1, \ldots, N–1$), where 1 is set to V_{cc}, the output (analog) voltage will be

$$V_o = -V_{cc} \frac{R_f}{R} \left\{ \frac{1}{2^0} b_{N-1} + \frac{1}{2^1} b_{N-2} + \cdots + \frac{1}{2^{N-3}} b_2 + \frac{1}{2^{N-2}} b_1 + \frac{1}{2^{N-1}} b_0 \right\} \tag{7.22}$$

where b_{N-1} and b_0 correspond to LSB and MSB, respectively. Like in the ADC conversion process, a reference voltage V_{ref} is preferred instead of V_{cc}. So, we can rewrite Equation (7.22) thus

$$V_o = -V_{\text{ref}} \frac{R_f}{R} \left\{ \frac{1}{2^0} b_{N-1} + \frac{1}{2^1} b_{N-2} + \cdots + \frac{1}{2^{N-3}} b_2 + \frac{1}{2^{N-2}} b_1 + \frac{1}{2^{N-1}} b_0 \right\} \tag{7.23}$$

With this method, and for full-scale range, the resolution can be expressed as

$$\Delta_{\text{res}} = \frac{V_{\text{ref}}}{2^N - 1} \tag{7.24}$$

Equation (7.24), the resolution, represents the quantization error that is inherent in the conversion of the signal to digital form.

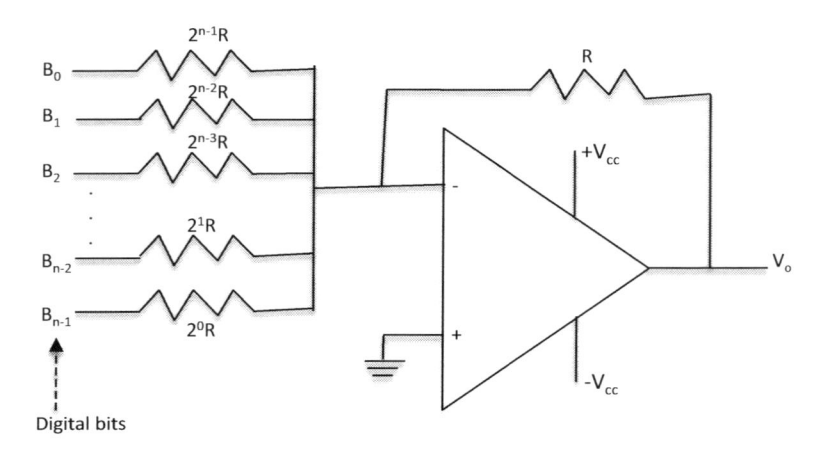

FIGURE 7.11 Summing amplifier as D/A converter with weighted binary input bits B_i.

Example 7.2

Suppose we reconfigure Figures 7.11 and 7.12 for 4-bit inputs. If we take $R_f = R$ and $V_{ref} = 4\,V$, estimate the analog output voltage.

Solution:

From Equation (7.23), we can write for any digital 4-bit input as

$$V_o = -V_{ref}\left\{\frac{1}{2^0}b_3 + \frac{1}{2^1}b_2 + \frac{1}{2^2}b_1 + \frac{1}{2^3}b_0\right\}$$

$$= -4\left\{\frac{1}{1}b_3 + \frac{1}{2}b_2 + \frac{1}{4}b_1 + \frac{1}{8}b_0\right\} = -4b_3 - 2b_2 - 1b_1 - 0.5b_0 \qquad (7.25)$$

So, from Equation (7.25) and for any digital input codes, we can estimate the corresponding output voltages, as shown in Table 7.1, given $V_{ref} = 4\,V$.

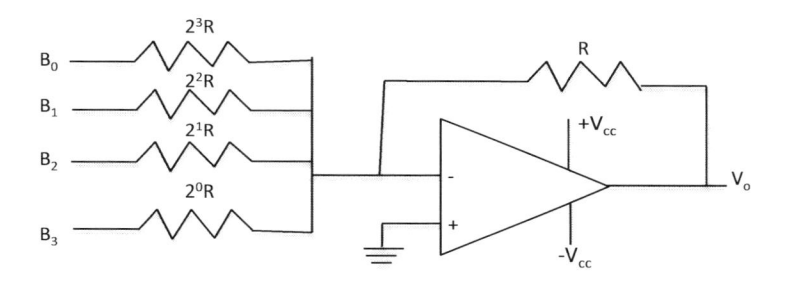

FIGURE 7.12 Summing amplifier as D/A converter with weighted 4-bit inputs.

TABLE 7.1
Analog Output Voltages Corresponding to the Digital Input Codes

Digital Input Code				Analog Output Voltage, V_o
b_3	b_2	b_1	b_0	(V)
0	0	0	0	0.0
0	0	0	1	−0.5
0	0	1	0	−1.0
0	0	1	1	−1.5
0	1	0	0	−2.0
0	1	0	1	−2.5
0	1	1	0	−3.0
0	1	1	1	−3.5
1	0	0	0	−4.0
1	0	0	1	−4.5
1	0	1	0	−5.0
1	0	1	1	−5.5
1	1	0	0	−6.0
1	1	0	1	−6.5
1	1	1	0	−7.0
1	1	1	1	−7.5

Although simple, this approach is not satisfactory for a large number of bits because it requires too much precision in the summing resistors. This problem is overcome in the R-2R resistor ladder network DAC.

R-2R ladder network: A symmetric network that provides a simple and inexpensive way to performing D/A conversion, using repetitive arrangements of precision resistor networks in a ladder-like configuration. Figure 7.13 shows a basic R-2R resistor ladder network. The switches are controlled by a digital logic with bits b_i (where $i = 0, 1, 2, \ldots, N-1$) and b_i is either 1 or 0. Ideally, the bits are switched between logic "0" (0 V) and logic "1" (V_{ref}). The 2R resistor is formed by connecting two resistors in series to obtain the best resistor matching, and can be connected either to the amplifier-inverting node, or to the ground. For instance, when connected to the amplifier-inverting node, the corresponding bit, b_i, assumes the high state. In the other case, when connected to the ground it is in the low state.

The basic principle of this R-2R ladder DAC is to split the reference currents equally through the switches. Accordingly, the currents I_i can be obtained as

$$I_0 = 2^1 I_1 = 2^2 I_2 = 2^3 I_3 = \cdots = 2^{N-1} I_{N-1} \tag{7.26}$$

where

$$I_i = \frac{V_{ref}}{2^i R} \tag{7.27}$$

The DAC's output voltage V_o is given by

$$V_o = -R_f \sum_{i=0}^{N-1} b_i I_i$$

$$= \frac{R_f}{R} V_{ref} \sum_{i=0}^{N-1} b_i 2^{-i} \tag{7.28}$$

where b_i is either 1 or 0. The output voltage V_o is proportional to the switched-on bits. To a large extent, the reference voltage V_{ref} defines the DAC characteristics.

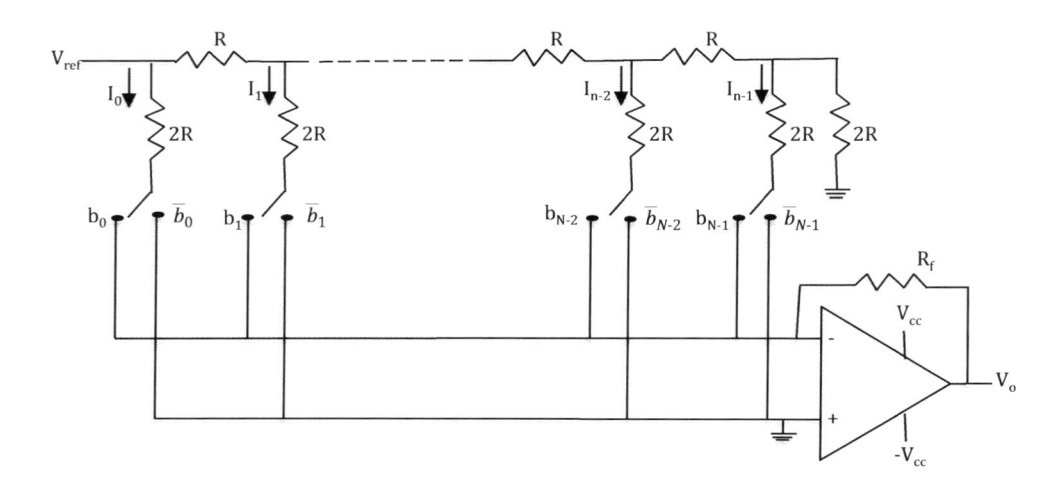

FIGURE 7.13 Block diagram of R-2R resistor ladder DAC with logic bits $b_i (i = 0,1,\ldots)$.

The R-2R DAC can exhibit a lower noise and nonlinearity errors of ±1 LSB. One LSB can be defined as

$$1\text{ LSB} = \frac{V_{\text{ref}}}{2^N} \tag{7.29}$$

However, without involving some amount of resistor trimming to reduce the matching errors, the resolution of an R-2R D/A ladder converter is limited to 12 bits [16]. However, it is less attractive for glitch-sensitive applications such as waveform generation.

Binary-weighted capacitor array DAC: A binary-weighted capacitor array is often used, as shown in Figure 7.14, by grouping unit capacitors C_j (where $j = 0, 1, ..., N-1$) in binary ratio values. C_p is the parasitic capacitance at the output node, V_x. Each capacitor in the DAC array has nominally the same unit capacitance value, C.

The capacitor-array DAC requires two-phase non-overlapping clocks for proper operation [17]. Initially, all capacitors are charged to ground. After initialization, depending on the digital input, the bottom plates are connected either to V_{ref} or to the ground.

If the top plate is floating without the feedback buffer, and the charge at the top finishes its redistribution—neglecting the top-plate parasitic effect—the output voltage becomes

$$V_o = \sum_{j=0}^{N-1} \frac{b_{N-1-j}}{2^j} V_{\text{ref}} \tag{7.30}$$

where b_j represents the input binary word. To consider the parasitic effect, for example, switching MSB capacitor bottom to V_{ref} changes the output voltage by

$$\frac{2^{N-1}C}{\sum_{j=0}^{N-1} C_j} V_{\text{ref}} = \frac{2^{N-1}C}{2^N C} V_{\text{ref}} = \frac{V_{\text{ref}}}{2} \tag{7.31}$$

where the capacitor C_i, for the jth bit is normally scaled to $2^{j-1}C$. As a result, the nonlinearity at the midpoint of the full range is limited by the ratio mismatch of the half sum of the capacitor array to

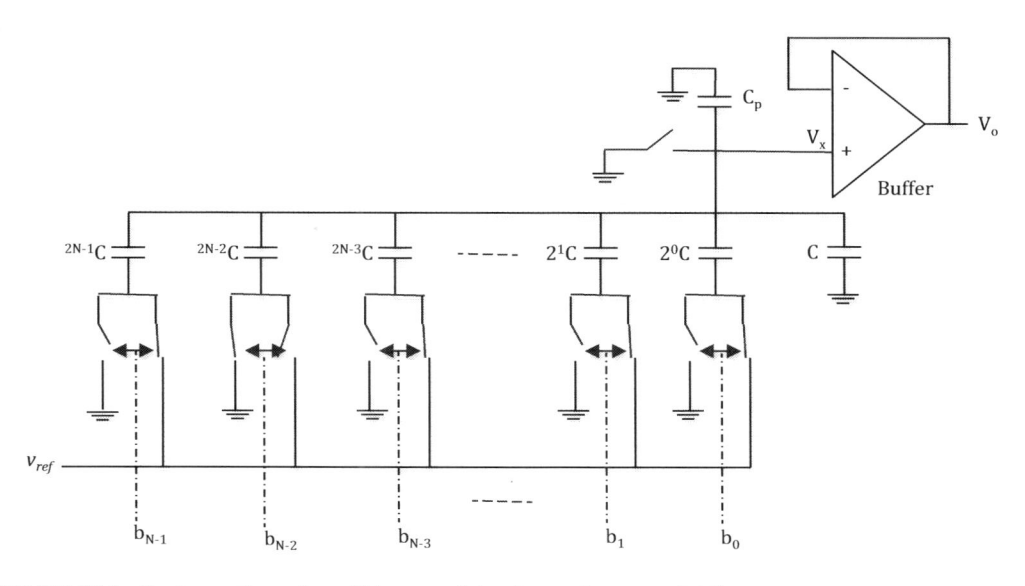

FIGURE 7.14 Basic configuration of binary-weighted capacitor array DAC.

the total sum of the array. Similarly, the nonlinearity at one-fourth of the full range is limited by the ratio mismatch of one-fourth of the capacitor array to the total sum of the array. Same argument applies to the eighth capacitor array, and subsequent capacitor array mismatch causes deviation from ideal output voltage step for 1 LSB leading to INL and DNL errors. Using the ground "0 V" and V_{ref} as endpoints, these errors are defined as follows.

$$\text{INL} = \frac{V_o(d_{i+1}) - i\left(V_{\text{ref}}/2^N\right)}{V_{\text{ref}}/2^N} \quad \text{for} \quad i = 0,1,\ldots,2^N - 1 \tag{7.32}$$

and

$$\text{DNL} = \frac{V_o(d_{i+1}) - V_o(d_i) - \left(V_{\text{ref}}/2^N\right)}{V_{\text{ref}}/2^N} \quad \text{for} \quad i = 0,1,\ldots,2^N - 1 \tag{7.33}$$

where d_i represents ith output voltage step.

The largest positive and negative numbers of Equations (7.32) and (7.33) are usually quoted to specify the DAC static performance.

If the bottom plates of the capacitor array are set at V_{ref} with $b_j = 1$ and at ground with $b_j = 0$, then the output voltage V_o can be expressed as

$$V_o = \left(\frac{\sum\limits_{j=0}^{N-1} b_{N-1-j} 2^{N-1-j} C}{C_p + C + \sum\limits_{j=0}^{N-1} 2^{N-1-j} C}\right) V_{\text{ref}} = \left(\frac{\sum\limits_{j=0}^{N-1} b_{N-1-j} 2^{N-1-j} C}{C_p + 2^N C}\right) V_{\text{ref}}$$

$$= \left(\frac{2^N C}{C_p + 2^N C}\right) V_{\text{ref}} \sum_{j=0}^{N-1} \frac{b_{N-1-j}}{2^j} \tag{7.34}$$

One important application of the capacitor-array DAC is as a reference DAC for ADCs.

If the DAC is used as a subblock of an ADC, the total capacitor should be $2^N C$, as shown in Figure 7.15, and the top plate of the array is usually connected to the input nodes of comparators or high-gain operational amplifiers, depending on the ADC architecture. As a result, the top plate has a parasitic capacitance, but its effect on the DAC performance is negligible. In the case of this architecture, a large buffer (gain A_b) would be needed to attenuate summing-node charge sharing. The output voltage V_o can be expressed as

$$V_o = \left(\frac{2^N C}{2^N C + \dfrac{C_p + C\left(2^{N+1} - 1\right)}{A_b}}\right) V_{\text{ref}} \sum_{j=0}^{N-1} \frac{b_{N-1-j}}{2^j} \tag{7.35}$$

Resolution: Resolution describes the minimum voltage or current a DAC can resolve. The fundamental limit of a DAC is the quantization noise due to finite resolution of the DAC. Suppose the input digital word of a DAC is N bits long, then the minimum step it can resolve approximates to Equation (7.4); that is,

$$\delta V_n = \frac{V_{\text{ref}}}{2^N} \tag{7.36}$$

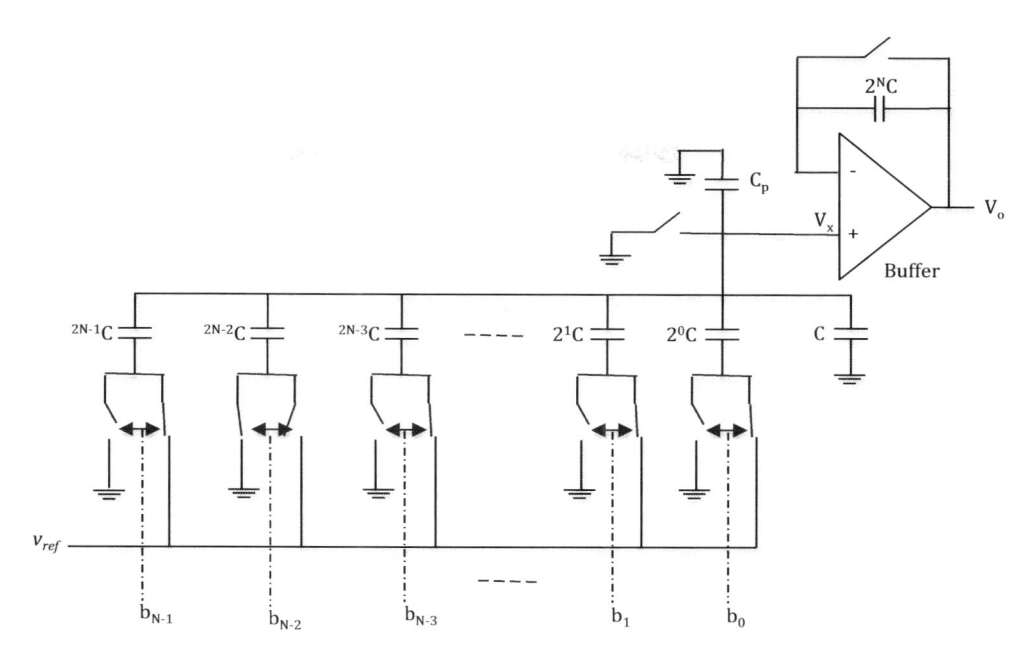

FIGURE 7.15   Basic configuration of stray-insensitive binary-weighted capacitor array DAC.

Usually, the resolution of a DAC is characterized in terms of signal-to-noise ratio (SNR), although the SNR only accounts for the uncorrelated noise. If output voltages were reproduced with δV_n of uncertainty, the SNR of an ideal DAC is

$$\text{SNR} = \frac{3}{2} 2^N \tag{7.37}$$

Whilst the conventional binary-weighted capacitor array DAC has a most efficient architecture because of its insensitivity to stray capacitance, it uses charge inefficiently during a conversion.

In essence, there is room for improvement as there are limitations to most basic techniques. The future trends tend to be in areas of switch-capacitor integration, electronic calibration of DAC nonlinearity, and oversampling interpolation techniques DACs, which trade speed with resolution. In particular, the oversampling interpolative DACs are used widely in stand-alone applications such as digital audio playback systems or digital communications due to their inherent monotonicity.

7.5   SUMMARY

Communication signals can be analog or digital. Information can be transmitted using analog or digital signals. Interchangeability of information transfer allows the development of conversion process without loss of detail. The challenge is to achieve a high sampling rate and high conversion accuracy in the presence of component mismatch, nonlinearity errors, and noise. Although the electronic circuits required to do this conversion processing can be quite complex, the basic idea is fairly simple. This chapter has explained the basic concepts of data conversion and their inherent errors, as well as the choice of the converter types that strongly influence the architecture of the overall system that are fundamental to the continuing revolution in information technology and communication systems.

PROBLEMS

7.1. A digital signal is the one which is sampled and quantized. Can a discrete signal be considered a digital signal?

7.2. In flash ADCs [18]:
   a. The high speed is made possible by simultaneous comparisons
   b. There are many comparators as there are bits.
   c. There are many reference voltages as there are bits.
   d. The basic string for comparisons is a capacitor charge-redistribution network.

7.3. An S/H circuit [18]:
   a. Has a momentary switch that connects the input voltage to a capacitor long enough for the capacitor to charge.
   b. Has a resistor in series with a capacitor in series with a switch.
   c. Has a capacitor that is charged to hold the value of the input voltage
   d. All of the above
   e. a and c above
   f. None of the above

7.4. Name two effects that determine the acquisition time of an S/H circuit.

7.5. What sort of distortion components does the quantization process generate when sinusoidal input signals are used for sampling?

REFERENCES

1. Kitayama, K., Notomi, M., Naruse, M., Inoue, K., Kawakami, S. and Uchida, A. (2019). Novel frontier of photonics for data processing—Photonic accelerator. *APL Photon*, 4, 090901-1-24.
2. Mishali, M. and Eldar, Y. C., (2010). From theory to practice: Sub-Nyquist sampling of sparse wideband analog signals. *IEEE Journal of Selected Topic in Signal Processing*, 4(2), 375–391.
3. Callahan, P.T., Dennis, M.L. and Clark, T.R. (2012). Photonic analog-to-digital conversion. *Johns Hopkins APL Technical Digest*, 30(4), 280–286.
4. Kolawole, M.O. (2013). *Satellite Communication Engineering*. Boca Raton, FL: CRC Press.
5. Gersho, A. and Gray, R.M. (2012). *Vector Quantization and Signal Compression*. New York: Springer.
6. Zjajo, A. and Pineda de Gyvez, J. (2011). *Low-Power High-Resolution Analog to Digital Converters, Analog Circuits and Signal Processing*. New York: Springer.
7. Sangwine, S. (2007). *Electronic Components and Technology*, 3rd edition. Boca Raton, FL: CRC Press.
8. McCreary, J.L. and Gray, P.R. (1975). All-MOS charge redistribution analog-to-digital conversion techniques. I. *IEEE Journal of Solid-State Circuits*, 10(6), 371–379.
9. Baker, R.J. (2008). *CMOS: Circuit Design, Layout, and Simulation, Volume 1*. Piscataway, NJ: Wiley.
10. Yuan, F. (2010). *CMOS Circuits for Passive Wireless Microsystems*. New York: Springer.
11. Jung, W. (2004). *Op Amp Applications Handbook*. New York: Newnes.
12. Zumbahlen, H. (2008). *Linear Circuit Design Handbook*. Norwood, MA: Newnes.
13. Allen, P.E. and Holberg, D.R. (2002). *CMOS Analog Circuit Design*, 2nd edition. Oxford: Oxford University Press.
14. Carbone, P., Kiaei, S. and Xu, F. (eds.) (2013). *Design, Modeling and Testing of Data Converters*. Berlin: Springer.
15. Dai, L. and Harjani, R. (2000). CMOS switched-op-amp-based sample-and-hold circuit. *IEEE Journal of Solid-State Circuits*, 35(1), 109–113.
16. Ndjountche, T. (2011). *CMOS Analog Integrated Circuits: High-Speed and Power-Efficient Design*. Boca Raton, FL: CRC Press.
17. Chen, W.-K. (ed.) (2009). *Analog and VLSI Circuits*. Boca Raton, FL: CRC Press
18. Luecke, G. (2005). *Analog and Digital Circuits for Electronic Control System Applications Using the TI MSP430 Microcontroller*. Amsterdam: Newnes.

8 Nanomaterials, Nanoelectronics, and Nanofabrication Technology

Progress in electronics has been driven by miniaturization. However, electronic devices are approaching molecular scale, thus exercising spurious electrons whose dynamic behavior is at the nanoscale. Nanostructured systems could offer electronic performance beyond what is possible with current very large-scale integrated circuits (IC). Moreover, by enabling the possibility of devices that are low cost and simultaneously flexible, transparent, and lightweight, nanostructured solutions are appealing to nearly any device system [1]. To fully understand nanostructured device dynamic behavior requires an understanding of quantum mechanics and quantization, the wave-particle duality, wavefunctions, and their computational limits, to name a few. This chapter does not dwell into device dynamic behavioral studies per se, but rather provides an overview of the electronic properties of nanomaterials from bulk to nanoscale, the crystallites' size reduction effect, and the principles of operating technology enabling fabrication of nanoscale structures.

8.1 INTRODUCTION

Nanotechnology is essential to the continuing advances in integrated electronics: increasing computational power, reducing device scale, and limiting energy consumption [1]. What is nanotechnology? Nanotechnology, in the simplest form, is the way discoveries made at the nanoscale are put to work. Technically, nanotechnology involves the manipulation of matter at the atomic level where conventional physics breaks down to impart new materials or devices with performance characteristics that far exceed those predicted for more orthodox approaches [2]. Actually, physical, chemical, electrical, magnetic, optical, as well as mechanical properties of materials change when going from conventional bulk to nanoscale. For instance, the melting point of metals can deviate from the bulk melting point as much as a couple of hundreds of degrees for a particle of 5 nanometers (nm) size. The unit of nanometer derives its prefix "nano" from a Greek word meaning "dwarf" or "extremely small." One nanometer spans 3–5 atoms lined up in a row. By comparison, the diameter of a human hair is about five orders of magnitude larger than a nanoscale particle. At the nanoscale level, some material properties are affected by the laws of atomic physics rather than behaving as traditional bulk materials do. For instance, the behavior a nanomaterial encounters may be quite different from that seen in relation to larger-scale materials. So also is the performance of the nanometer devices. As noted in Chapters 3 and 5, there is an energy range in the solid semiconductor where no electron states can exist. This range is called a *bandgap*. The bandgap of a silicon nanowire can increase appreciably for a handful of nanodiameter. As such, nanotechnology is not just about size alone but more about how to harness the change in properties and produce useful functionalities. Their extremely small feature size is of the same scale as the critical size for physical phenomena. Fundamental electronic, magnetic, optical, chemical, and biological processes are also different at this level.

8.2 NANOMATERIALS

Nanomaterials are materials that have particles or constituents of nanoscale dimensions, and can be metals, ceramics, polymeric materials, or composite materials. Nanomaterials properties—e.g. chemical reactivity, strength, and electrical and magnetic behaviors—can differ from those of the same materials with micron- or millimeter-scale dimensions. Surface properties such as energy levels, electronic structure, interfaces, and reactivity can be quite different from interior states and can give rise to quite different material properties. For instance, the small feature size ensures that many atoms, perhaps half or more in some cases, will be near interfaces, which can be used to physically and chemically manipulate the nanomaterials for specific applications.

Nanoparticles are minute substances in the size range of 1–100 nm (1 nm = 10^{-9} m) in all directions. Isolated nanoparticles exhibit amazing properties; however, when mixed with other materials the new compound becomes improved in a variety of ways. The size of the nanoparticle grains (or crystallites) and the overlapping of different grain sizes affect the physical strength of the nanomaterial. Also, when the crystallites of a material are reduced to the nanometer scale, there is an increase in the role of interfacial defects: grain boundaries, triple junctions, and elastically distorted layers [3]. The expression that shows the relationship between grain size (d_g) and material yield strength σ_o, known as Hall–Petch model [4,5]:

$$\sigma_o = \sigma_i + \frac{k_y}{\sqrt{d_g}} \tag{8.1}$$

has been proven to hold with nanoparticles as well [3], where σ_i is the material constant for the starting stress for dislocation movement (or the resistance of the lattice to dislocation motion) and k_y is the strengthening coefficient (a constant specific to each material which depends on interatomic bonds' strength as well as on a crystallography of the material). In some nanocrystalline materials, grain boundary weakened. Of course, a series of quantum confinement effects arises that could significantly change the way the nanoparticles behave, such as conductivity, specific heat, increases in energy bandgap, and different wavelengths of emitted light.

There are different ways nanoparticles are classified: for instance, whether they are carbon-based (e.g. graphene), ceramic, semiconducting (e.g. quantum dots), and polymeric (e.g. colloidal particles); or as inorganic or hard (e.g. titanium dioxide (TiO_2), quartz–silica dioxide (SiO_2), and fullerenes); or as organic or soft (e.g. liposomes, vesicles, and nanodroplets [6]). Spherical and ellipsoidal carbon-based nanomaterials are referred to as fullerenes—a class of allotropes of carbon, while cylindrical ones are called nanotubes. An allotrope is a variant of a substance consisting of only one type of atom. Besides carbon, oxygen and sulfur are also substances that have allotropes. A *quantum dot* is a closely packed semiconductor crystal composed of hundreds, or if not thousands, of atoms, and whose size is on the order of a few nanometers to a few hundred nanometers. Overall, nanoparticles have a vast range of compositions. Despite the diverse compositional formations, they have three major physical properties; namely, (i) they are highly mobile in the free state, (ii) they have enormous specific surface areas, and (iii) they may exhibit subtle quantum effects; that is, consequences of wave-particle duality.

Carbon is a group 14 element that resides above silicon in the *periodic table*, as seen in Figure 1.2, Chapter 1. Like silicon and germanium, carbon has four electrons in its valence shell. In its most common state, carbon is an amorphous non-metal like coal or soot. In a crystalline state, that is, in tetrahedrally bonded state, carbon becomes diamond that is an insulator with a relatively large bandgap. However, when carbon atoms are arranged in crystalline structures composed of hexagonal benzene-like rings, they form a number of allotropes that offer exceptional electrical properties. Carbon is the substance with the greatest number of allotropes. Graphite, charcoal, and diamond are all allotropes of carbon. Carbon possesses allotropes most radically different from one another, ranging from soft to hard, opaque to transparent, abrasive to smooth. These allotropes include the amorphous carbon allotrope, carbon nanofoam, carbon nanotube, graphite, and ceraphite allotrope, to name a few. For the replacement of silicon in future transistor channels, the two most promising of these allotropes to date, as noted in the literature, are graphene and carbon nanotubes.

Carbon nanotubes are not the same as carbon fibers. Carbon nanotubes are remarkable materials, the most rigid materials known to date, with excellent charge transport properties.

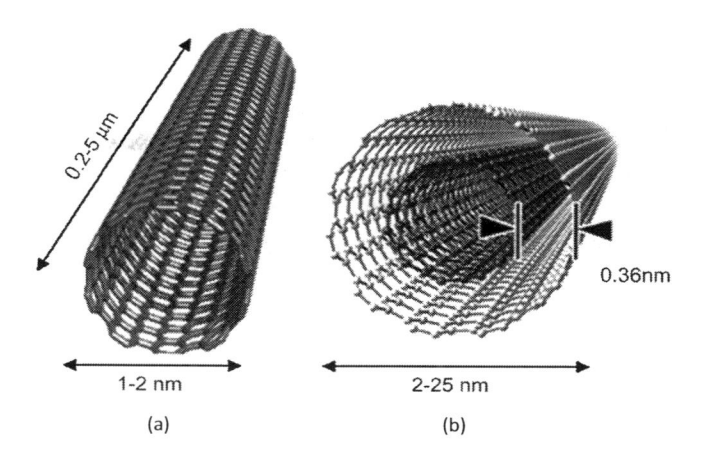

FIGURE 8.1 Carbon nanotube molecular structures: (a) SWCNT [7] and (b) MWCNT [8].

Carbon nanotubes can be divided into two groups, namely: single-walled carbon nanotubes (SWCNTs) and multiwalled carbon nanotubes (MWCNTs). Figure 8.1 shows the molecular structures of SWCNT and MWCNT.

An SWCNT is a hollow and cylindrically shaped one-dimensional graphitic carbon allotrope that appears as a rolled-up sheet of monolayer graphene. An MWCNT is composed of a number of SWCNTs nested inside one another of 3–30 nm diameter, which can grow several centimeters (cm) long and can be thought of as a rolled-up sheet of multilayer graphene. SWCNTs are a few nanometers in diameter roughly 1–4 nm and up to several microns long. MWCNTs have dimensions greater than SWCNTs and can be distinguished from SWCNTs on the basis of their multiwalled Russian-doll structure and rigidity, and from carbon nanofibers on the basis of their different wall structure, smaller outer diameter, and hollow interior [9]. With diameters less than 10 nm, this allows for exceptionally high aspect ratios making nanotubes an essentially one-dimensional material [10]. At low temperature, an SWCNT is a quantum wire in which the electrons in the wire move without being scattered [11].

SWCNTs can behave as a metal or a semiconductor depending on the intrinsic bandgap and chirality. Chirality is the noun of chiral: an asymmetric nature in which the structure and its mirror image are not superimposable. A biological example of chiral is a human hand in which both right and left hands are structurally the same but the right hand is a mirror image of the left hand. Chemically, chiral compounds are typically optically active; large organic molecules often have one or more chiral centers where four different groups are attached to a carbon atom. For an SWCNT, the chirality is the amount of "twist" present in the tube. If you think of SWCNTs as a rolled-up graphene sheet made up of hexagons (see Figure 8.2), the chirality is how far the axis of the tube (line down center of tube) is from being parallel to one side of the hexagons (y-axis in Figure 8.2) [12]. If the tube axis is parallel, the SWCNT will be semiconducting. The chirality of nanotubes has a significant impact on their transport properties, particularly the electronic properties [13].

Carbon forms very strong carbon–carbon bonds, so it has a very large bandgap. The tight-binding structure allows us to predict the bandgap E_G as decreasing inversely with an increase in the single-wall carbon nanotube diameter, thus

$$E_g = = \frac{0.894}{d_s} \quad \text{eV (electron-volt)} \tag{8.2}$$

where d_s is the diameter (in nm).

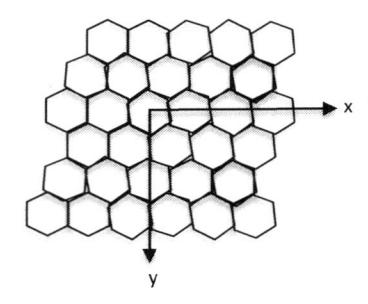

FIGURE 8.2   An illustration of the chirality of SWCNT.

In other words, the lower the bandgap, the better the conducting material. The diameter of single-wall nanotubes puts their bandgap energy at levels that are good for transistor or diode applications. The larger diameter of MWCNTs decreases their bandgap energy so low that they behave like metals regardless of their chirality. In essence, the structure and bandgap of nanosamples determine whether the samples are metallic or semiconducting.

Graphene is simply a single atomic layer of graphite; the material that makes up a pencil lead. Monolayer graphene is made from a sheet of carbon exactly one atom thick, making it a pure two-dimensional material. Unlike silicon, which has a bandgap that can be bridged to have electrical transport, graphene does not have a gap; this makes graphene difficult to performing the ordinary function of a transistor, like turning it ON and OFF. New research has shown that by confining the electrons in a very narrow ribbon of material, a gap can be opened up in graphene in a manner similar to a nanotube that produces a quantum wire [14]. Graphene may have the advantages of placement and uniformity over carbon nanotubes because it is a planar sheet rather than a tube: it could also be rolled up into cylinders to form carbon nanotubes.

The regular crystalline structure of both graphene and carbon nanotubes gives them robust physical properties. Also, the strength and stability of carbon nanomaterials make them attractive for use in nanoelectronic devices that require stiffness, lightness, and robustness, such as nanoelectrical mechanical systems (NEMS). Building transistor from carbon nanotubes will enable minimum transistor dimensions of a few nanometers as well as enabling development techniques to manufacture ICs built with nanotube transistors. Both graphene and embedded nanowires allow development of flexible and stretchable electronics. Graphene with its breaking strength of 100 GPa and capability to be a single atomic thickness sheet offers an extraordinary material choice to develop flexible, transparent electronics [2]. In their analysis, Meador et al. [2] noted that the highlight features of nanomaterial-based electronics are in many cases highly radiation resistant (due to their small target cross-section), or can be made radiation tolerant (tens of giga rads) without special processing/fabrication methods. Consequently, carbon nanotubes could be the basis for a new era of electronic devices smaller and more powerful than any previously envisioned.

8.3   NANOELECTRONICS

There is hardly a field where the links between basic science and application are tighter than in nanoelectronics and information technology. Nanoelectronics offers a broad set of opportunities by focusing on quantum devices and addressing their potential for high performance through increases in density, speed, and reduced power. Nanoelectronics generally refers to semiconductor or microelectronic devices that have been shrunk to nanoscale. Nanoelectronics is a critical enabling technology for a range of existing and developing industries. Figure 8.3 shows the different physical levels of existing technology for semiconductors, their effects as well as for the region of nanostructures if the feature sizes are scaled down. The ICs are simply *integrated circuits*. As seen in Figure 8.3,

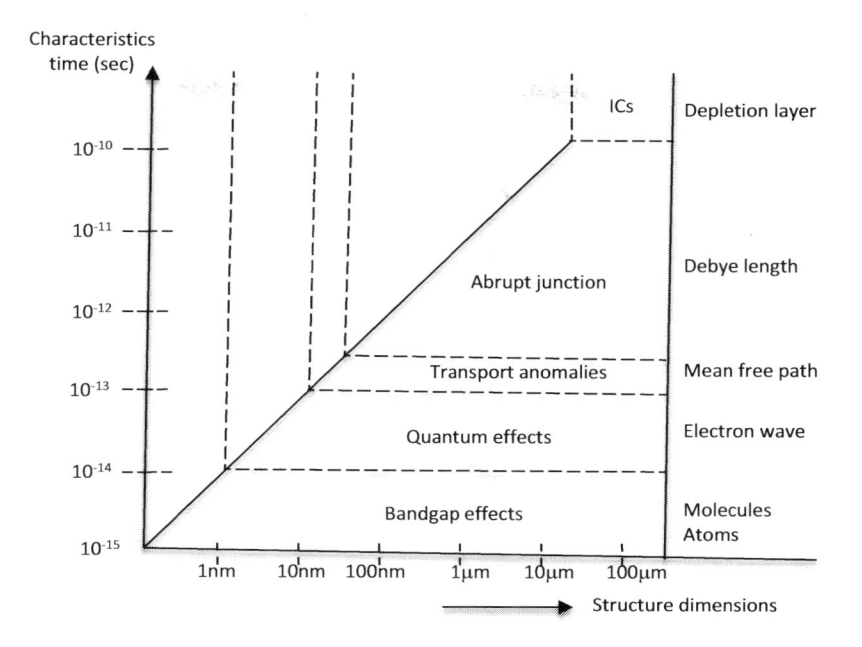

FIGURE 8.3 Device region characteristic times and physical effects.

the characteristic times are plotted against the structure dimensions of the devices. These times are correlated with different physical effects. The domain of nanoelectronics and molecular electronics is reached, respectively, when the characteristic times and dimensions are less than 1 picosecond (1 ps) and less than 10 nm.

In the nanodomain, the quantum and bandgap effects are present. Besides bandgap defined by Equation (8.2), the next few paragraphs define how other effects are quantified.

It is noted in [15] that the depletion-layer width limits the reduction in size of the *pn* junction in a diode. For a *pn* junction and assuming an abrupt junction, the depletion width shows a square-root dependency approximated by

$$l_n = \sqrt{\frac{2\varepsilon(V_d - V)}{qN_D}} \tag{8.3}$$

where ε is the semiconductor dielectric permittivity, V_d is the built-in potential (or *turn-on* voltage as defined in Chapter 2), and q is the elementary charge. The width l_n of the depletion layer increases with the voltage V and decreases with the doping density N_D of the region.

Another important dimension is Debye length, which is similar to that of the depletion-layer width. For example, the voltage term $(V_d - V)$ in Equation (8.3) is replaced by the thermal voltage V_T (as defined by Equation (3.4) in Chapter 3), and the doping density N_D is replaced by the intrinsic doping n_i of a semiconductor. The Debye length can be expressed as

$$L_D = \sqrt{\frac{\varepsilon V_T}{qn_i}} \tag{8.4}$$

Note that for a doped semiconductor n_i is substituted by the density of majority carriers. The Debye length describes the spatial extension of a perturbation inside a semiconductor.

Particles moving due to thermal energy have a mean free path $\langle l \rangle$ between two collisions

$$\langle l \rangle = v_{\text{th}} \tag{8.5}$$

This expression (Equation 8.5) depends on the thermal velocity v_{th} and on the mean free time τ_c of the particles, which determines the mobility μ of the charge carriers.

Quantum effects become relevant in devices if the wavelength λ of the electrons is in the range of the feature size of the devices.

$$\lambda = \frac{h}{m_c v} \tag{8.6}$$

where m_c, v, and h are mass of the particle, velocity of the particle, and Planck's constant, respectively ($h \approx 6.63 \times 10^{-34}$ Js).

Most of the nanoelectronic devices function in this range; so, the wave behavior of the electrons has to be considered. If one continues to decrease the structural dimension, the domain of atoms and molecules is reached. If a nanoelectronic device has only one structural dimension in this range, we call it a *mesoscopic* device (*meso* means "in between"). Due to the disturbing influence of the thermal energy, quantum devices are operated at low temperatures. However, the quantum effects increase with a decrease in the feature size of the devices. Therefore, the devices must be very small if operated at room temperature. The latter is also an important point for nanoelectronics [16]. The structural smallness of the devices introduces other complexity problem; such as, the high complexity of integrated nanoelectronic systems. Modern microelectronic systems contain up to 100 million devices on a single chip. Nanoelectronics will push this number up to 1 billion devices or even more. The main problem is not only the large number of devices but also the development time and the time for testing such systems. Another important point of view is the choice of architecture for an efficient interaction between the subsystems.

The development of higher integration levels forces two conclusions [16]: First, new physical effects, for example quantum effects due to small dimensions, replace the classical transistor. Second, the huge number of devices on a single chip demands new system concepts due to the reduced efficiency of the classical architectures. The architectures must consider fault tolerance, self-organization, and they must satisfy a large variety of applications. If silicon technology will eventually reach its limits other technologies must be applied unless there is no demand from the market to introduce other technologies with improved characteristics. However, this is unlikely because demands for information technology will be tremendous.

Whilst we are talking of nanomaterials and nanoelectronics, advances in technology might propel us to consider further reduction in device structures and sizes. If this trend of reduction in size of electronic devices continues at its present exponential pace, the size of entire devices will approach that of molecules within a few decades. This means our paradigm, both the physics upon which electronic devices are based and the manufacturing procedures used to produce them, will have to change dramatically. This is because current electronics are based primarily on classical mechanics, but at the scale of molecules, electrons are quantum mechanical objects. This means attention will shift to *molecular electronics* where molecules that are quantum electronic devices are designed and synthesized using the batch processes of chemistry and then assembled into useful circuits through the processes of self-organization and self-alignment. As a result, individual molecules will be used as switches and carbon nanotubes as wires in the electronic circuits. If this goal can be achieved, we can anticipate non-volatile memories with 1 million times the bit area density of today's *dynamic random access memories* (DRAM) and power efficiency 1 billion times better than conventional complementary metal oxide semiconductor (CMOS) circuitry [17]. Also, the cost of building the factories for fabricating electronic devices is increasing at a rate that is much larger than the market for electronics; therefore, much less expensive manufacturing process will need to be invented. New electronic systems-on-a-chip in the future will involve combining different technologies, possibly on a CMOS fabric or hybrid. The physical nature of the nanoscale materials is complicated, and because of its novelty, only simple circuit structures have so far been formulated.

8.4 NANOFABRICATION

The major traction in electronic devices development is the method of fabrication, which is the shift to nanofabrication where individual wires, diodes, field-effect transistors (FETs), and switches can be created abundantly, and, hopefully, reliably and cheaply.

Nanofabrication manufacturing involves making devices at the smallest dimensions. It is generally made by "top-down" fabrication technique involving steps such as deposition of thin films, patterning, and etching discussed in Chapter 1. This approach used microfabrication technologies to integrate microstructural devices on a single chip. Semiconductor memory chips, such as DRAM or *static random access memory* (SRAM), are examples. They have different types of device structures for storing charges, which are arranged in well-organized patterns of *rows* and *columns*. This microfabrication approach may prove inadequate for nanofabrication. For instance, *IC* fabrication involves deposition or growth of different types of thin films of electrical insulators (which do not conduct electricity), semiconductors (where electrical conduction can be reasonably controlled), and metals (which are always excellent conductors of current). Nanoelectronic devices need all three categories of solids to function properly. For example, insulators are used to electrically isolate the billions of nanoelectronic devices on a chip, and metals can be used to connect them up as needed by the circuitry. The deposition and growth of such thin films are coupled with patterning using photolithography and etching according to the circuit patterns one is trying to form. Vertical furnaces such as this allow a whole batch of wafers to be processed at a time with excellent uniformity and low particle defects [18]. This suggests that existing microfabrication technologies may need further refinement or development in order to meet reliability, repeatability, and yield requirements. Assembly of new nanomaterials-based composites, critical to advanced systems, will also rely on such technologies for high yield production and processing of nanomaterials, drawing upon expertise from a broad range of traditional disciplines.

Top-down technique is in contrast to "bottom-up" or "self-assembly" schemes where assembling is done as nature does: atom-by-atom, molecule-by-molecule. Results so far have been encouraging with the "bottom-up" techniques; they may not generally be enough by themselves to make useful nanoelectronic devices, which have to be designed in complex, irregular geometries to satisfy circuit functions. There is the suspicion that with the bottom-up approach, if successful, new properties may be observed in the process, which, again, would need new techniques into making smart materials and working devices.

New electronic systems-on-a-chip in the future will involve combining different technologies. New nanotechnologies are actively being investigated by tunneling of electrons. Examples include that based on (i) Josephson tunnel junction, which uses electron beam lithography and shadow evaporation [19,20] and (ii) *quantum-dot cellular automata* (QCA) [21,22]. A *Josephson tunnel junction* is a quantum mechanical device that is made of two superconducting electrodes separated by a barrier (e.g. insulating tunnel barrier, thin normal metal, etc.). QCA works on the intercell interaction mechanism where Coulombic effects dominate over tunneling. Small metal pieces can also behave as quantum dots, if the energy states an electron can occupy are distinguishable, instead of the usual energy band. This means that the difference between two consecutive energy states must be well above the thermal noise energy, discussed in Chapter 6, Section 6.1. QCA is not a physical implementation yet, since there are several ways to build quantum dots and connect them. (More is said about *tunneling* and *Josephson tunnel junction,* as well as QCA in Chapter 9, Section 9.5.) The impact of nanotechnology will be felt in different fields including communications, information technology, medicine, defense, space, power, and energy systems to name a few. Whilst there have been many nanoelectronics device proposals and/or demonstrations, there are still major problems to overcome either at high functionality level or simple fabrication and integration ensuring efficient interaction between the subsystems. We are entering technologically exciting period with equally exciting advances in store and new frontiers.

8.5 SUMMARY

As a result of recent improvement in technologies, engineers and research scientists have been able to manipulate existing materials to nanoscale, leading to an array of nanomaterial fields. These fields include nanoelectronics, nano-optics, nanosensors, and actuators (e.g. intelligent chembio-rad sensing systems), NEMS, nanorobotics, nanobiofusion, further nanofabrication (e.g. small satellites), nano-optoelectronics, lightweight electric motor, and the list goes on. This chapter has briefly discussed the electronic properties of nanomaterials from bulk to nanoscale, the crystallites' size reduction effect, and the principles of operating technology enabling nanofabrication of structures. Whilst these rapid technological advances are exciting, the structural smallness of the devices could introduce other complexity problems, for example, the high complexity of nanoelectronic subsystems integration and efficient interaction between systems.

PROBLEMS

8.1.   Why is channel doping unnecessary at nanoscale?
8.2.   Is carbon nanotube suitable for FET manufacturing?
8.3.   Is tunneling criterion the only obstacle to factor in when considering manufacturing electronic?
8.4.   If we reduce the grain size of a nanoparticle, what will happen to the nanomaterial being formed?
8.5.   What does Coulomb blockade mean?

REFERENCES

1. Gramling, H.M., Kiziroglou, M.E. and Yeatman, E.M. (2017). Nanotechnology for consumer electronics. In *Nanoelectronics: Materials, Devices, Applications*. Van de Voorde, h.c.M., Puers, R., Livio Baldi, L., and van Nooten, S.E. (eds). Weinheim, Germany: Wiley-VCH Verlag GmbH & Co, pp. 501–526.
2. Meador, M.A., Files, B., Li, J., Manohara, H., Powell, D. and Siochi, E.J. (2010). NASA Draft nanotechnology roadmap: Technology area 10. *Technical report*, TA10-1-TA10-23.
3. Andrievski, R.A. and Glezer, A.M. (2000). Size effects in properties of nanomaterials. *Scripta Materialia*, 44(8/9), 1621–1623.
4. Hall, E.O. (1951). The deformation and ageing of mild steel: III discussion of results. *Proceedings of the Physical Society London*, 64(9), 747–753.
5. Petch, N.J. (1953). The cleavage strength of polycrystals. *Journal of Iron Steel Institute London*, 173, 25–28.
6. U.S. Environmental Protection Agency Nanotechnology White Paper (2007). EPA 100/B-07/001.
7. Iijima, S. (2002). Carbon nanotubes: Past, present, and future. *Physica B: Physics of Condensed Matter*, 323, 1–5.
8. Hirsch, A. (2002). Funktionalisierung von einwandigen Kohlenstoffnanoröhern. *Angewandte Chemie*, 114, 1933–1939.
9. Kukovecz, A., Kozma, G. and Konya, Z (2013). Multi-walled carbon nanotubes. In *Handbook of Nanomaterials*. Vajtai, R. (ed). Berlin: Springer, pp. 147–188.
10. Chen, D., Chilstedt, S., Dong, C. and Pop, E. (2010). What everyone needs to know about carbon-based nanocircuits. *Design Automation Conference Proceedings*, pp. 2–18.
11. Petty, M.C. (2008). *Molecular Electronics: From Principles to Practice*. Chichester: Wiley.
12. Raja, T., Agrawal, V.D. and Bushnell, M.L. (2004). A tutorial on the emerging nanotechnology devices. *Proceedings of the International Conference on VLSI Design*, pp. 343–360.

13. Reich, S., Thomsen C. and Maultzsch, J. (2004). *Carbon Nanotubes: Basic Concepts and Physical Properties*. New York: Wiley-VCH, pp. 31–40.
14. IBM (2011). Working on wafer scale grapheme RF nanoelectronics. *Defence Industry Daily*. http://www.defenseindustrydaily.com/IBM-Working-on-Wafer-Scale-Graphene-RF-Nanoelectronics-04944/. Accessed 22 December 2019.
15. Hayden, O., Zheng, G., Agarwal, P. and Lieber, C.M. (2007). Visualization of carrier depletion in semiconducting nanowires. *Small*, 3(12), 2048–2052.
16. Goser, K. Glösekötter, P. and Dienstuhl, J. (2004). *Nanoelectronics and Nanosystems: From Transistors to Molecular and Quantum Devices*. Berlin: Springer-Verlag.
17. Jasinski, J and Petroff, P. (2001). *Applications: Nanodevices, Nanoelectronics, and Nanosensors in Nanotechnology Research Directions: IWGN Workshop Report: Vision for Nanotechnology in the Next Decade*. Dordrecht: Kluwer Academic Publisher.
18. NIST (National Institute of Standards and Technology) (2006). Competing for the future: A historical review of NIST ATP investments in semiconductor and micro/nano-electronics, June 2006, Report. http://www.atp.nist.gov/iteo/semi-nanoelec.htm. Accessed 15 December 2018.
19. Göppl, M.V. (2009). *Engineering quantum electronic chips—Realization and characterization of circuit quantum electrodynamics systems*. PhD Thesis, Diplom-Physiker (University), TU München.
20. Kielpinski, D., Kafri, D., Woolley, M., Milburn, G. and Taylor, J. (2012). Quantum interface between an electrical circuit and a single atom. *Physical Review Letters*, 108, 130504.
21. Tougaw, P.D. and Lent, C.S. (1996). Dynamic behavior of quantum cellular automata. *Journal of Applied Physics*, 80, 4722–4736.
22. Haselman, M. and Hauck, S. (2010). The future of integrated circuits: A survey of nanoelectronics. *Proceedings of the IEEE*, 98(1), 11–38.

9 Elements of Quantum Electronics

As noted in Chapter 8, electronic states of nanostructures are determined by quantum mechanics. Quantum mechanics is a fundamental theory in physics that describes the underlying theory concerning the world around us, from atoms, electrons, and molecules to materials, lasers, and other functioning circuits. The consequences of the quantization of energy states and their interaction with electromagnetic radiation and matter, and on the effects of quantum mechanics on the behavior of electrons lead to the field of *quantum electronics*. In this chapter, the following topics are explained: (a) an overview of the physics governing quantum mechanics; (b) the fundamental quantum electronics principles leading to the development of lasers, masers, and reversible quantum gates and circuits; (c) basic quantum gates developed based on quantum cells; (d) fabrication methods include self-assembly—as occurs for *quantum* dots (e.g. *quantum*-dot *cellular automata (QCA)*) and tunneling—as occurs for superconducting circuit techniques (e.g. Josephson tunnel junction); and (e) the protocols required to ensure uninfected information transmit through quantum communication channel and quantum error-correction (QEC) techniques to clean up erroneous pairs and/or errors (or decoherence) in the quantum channel.

9.1 INTRODUCTION TO BASIC QUANTUM PRINCIPLES

The ability to control individual electrons in an electronic conductor, long considered in quantum physics, has paved the way for novel quantum technologies. For instance, in analog or digital circuit, a short voltage pulse $V(t)$ on the contact of resistive conductor R (in Ω) produces an elementary current pulse.

$$I(t) = \frac{V(t)}{R} \tag{9.1}$$

While this solution seems trivial and standard in analog and digital systems, this is different in quantum conductor. Theoretically, for example, when a short voltage pulse $V(t)$ is applied on a contact of a quantum conductor carrying a single electronic mode, an elementary current pulse $I(t)$ is then generated [1]; that is,

$$I(t) = e^2 \frac{V(t)}{h} \tag{9.2}$$

where
    h = the Planck's constant $\approx 6.62608 \times 10^{-34} \text{J s}^{-1}$ (which is the currently accepted value). This constant relates "wave-like" quantities to "particle-like" quantities.
    e = the charge of an electron = $1.6 \times 10^{-19} \text{C}$ (Coulombs).

These two expressions (Equations 9.1 and 9.2) explain the difference in the behavior of energy levels of atoms and subatomic particles and, perhaps, the interactions of the particles with one another. The interactions can be (a) strong—binding nucleus together, (b) weak—causing beta decay: a process where unstable atoms become more stable and vice versa, (c) electrodynamics—between charged particles with the electromagnetic field, and/or (d) gravitational—between states and energy.

In any given instance, a particle may not necessarily be subject to all four interactions. The science dealing with the nature and behavior of energy levels of atoms and their constituents—electrons, protons, neutrons, and other obscure particles (e.g. quarks)—is referred to as *quantum physics* (also called *quantum mechanics*). Actual physical theories, such as quantum electrodynamics—that is, the quantized theory of interacting electrons and photons—are built upon a foundation of quantum mechanics. *Quantum electrodynamics* is often abbreviated QED in literature. The structural properties of an atom, electron, and spin have been discussed in Chapter 1, Section 1.1. Particles can behave like waves, and waves behave as though they are particles. The term *quantum* means *discreteness*; implying the existence of individual lumps as opposed to a continuum. As such, the quantum physics thus requires fundamental changes in how we view nature.

A lot of research is going on in the field of quantum physics regarding quantum computation and understanding quantum information; particularly of comparable importance to energy and information, as well as their exploitation to fabricating physical electronic devices. One of the messages of quantum computation and quantum information is how new tools would be required to traverse the gulf between small and relatively complex systems, as well as what computational algorithms would be needed to provide a systematic means for constructing and understanding such systems. As a starter, we look at what quantum computation and quantum information are built on. Basically, quantum computation and quantum information are built upon an analogous concept: the *quantum bit*, or *qubit* for short.

9.1.1 WHAT IS A QUBIT?

Qubit is the fundamental carrier of quantum information, which may take many physical forms, such as trapped ions, neutral atoms, photons, superconducting devices, to name a few. Classical computation and classical information work by manipulating *binary digital bits* that exist in one of two states: zero "0," or one "1," and there is nothing in between. A qubit state is not restricted to these two states: "0" {green} and "1" {yellow} but can also exist in a superposition of both states, depicted as {red} in Figure 9.1.

Just as a classical bit can have a state of either 0 or 1, the two most common states for a qubit are the states $|0\rangle$ and $|1\rangle$ which arise from *Dirac* (also called *Bra-ket*) notational representation of vector space $|\psi\rangle$. (More is said about the quantum states' notations in Section 9.2.) Technically, qubits are represented in a similar way as classical bit as base-2 numbers, and they take on the value 1 or 0 when measured and thus collapsed to a classical state. The actual difference though is that a qubit can be in a state other than $|0\rangle$ and $|1\rangle$. It is also possible to form *linear combination* of states, also known as *superpositions*, as

$$|\psi\rangle = c_0|0\rangle + c_1|1\rangle \qquad (9.3)$$

where the coefficients c_0 and c_1 are complex numbers called (probabilities) amplitudes of measuring or observing $|0\rangle$ and $|1\rangle$, respectively; each with respective *quantum* probabilities $|c_0|^2$ and $|c_1|^2$. These *quantum probabilities* represent the chance that a given quantum state $|0\rangle$ or $|1\rangle$ will be

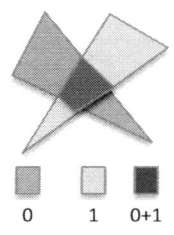

0   1   0+1

FIGURE 9.1 Quantum information representation.

observed when the superposition is collapsed. Surmising therefore, a qubit (in its uncollapsed quantum state) can exist in a continuum of states between $|0\rangle$ and $|1\rangle$—"ground" and "excited" states, respectively—until it is observed (or measured). From this expression (Equation 9.3), we can say that the state of a qubit is a vector in a two-dimensional vector space. The special states $|0\rangle$ and $|1\rangle$ are therefore known as *computational basis states* and form an orthonormal basis for this vector space.

Note that there is a fundamental difference between traditional (classical) probabilities (amplitudes) and that of the quantum states: that of the quantum states are complex numbers, while that of the traditional probabilities are real numbers. In the quantum states, complex numbers are required to fully describe the superposition of states as well as the interference and/or entanglement inherent in quantum systems. More is said about interference and its consequence later.

Unlike a classical system whose register contains only one value at any given time, a quantum register exists in a superposition of all its possible configurations of 0's and 1's at the same time and encodes information as *qubits*. More generally, we may consider a system of n qubits. The good news is that n qubits can store 2^n binary numbers simultaneously, which suggests a massive parallelism. As an example, say $n = 2$, we write

$$|\psi_n\rangle = c_0|00\rangle + c_1|01\rangle + c_2|10\rangle + c_3|11\rangle \tag{9.4}$$

In general, for n states, we write

$$|\psi_n\rangle = \sum_{i=0}^{2^n-1} c_i \left| b_{i,n-1} \quad b_{i,n-2}\dots \quad b_{i,0} \right\rangle \tag{9.5}$$

The bad news is that measurement yields just one of the 2^n superimposed numbers $|b_{i,n-1} \quad b_{i,n-2}\dots \quad b_{i,0}$ and destroys the superposition. This dichotomy between the unobservable state of a qubit and the observed states lies at the heart of quantum computation and quantum information. Since there is no direct correspondence, in quantum mechanics, it makes it difficult to intuit the behavior of quantum systems. In essence, a collection of n qubits is called a *quantum register* of size n of 2^n states. Quantum states have wave-like properties that allow powerful non-classical operations such as *interference* and *entanglement*.

Quantum *entanglement* is simply a sort of theoretical teleportation: it suggests that when pairs of particles, or qubits, interact or share spatial proximity, the quantum state of each particle, or qubit, cannot be described independently of the state of the other—meaning, whatever happens to one particle immediately affects the other particle regardless of the distance between them: the connection between the particles is known as Bell Entanglement [2]. As a result of such interaction, the two representatives (or ψ-functions) have become entangled. Measurements performed on entangled particles, or qubits, are correlated, which means that measurements performed on one particle or qubit seem to be instantaneously influencing other particles or qubits entangled with it. This leads to correlations between observable physical properties of the systems. Despite this conundrum, quantum entanglement has applications in the emerging technologies of quantum computing and quantum cryptography. Quantum entanglement and quantum coherence are both rooted in the superposition principle.

9.1.2 What Is Quantum Coherence?

Quantum *coherence* deals with the idea that all objects have wave-like properties. If an object's wave-like nature is split into two, then the two waves may coherently interfere with each other in such a way as to form a single state; that is, a superposition of the two states. Iteratively, quantum coherence is the ability of a quantum system to build interferences. What is special about quantum *interference* is that it is associated with the single-particle probability distribution. Of course, when more than one particle interacts, the corresponding multiparticle coherent superpositions are described by entangled states, which give rise to interesting (multiparticle) interference processes. In the real world, the interaction of qubits with noise in its processing environment or quantum communications channels occurs. When interaction with noise happens, coherency is lost; that is, the process becomes disturbed. The process by which the quantum systems is disturbed is called *decoherence*. In this situation, the qubit's superposition of states is awfully fragile and vulnerable. Of course, the qubit states' vulnerability to channel's noise would lead to errors, which can be overcome by *QEC*. More is said about *QEC* in Section 9.4.3. Error-correction techniques are themselves susceptible to noise. As such, it is crucial to develop fault-tolerant correction. QEC of *decoherence* and faulty control operations forms the backbone of all of quantum information processing.

Superposition principle, an explanation: The superposition principle—also called the principle of linear superposition of states—is one of the most fundamental principles in quantum mechanics. The superposition principle states that a physical system—such as an atom, a photon or electron, molecule, or particle—simultaneously exists partly in all theoretically possible states. For clarity, the concept of *superposition principle* is explained thus by drawing from the Schrödinger's theory of probability waves, which permits the existence of two or more waves [3]. Consider light waves $\varphi_1(x, t)$ and $\varphi_2(x, t)$, as shown in Figure 9.2, shining on an imperfectly transparent glass or mirror.

One wave would correspond to a photon passing through the mirror and another wave would correspond to the photon bouncing back. It is also possible for both waves to have superposed waves. It is also possible that the photon is being both transmitted and reflected, and, conceivably, appearing simultaneously on both sides of the mirror. This means, within this context, that any wavefunction solution, $\psi(x, t)$, can be linearly decomposed (at any time, t) as a sum of a series of other wave solutions of the same equation (for the same potential function), described by

$$\psi(x,t) = a_1\psi_1(x,t) + a_2\psi_2(x,t) \tag{9.6}$$

where a_1 and a_2 are weighting factors, or expansion coefficients, of the arbitrary wavefunction ψ.

What quantum states allow for is much more complex information to be encoded into a single bit. As such, *quantum electronics* are seen as culminating from the new qubits. Building an entangled network of such qubits will allow multiple calculations to be performed simultaneously that could produce a completely different form of quantum systems and computing power, enabling a whole new technology to be harnessed. Of course, classical supercomputers can perform linear calculations with astonishing speed if we are able to fabricate chips capable of hosting or imbedding billions of transistors on a chip. If the state of a quantum system is a kind of information, then the dynamics of that system is a kind of information processing. This is the basis of *quantum computing*, which seeks to exploit the physics of quantum systems to do useful information processing. Analogous to the way a classical computing system is built from logic gates and wires, a quantum *computing* device (or computer) would be built from quantum electronics comprising of circuits, quantum

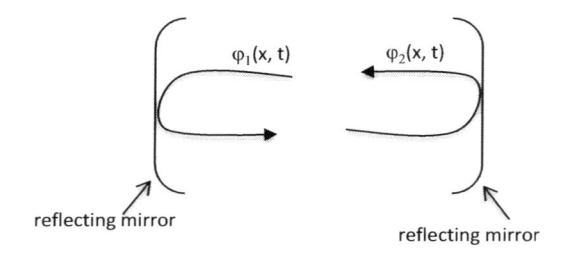

FIGURE 9.2   Superposition of light waves.

logic gates, and wires to carry around and manipulate the quantum information [4]. As a result, a quantum computing system would make use of the quantum states of subatomic particles to store information. Quantum computers are still largely hypothetical at this stage because the engineering required to construct quantum electronic components and computers is very difficult.

Currently, four main technological approaches have been suggested to building a quantum computer; they are trapped ions, superconductors, semiconductors, and majorana fermions—the braiding or topological knotting of quasiparticles in solid-state systems that not only change the phase but lead to different quantum states. It is envisaged that the application of qubits superposition would give quantum computers their inherent parallelism. More advanced research is continuing to building viable states; however, simple quantum logic gates are considered in this chapter.

9.2 ASSOCIATED OPERATORS OF SPINS AND ANGULAR MOMENTUM: AN OVERVIEW

The operator associated with the measurement of a spin has a vector character of its own. The operators representing the components of spin—denoted as S_x, S_y, S_z—are defined by Pauli spin matrices [5]:

$$S_x = \frac{h}{2} \begin{pmatrix} 0 & 1 \\ 1 & 0 \end{pmatrix} \tag{9.7a}$$

$$S_y = \frac{h}{2} \begin{pmatrix} 0 & -i \\ i & 0 \end{pmatrix} \tag{9.7b}$$

$$S_z = \frac{h}{2} \begin{pmatrix} 1 & 0 \\ 0 & -1 \end{pmatrix} \tag{9.7c}$$

from where "spin-1/2" particle is derived. By definition, h-bar ($\hbar$) is the quantization of angular momentum, defined as

$$h = \frac{h}{2\pi} \tag{9.8}$$

where h is the Planck's constant—as defined in Equation (9.2), and i is an imaginary unit.

For clarification, spin-1/2 is used to represent the internal angular momentum of some particles with just two possible states with different z components of spin:

i. Spin up $\begin{pmatrix} 1 \\ 0 \end{pmatrix}$ having $+\frac{h}{2}$ angular momentum in z-direction; and

ii. Spin down $\begin{pmatrix} 0 \\ 1 \end{pmatrix}$ having $-\frac{h}{2}$ angular momentum in z-direction;

leading to the derived properties of spin-operators expressed by Equation (9.7a–c).

A very important special property of spin-1/2 particle (or system) is

$$S^2 = S_x^2 + S_y^2 + S_z^2 = \frac{3}{4}h^2 \begin{pmatrix} 1 & 0 \\ 0 & 1 \end{pmatrix} \tag{9.9}$$

which means that the operator is a multiple of the identity operator.

We can work out the commutators of the spin operators by doing some linear algebra. Suppose, for example, we want to work out the commutator of S_x and S_y, which translates to

$$[S_x, S_y] = S_x S_y - S_y S_x = \frac{\hbar}{2} \begin{pmatrix} 0 & 1 \\ 1 & 0 \end{pmatrix} \frac{\hbar}{2} \begin{pmatrix} 0 & -i \\ i & 0 \end{pmatrix} - \frac{\hbar}{2} \begin{pmatrix} 0 & -i \\ i & 0 \end{pmatrix} \frac{\hbar}{2} \begin{pmatrix} 0 & 1 \\ 1 & 0 \end{pmatrix}$$

$$= \frac{\hbar^2}{4} \left[\begin{pmatrix} 0 & 1 \\ 1 & 0 \end{pmatrix} \begin{pmatrix} i & 0 \\ 0 & -i \end{pmatrix} - \begin{pmatrix} -i & 0 \\ 0 & i \end{pmatrix} \begin{pmatrix} 0 & 1 \\ 1 & 0 \end{pmatrix} \right]$$

$$= \frac{\hbar^2}{4} \left[\begin{pmatrix} i & 0 \\ 0 & -i \end{pmatrix} - \begin{pmatrix} -i & 0 \\ 0 & i \end{pmatrix} \right] = \frac{\hbar^2}{4} \begin{pmatrix} i & 0 \\ 0 & -i \end{pmatrix} \tag{9.10}$$

which turns out as

$$[S_x, S_y] = i\hbar S_z \tag{9.11}$$

In essence, any operator in the 2×2 space can be written as a linear combination of S_x, S_y, S_z, and S^2.

Often, the unitless equivalents (or versions) of S_x, S_y, and S_z (i.e. matrices without the leading factor $\hbar/2$ shown in Equation 9.7) are defined as Pauli matrices, thus

$$\sigma_x \equiv \begin{pmatrix} 0 & 1 \\ 1 & 0 \end{pmatrix} \quad \sigma_y \equiv \begin{pmatrix} 0 & -i \\ i & 0 \end{pmatrix} \quad \sigma_z \equiv \begin{pmatrix} 1 & 0 \\ 0 & -1 \end{pmatrix} \tag{9.12}$$

Following the rules of matrix multiplication, we can find the multiplication table for the Pauli operators as follows:

$$\sigma_x \sigma_y = -\sigma_y \sigma_x = i\sigma_z \tag{9.13a}$$

$$\sigma_y \sigma_z = -\sigma_z \sigma_y = i\sigma_x \tag{9.13b}$$

$$\sigma_z \sigma_x = -\sigma_x \sigma_z = i\sigma_y \tag{9.13c}$$

$$\sigma_a^2 = I \tag{9.13d}$$

where σ_a and I are just a certain operator and identity matrix, respectively. And,

$$\sigma_x^2 = 1$$

$$\sigma_y^2 = 1 \tag{9.14}$$

$$\sigma_z^2 = 1$$

Furthermore, we can express the algebra of Pauli operators as follows:

$$\sigma_y \sigma_x \sigma_y = \sigma_z \sigma_x \sigma_z = -\sigma_x \tag{9.15a}$$

$$\sigma_z \sigma_y \sigma_z = \sigma_x \sigma_y \sigma_x = -\sigma_y \tag{9.15b}$$

$$\sigma_x \sigma_z \sigma_x = \sigma_y \sigma_z \sigma_y = -\sigma_z \tag{9.15c}$$

In some textbooks, the Pauli matrices σ_x, σ_y, and σ_z are simply labeled matrices X, Y, and Z, respectively. (More is said about the Pauli matrices later in this chapter.)

TABLE 9.1

Table of Elementary Particle Masses and Charges

Particle	Mass (kg)	Charge (*Coulomb*, C)
Electron	$9.1013897 \times 10^{-31}$	$-1.60217733 \times 10^{-19}$
Proton	$\approx 1.6726 \times 10^{-27}$	$+1.60217733 \times 10^{-19}$
Neutron	$\approx 1.6749 \times 10^{-27}$	0

As noted in Chapter 1, Section 1.1, protons and neutrons are composite particles. Unlike neutron which is electrically neutral, protons and electrons are electrically charged; for instance, protons have a relative positive charge $+e$ while electrons have a relative negative charge $-e$ (see Table 9.1). The size of the charge is in *Coulomb*, C; i.e. $\pm e = \pm 1.60217733 \times 10^{-19}$ C. Since an atom contains equal numbers of positively charged protons and negatively charged electrons, overall it is neutral; meaning that the atomic number Z is the same as the number of electrons in an atom. So, the atomic radius R_n is empirically related to the particle's atomic number; thus

$$R_n \cong r_o \sqrt[3]{Z} \tag{9.16}$$

where $r_o = 1.3 \times 10^{-15}$ m = 1.3 fm.

(Note: 1 fm = 10^{-15} m. Nuclei are so small that the fermi (fm) is an appropriate unit of length.)

The quarks carry electric charges but are smaller in magnitude than e, and *spin*. The smaller charge magnitude enables quarks to combine together with other particles to give the correct electric charges for the observed particles. Quarks are quantified as "up" quark—with a charge of $+\dfrac{2}{3}e$ and "down" quark—with a charge of $-\dfrac{1}{3}e$. The way we build particles—protons and neutrons—depends on how we manipulate the number of quarks constituting each particle. For example, for a total charge of $+e$ for a proton, we need two "up" quarks and one "down" quark; that is, $\left\{ +e = +\dfrac{2}{3}e + \dfrac{2}{3}e - \dfrac{1}{3}e \right\}$. On the other hand, a neutron is built from one "up" quark and two "down" quarks to have a net charge of zero; that is, $\left\{ 0 = +\dfrac{2}{3}e - \dfrac{1}{3}e - \dfrac{1}{3}e \right\}$. We can also form a proton (or a neutron) by aligning quarks in such a way that two of the three spins cancel each other, leaving a net value of spin. Given the neutrality-charge status of the neutron, the proton's mass is an important parameter: one of the factors that affect how the electrons move around the atomic nucleus.

Spin angular momentum: Both the proton and neutron, like the electron, have an intrinsic spin. The *spin angular momentum* can be expressed as

$$L_s = \frac{h}{2\pi} \tag{9.17}$$

where the quantum number l, commonly called the spin, has a magnitude ½. In view of Equation (9.17), the *spin angular momentum* has a value

$$L_s = \frac{\sqrt{3}}{2} \frac{h}{2\pi} = \frac{\sqrt{3}}{2} \hbar \tag{9.18}$$

where h-bar ($\hbar$) is the quantization of angular momentum, as previously defined.

In addition to the *spin angular momentum*, the protons and neutrons in the nucleus have an *orbital angular momentum*. The resultant angular momentum of the nucleus is obtained by adding the spin and orbital angular momenta of all the nucleons within the nucleus. So, the total angular momenta, L_N, is given as

$$L_N = \hbar \sqrt{l_N (l_N + 1)} \tag{9.19}$$

where l_N is the nucleons spin number. Equation (9.19) is the *total angular momenta*, which is also called *nuclear spin*.

We know that neutron is a particle with spin but has a net charge of zero. Because the neutron has spin (and therefore a magnetic moment that points in the same direction as the spin), the neutron feels a torque in a uniform magnetic field [6]. If the spin direction is perpendicular to the magnetic field, the neutron will process about the magnetic field direction. In the same way, if the neutron has an electric dipole moment (EDM), it will process about an electric field. The difference is that the spatial transformation properties of magnetic and electric fields are different so that the existence of an EDM violates time reversal symmetry. We know that the spinning electron μ_e has an associated magnetic dipole moment of 1 Bohr magneton, i.e.

$$\mu_e = \frac{1}{2} \left(\frac{e\hbar}{m_e} \right) \tag{9.20}$$

where m_e is the mass of electron. Since a proton has a positive elementary charge and due to its spin, it should have a magnetic dipole moment. Dirac's relativistic quantum theory [7] showed that the electron must have a spin quantum number of ½ and a magnetic moment. By Dirac's theory, for nucleons, magnetic dipole moment μ_N is measured in nuclear magneton, defined using

$$\mu_N = \frac{1}{2} \left(\frac{e\hbar}{m_p} \right) \tag{9.21a}$$

where m_p = mass of proton (g).

Using data in Table 9.1 in Equation (9.21a), magnetic moment $\mu_N = 5.05 \times 10^{-27}$ J T^{-1}.

For electrons, $\mu_e \approx \mu_B$ where μ_B is Bohr magneton of a point-like particle defined as

$$\mu_B = \frac{1}{2} \left(\frac{e\hbar}{m_e c} \right) \tag{9.21b}$$

For nucleons, however, measurements give the magnetic moment of proton, $\mu_p = 2.7925\mu_N$, and neutron, $\mu_n = -1.9128\mu_N$. Physicists have found that it is especially hard to understand how the neutron, which is a neutral particle, can have a magnetic moment.

9.2.1   AXIOMS AND VECTOR SPACES

Before we delve into qubits in state space, it is important to define some notations used in quantum mechanics as compared to classical system. The space of states in classical system is Boolean. Analogous to its Boolean counterpart, the space of states in quantum system is *not* a mechanical set: it is a *vector space* (also called *Hilbert space*). A *vector space*, say V, over the field of complex numbers $\mathbb{C}$ is a non-empty set of elements called vectors. As such, in V, it is defined as the operations of vector addition and multiplication of a vector by a scalar in $\mathbb{C}$. A vector space may have either a finite or an infinite number of dimensions, but in most applications in quantum computation, *finite vector spaces* are used for completeness; denoted by $\mathbb{C}^n$. The *dimension* of a vector space is the number of *basis* vectors.

In quantum mechanics, *Dirac (or Bra-ket) notation* is used for vector spaces. A notation for *bra* is $\langle ... |$ and for *ket* is notation $| ... \rangle$.

How do we relate the *bra-ket notation* to a wavefunction? Simple. Let a wavefunction $\psi(x,t)$ be represented by a *ket*. As such,

$$\psi(x,t) \text{ is written as } |\psi\rangle.$$

The complex conjugate of the wavefunction is called its *bra*. For example, the bra

$$\left(\psi(x,t)\right)^* \text{ is written as } \langle\psi'| \text{ or } \langle\psi^*|$$

The Dirac notation of a generic vector **v** can be written as

$$\mathbf{v} \equiv |v\rangle \tag{9.22}$$

Often, in the first basis vector, denoted by $\mathbf{v}_0$, which we write in the Dirac notation as

$$\mathbf{v}_0 \equiv |0\rangle \tag{9.23}$$

Note that vector $\mathbf{v}_0$ is **not** the zero vector. The zero vector is an exception, whose notation is not modified; usually the notation is just 0.

In quantum mechanics, a wavefunction can be expressed in terms of a set of other functions. For example, $|\psi\rangle$ may be expressed as a linear combination of other functions like

$$|\psi\rangle = a_1|1\rangle + a_2|2\rangle + a_3|3\rangle + \cdots + a_n|n\rangle \tag{9.24}$$

where the $|i\rangle$ are called the "basis functions" and a_i are coefficients (numbers). If we have basis $\{v_1, v_2, \ldots, v_n\}$, we can express it like $\{|v_1\rangle, |v_2\rangle, \ldots, |v_n\rangle\}$ or $\{|1\rangle, |2\rangle, \ldots |n\rangle\}$. Therefore, the basis for $\mathbb{C}^n$ consists of exactly n linearly independent vectors.

Axioms: The axioms associated with the vector space of states of a quantum system, described in [8] and other articles, can be explained as follows:

1. The sum of any two kets is also a ket:

$$|A\rangle + |B\rangle = |D\rangle \tag{9.25a}$$

2. Vector addition is commutative:

$$|A\rangle + |B\rangle = |B\rangle + |A\rangle \tag{9.25b}$$

$$\{|A\rangle + |B\rangle\} + |D\rangle = |A\rangle + \{|B\rangle + |D\rangle\} \tag{9.25c}$$

3. For a unique vector 0 being added to any ket, the result gives the same ket:

$$|A\rangle + 0 = |A\rangle \tag{9.25d}$$

4. Given any ket, say $|A\rangle$ and complex number, say z or ω, linearity properties hold that allow formation of new ket, as follows:

$$|zA\rangle = z|A\rangle = |B\rangle \tag{9.25e}$$

$$z\{|A\rangle + |B\rangle\} = z|A\rangle + z|B\rangle \tag{9.25f}$$

$$\{z+\omega\}|A\rangle = z|A\rangle + \omega|A\rangle \tag{9.25g}$$

5. *Bra*-vectors satisfy the same axioms as the ket-vectors, except in the case of multiplication of complex numbers; i.e.

    i.

$$z|A\rangle = \langle A|z^*|$$ (9.26)

   where z^* is a complex conjugate of z.

   ii. If *kets* are represented by column vectors, the *dual bras* are represented by row vectors. For example, suppose vector **v** has basis $|v\rangle$; that is,

$$|v\rangle = \begin{bmatrix} a_1 \\ \vdots \\ a_n \end{bmatrix}$$ (9.27a)

   then the corresponding *bra* is

$$\langle v| = \begin{bmatrix} a_1^*, \ldots, a_n^* \end{bmatrix}$$ (9.27b)

   where a^* denotes the *complex conjugate* of a.

   Note that vectors are sometimes written in column format, as in Equation (9.27a) for example, and sometimes for readability in the format $(a_1, \ldots, a_n)$. As it will become obvious later in this chapter that for two-level quantum systems used as qubits; that is, the state $|0\rangle$ is usually identified with the vector $(1, 0)$, and similarly $|1\rangle$ with $(0, 1)$.

   The key to the Dirac notation is to always view *kets* as column matrices, and *bras* as row matrices. The significance of the *bra-ket* can be viewed as the inner (or dot) product of two vectors. [The word "dot" here is an algebraic operation and should not be confused with "quantum dot." A quantum dot is basically a region of space with energy barriers surrounding it. More is said about quantum dot in Section 9.5.]

6. We also have *products* which are analogous operations of *kets* and *bras*. These products are called (a) *inner product*, (b) *outer product*, and (c) *tensor product*. These *products* are described in the next pages under appropriate headings.

9.2.2 Product Spaces

Inner product: We write inner product of bra B and ket A as $\langle B|A\rangle$ or $\langle B, A\rangle$.

We can interchange bras and kets to have

$$\langle B|A\rangle = \langle A|B\rangle^*$$ (9.28a)

In terms of row and column vectors, we write

$$\langle B|A\rangle = \begin{pmatrix} b_1^* & b_2^* & b_3^* \end{pmatrix} \begin{pmatrix} a_1 \\ a_2 \\ a_3 \end{pmatrix} = b_1^* a_1 + b_2^* a_2 + b_3^*$$ (9.28b)

We can linearize as

$$\langle D|\{|A\rangle + |B\rangle\} = \langle D|A\rangle + \langle D|B\rangle$$ (9.28c)

The inner product of a ket with itself:

$$\langle v|v \rangle = \|v\|^2 \tag{9.28d}$$

is a positive real number (greater than zero) unless $|v\rangle$ is the zero vector, in which case $\langle v|v \rangle = 0$. The positive square root of $\langle v|v \rangle$ is called norm of $|v\rangle$; that is,

$$\text{norm} = \| v\| = \sqrt{\langle v|v \rangle} \tag{9.28e}$$

Multiplying any ket by the number 0 yields the unique *zero vector* or *zero ket*, which is denoted by 0. An *inner product* must satisfy the following conditions:

1. Interchanging the two arguments $|v\rangle$, $|\varphi\rangle$ results in the complex conjugate of the original expression:

$$\left(|v\rangle, \ |\varphi\rangle \right) = \left(|\varphi\rangle, \ |v\rangle \right)^* \tag{9.29a}$$

2. The inner product is a linear function of its second argument:

$$\left(|\varphi\rangle, \alpha|\phi\rangle + \beta|v\rangle \right) = \alpha \left(|\varphi\rangle, |\phi\rangle \right) + \beta \left(|\varphi\rangle, |v\rangle \right) \tag{9.29b}$$

3. The inner product is an antilinear function of its first argument:

$$\left(\alpha|\phi\rangle + \beta|v\rangle, |\varphi\rangle \right) = \alpha^* \left(|\varphi\rangle, |\phi\rangle \right) + \beta^* \left(|v\rangle, |\varphi\rangle \right) \tag{9.29c}$$

The term "antilinear" in this expression refers to the fact that the complex conjugates of α and β appear on the right side, rather than α and β themselves, as would be the case for a linear function.

4. The normalization and orthogonality conditions, for example, can be written as

$$\langle \psi|\psi \rangle = 1 \tag{9.29d}$$

$$\langle \psi'|\psi \rangle = 0 \tag{9.29e}$$

Outer product: The outer product, denoted by $|v\rangle\langle u|$ or $|v\rangle \otimes \langle u|$—the product of ket and bra—is the *tensor* or *Kronecker product* $|v\rangle$ with the conjugate transpose of $|u\rangle$. As with the inner product, the result is a matrix not scalar. As an example, consider the product of $|v\rangle$ by $\langle u|$ of length n and m, respectively:

$$|v\rangle\langle u| = \begin{bmatrix} v_0 \\ v_1 \\ \vdots \\ v_n \end{bmatrix} \begin{bmatrix} u_0^* & u_1^* & \cdots & u_m^* \end{bmatrix} = \begin{bmatrix} v_0 u_0^* & v_0 u_1^* & \cdots & v_0 u_m^* \\ v_1 u_0^* & v_1 u_1^* & \cdots & v_1 u_m^* \\ \vdots & \vdots & \ddots & \vdots \\ v_n u_0^* & v_n u_1^* & \cdots & v_n u_m^* \end{bmatrix} sw \tag{9.30}$$

This sequence of *bras* and *kets* is a matrix product that is associative but non-commutative. Suppose

$$|v\rangle = \begin{bmatrix} 1 \\ 0 \end{bmatrix}$$

then, its outer product is

$$|v\rangle\langle v| = \begin{bmatrix} 1 \\ 0 \end{bmatrix}\begin{bmatrix} 1^* & 0^* \end{bmatrix} = \begin{bmatrix} 1 & 0 \\ 0 & 0 \end{bmatrix} \tag{9.31}$$

Outer product can also act on bras; for instance,

$$\langle B||\psi\rangle\langle v| \equiv \langle B|\psi\rangle\langle v| \tag{9.32}$$

Tensor product: In mathematics, a tensor refers to objects that have multiple indices. Roughly speaking, this can be thought of as a multidimensional array. Consider a pair of vector spaces V and W with basis $\{|v_1\rangle,|v_2\rangle,\ldots,|v_n\rangle\}$ and $\{|w_1\rangle,|w_2\rangle,\ldots,|w_n\rangle\}$, respectively. The tensor product of V by W, denoted by $V \otimes W$, is an nm-dimensional vector space, whose set basis is

$$\{|v_1\rangle\otimes|w_1\rangle, |v_2\rangle\otimes|w_2\rangle,\ldots,|v_{n-1}\rangle\otimes|w_{m-1}\rangle,|v_n\rangle\otimes|w_m\rangle\}$$

A tensor product $V \otimes W$ can also be denoted by $|v_i\rangle|w_j\rangle$, or $|v_i, w_j\rangle$, or $|v_iw_j\rangle$. As an illustration, the tensor product of two 2×1 column vectors is a 4×1 column vector. Suppose we express a and b as

$$a = \begin{bmatrix} a_{11} \\ a_{21} \end{bmatrix} \tag{9.33a}$$

$$b = \begin{bmatrix} b_{11} \\ b_{21} \end{bmatrix} \tag{9.33b}$$

then their tensor product is

$$a \otimes b = \begin{bmatrix} a_{11} \\ a_{21} \end{bmatrix} \otimes \begin{bmatrix} b_{11} \\ b_{21} \end{bmatrix} = \begin{bmatrix} a_{11}b_{11} \\ a_{11}b_{21} \\ a_{21}b_{11} \\ a_{21}b_{21} \end{bmatrix} \tag{9.34}$$

We can extend one-dimensional scheme to two (and higher) dimensions. For example, suppose

$$a = \begin{bmatrix} a_{11} & a_{12} \\ a_{21} & a_{22} \end{bmatrix} \tag{9.35a}$$

$$b = \begin{bmatrix} b_{11} & b_{12} \\ b_{21} & b_{22} \end{bmatrix} \tag{9.35b}$$

we can compute the tensor product $a \otimes b$ thus

$$a \otimes b = \begin{bmatrix} a_{11} & a_{12} \\ a_{21} & a_{22} \end{bmatrix} \otimes \begin{bmatrix} b_{11} & b_{12} \\ b_{21} & b_{22} \end{bmatrix} = \begin{bmatrix} a_{11}\begin{bmatrix} b_{11} & b_{12} \\ b_{21} & b_{22} \end{bmatrix} & a_{12}\begin{bmatrix} b_{11} & b_{12} \\ b_{21} & b_{22} \end{bmatrix} \\ a_{21}\begin{bmatrix} b_{11} & b_{12} \\ b_{21} & b_{22} \end{bmatrix} & a_{22}\begin{bmatrix} b_{11} & b_{12} \\ b_{21} & b_{22} \end{bmatrix} \end{bmatrix} \tag{9.36}$$

If the operators are reversed, their tensor product becomes

$$b \otimes a = \begin{bmatrix} b_{11} & b_{12} \\ b_{21} & b_{22} \end{bmatrix} \otimes \begin{bmatrix} a_{11} & a_{12} \\ a_{21} & a_{22} \end{bmatrix} = \begin{bmatrix} b_{11} \begin{bmatrix} a_{11} & a_{12} \\ a_{21} & a_{22} \end{bmatrix} & b_{12} \begin{bmatrix} a_{11} & a_{12} \\ a_{21} & a_{22} \end{bmatrix} \\ b_{21} \begin{bmatrix} a_{11} & a_{12} \\ a_{21} & a_{22} \end{bmatrix} & b_{22} \begin{bmatrix} a_{11} & a_{12} \\ a_{21} & a_{22} \end{bmatrix} \end{bmatrix} \tag{9.37}$$

Two column vectors $|u\rangle$ and $|v\rangle$ of lengths m and n, respectively, yield a column vector of length mn when tensored; for instance,

$$|u\rangle |v\rangle = |uv\rangle = \begin{bmatrix} u_0 \\ u_1 \\ \vdots \\ u_{m-1} \\ u_m \end{bmatrix} \otimes \begin{bmatrix} v_0 \\ v_1 \\ \vdots \\ v_{n-1} \\ v_n \end{bmatrix} = \begin{bmatrix} u_0 \cdot v_0 \\ u_0 \cdot v_1 \\ \vdots \\ u_0 \cdot v_n \\ u_1 \cdot v_0 \\ \vdots \\ u_{m-1} \cdot v_n \\ u_m \cdot v_0 \\ \vdots \\ u_m \cdot v_{n-1} \\ u_m \cdot v_n \end{bmatrix} \tag{9.38}$$

We can use the tensor products to obtain the appropriate combinations of *kets* and *bras*, as follows. Suppose two kets, $|u\rangle$ and $|v\rangle$, are mutually orthogonal and have unit length, and are defined as

$$|u\rangle = \begin{bmatrix} 1 \\ 0 \end{bmatrix} \tag{9.39a}$$

$$|v\rangle = \begin{bmatrix} 0 \\ 1 \end{bmatrix} \tag{9.39b}$$

then, the tensor products of the *kets'* appropriate combinations are

$$|uu\rangle = \begin{bmatrix} 1 \\ 0 \end{bmatrix} \otimes \begin{bmatrix} 1 \\ 0 \end{bmatrix} = \begin{bmatrix} 1 \\ 0 \\ 0 \\ 0 \end{bmatrix} \tag{9.40a}$$

$$|uv\rangle = \begin{bmatrix} 1 \\ 0 \end{bmatrix} \otimes \begin{bmatrix} 0 \\ 1 \end{bmatrix} = \begin{bmatrix} 0 \\ 1 \\ 0 \\ 0 \end{bmatrix} \tag{9.40b}$$

$$|vu\rangle = \begin{bmatrix} 0 \\ 1 \end{bmatrix} \otimes \begin{bmatrix} 1 \\ 0 \end{bmatrix} = \begin{bmatrix} 0 \\ 0 \\ 1 \\ 0 \end{bmatrix} \tag{9.40c}$$

$$|vv\rangle = \begin{bmatrix} 0 \\ 1 \end{bmatrix} \otimes \begin{bmatrix} 0 \\ 1 \end{bmatrix} = \begin{bmatrix} 0 \\ 0 \\ 0 \\ 1 \end{bmatrix} \tag{9.40d}$$

Another example of tensor product is considering spin operators (from Equation 9.12):

$$\sigma_x \otimes \sigma_z = \begin{bmatrix} 0 & 1 \\ 1 & 0 \end{bmatrix} \otimes \begin{bmatrix} 1 & 0 \\ 0 & -1 \end{bmatrix} = \begin{bmatrix} 0 & 0 & 1 & 0 \\ 0 & 0 & 0 & -1 \\ 1 & 0 & 0 & 0 \\ 0 & -1 & 0 & 0 \end{bmatrix} \tag{9.41}$$

$$\sigma_z \otimes \sigma_x = \begin{bmatrix} 1 & 0 \\ 0 & -1 \end{bmatrix} \otimes \begin{bmatrix} 0 & 1 \\ 1 & 0 \end{bmatrix} = \begin{bmatrix} 0 & 1 & 0 & 0 \\ 1 & 0 & 0 & 0 \\ 0 & 0 & 0 & -1 \\ 0 & 0 & -1 & 0 \end{bmatrix} \tag{9.42}$$

The two products $\sigma_x \otimes \sigma_z$ (Equation 9.41) and $\sigma_z \otimes \sigma_x$ (Equation 9.42) are not the same because they represent different observables. Realistically, *tensor product* is a special case of the *Kronecker product* of matrices. It is therefore conceivable that matrices of abstract operators and state vectors can replicate a known behavior. For instance,

$$(\sigma_z \otimes \sigma_x)|uv\rangle = \begin{bmatrix} 0 & 1 & 0 & 0 \\ 1 & 0 & 0 & 0 \\ 0 & 0 & 0 & -1 \\ 0 & 0 & -1 & 0 \end{bmatrix} \begin{bmatrix} 0 \\ 1 \\ 0 \\ 0 \end{bmatrix} = \begin{bmatrix} 1 \\ 0 \\ 0 \\ 0 \end{bmatrix} \tag{9.43}$$

This solution somewhat replicates the known behavior of Equation (9.40a); i.e. $|uu\rangle$ the tensor product of *kets*.

9.2.3   QUBITS IN *STATE SPACE*

Kets can be transformed into any two vectors commonly used in quantum computational basis; that is,

$$|0\rangle = \begin{bmatrix} 1 \\ 0 \end{bmatrix}$$
$$|1\rangle = \begin{bmatrix} 0 \\ 1 \end{bmatrix} \tag{9.44}$$

As such, the state of a qubit at any given time can be represented as a two-dimensional *state space* with orthonormal basis vectors $|0\rangle$ and $|1\rangle$.

The question is: When is a set, or subset, considered orthonormal?

Simple. A subset $\{v_1, \ldots, v_k\}$ of a vector space V, with inner product $\langle,\rangle$, is called orthonormal if $\langle v_i, v_j \rangle = 0$ when $i \neq j$; that is, when the vectors are mutually perpendicular and have unit length, meaning $\langle v_i, v_i \rangle = 1$.

As a consequence of space vector representation, other orthonormal basis vectors can be used in quantum computational basis. For instance,

$$|+\rangle = \frac{|0\rangle + |1\rangle}{\sqrt{2}} = \frac{1}{\sqrt{2}} \begin{bmatrix} 1 \\ 1 \end{bmatrix}$$

$$|-\rangle = \frac{|0\rangle - |1\rangle}{\sqrt{2}} = \frac{1}{\sqrt{2}} \begin{bmatrix} 1 \\ -1 \end{bmatrix}$$

(9.45)

and, when measured it gives the *quantum* probability $\left|\dfrac{1}{\sqrt{2}}\right|^2 = \dfrac{1}{2}$ of either state. This state is often denoted as $|+\rangle$.

In view of Equations (9.44) and (9.45), a qubit can be expressed slightly different from that defined in Equation (9.3), thus

$$|\psi\rangle = c_0|0\rangle + c_1|1\rangle = c_0 \frac{|+\rangle + |-\rangle}{\sqrt{2}} + c_1 \frac{|+\rangle - |-\rangle}{\sqrt{2}} = \frac{c_0 + c_1}{\sqrt{2}}|+\rangle + \frac{c_0 - c_1}{\sqrt{2}}|-\rangle$$

(9.46)

In this instance, instead of measuring the states $|0\rangle$ and $|1\rangle$ each with respective probabilities $|c_0|^2$ and $|c_1|^2$, the states $|+\rangle$ and $|-\rangle$ would be measured with probabilities $\left|\dfrac{c_0 + c_1}{\sqrt{2}}\right|^2$ and $\left|\dfrac{c_0 - c_1}{\sqrt{2}}\right|^2$, respectively.

9.3 QUANTUM ELECTRONICS: WHAT IS IT?

Preface: Quantum electronics is the area of applied physics that deals with the effects of quantum mechanics on the behavior of electronics in matter. Quantum mechanics explain that there are elementary excitations (in energy) of electromagnetic waves, with particle-like properties. Electromagnetic radiation consists of "particles" or "packets" of energy known as photons—a stream of massless particles that travel in a wave-like pattern at the speed of light when the particles propagate in a vacuum. The photon energy (also called *quantum energy*), E_p, can be estimated using

$$E_p = hf = \frac{hc}{\lambda} = 2\pi \frac{hc}{\lambda} \ \text{(electron Volts, eV)}$$

(9.47)

where
h and $\hbar$ are, respectively, the Planck's constant and quantization of angular momentum; as defined previously in Equations (9.2) and (9.8).
f = frequency of photon or electromagnetic radiation, Hz.
λ = wavelength, m.
c = speed of light in vacuum $\approx 3 \times 10^8 \, \text{m s}^{-1}$.

The photon energies vary so considerably that the electromagnetic spectrum is divided up into convenient regions which are connected with their effects on matter with which they might interact.

The regions range from cosmic rays, gamma γ-rays, through X-rays, ultraviolet (UV) rays, visible light waves (comprising violet, blue, green, yellow, orange, and red light), infrared waves (comprising near and thermal), microwave, and radar waves to extremely low frequency (ELF) waves, as shown in Figure 9.3. The various regions are distinguished by their different wavelength ranges. The broad range of the electromagnetic radiations is known as the electromagnetic spectrum.

It follows from Equation (9.47) that the smaller the wavelength, λ, the bigger the photon energy, E_p. Visible, optics infrared, and optical communications are carried out primarily within the 400 and 750 nm ranges. Lower temperatures are important in the near infrared and visible spectrum where, currently, the bulk of the photonic activities are carried out. Note that the particle properties of electromagnetic waves become of importance when the quantum (or photon) energy, E_p, is larger than the thermal energy, E_{th}, stored in an electromagnetic (EM) mode. As defined by Equation (1.6) in Chapter 1, thermal energy

$$E_{th} = kT \tag{9.48}$$

where k is the Boltzmann's constant (= 1.3805×10^{-23} J K^{-1}) and T is the absolute temperature (in degree kelvin, K). [Note that absolute temperature, $T = 273.15 + t$, where t = measured temperature in degree Celsius.]

It is worth noting that "quantum mechanics" equally assigns to both electrons and protons wave properties, and matter waves, just like it assigns to the electromagnetic waves' particle-like properties. Electrons are elementary excitations of matter waves just as photons are elementary excitations of electromagnetic waves. Basically, electronics uses primarily electrons or other charged particles.

Quantum electronics became a field of physics and of engineering with the development of the devices known as *maser* and *laser*. The acronyms "maser" and "laser" stand, respectively, for "microwave amplification by stimulated emission of radiation" and "light amplification by stimulated emission of radiation."

Maser is the microwave-wavelength version of the laser. Stimulated emission is a form of spectral line emission. A spectral line relates to specific wavelengths of light emitted by an object. It may result from an excess or deficiency of photons in a narrow frequency range when compared with the nearby frequencies. Both devices (maser and laser) are of the same general class: maser produces and amplifies electromagnetic radiation mainly in the microwave region of the spectrum. Maser operates according to the same basic principle as the laser and shares many of its characteristics. As such, we concentrate on addressing/defining laser and its attributes.

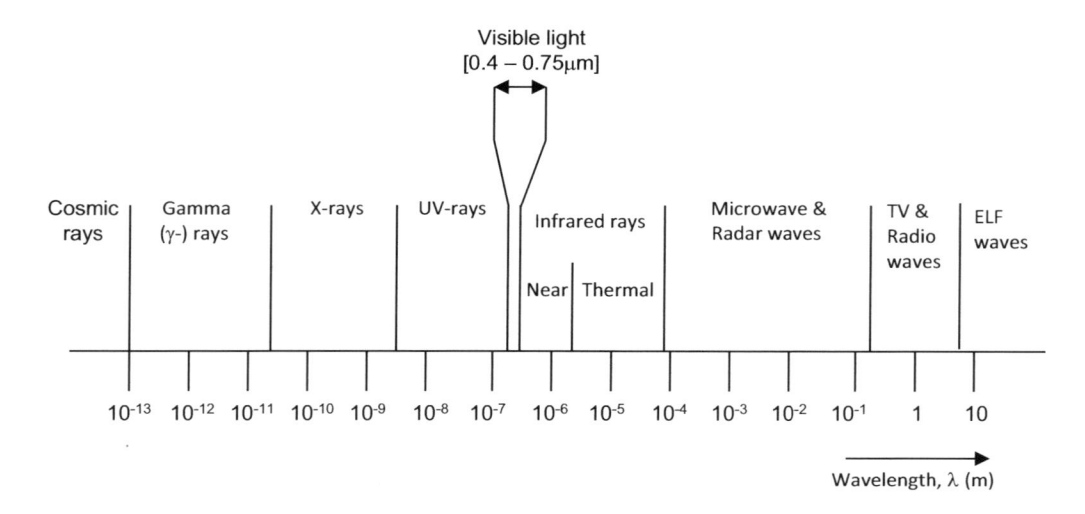

FIGURE 9.3   Wavelength ranges of electromagnetic radiation and its use.

Lasers, broadly speaking, are devices that generate or amplify light, just as transistors and vacuum tubes generate and amplify electronic signals at audio, radio, or microwave frequencies. Here "light" must be understood broadly, since different kinds of lasers can amplify radiation at wavelengths ranging from the very long infrared region, merging with millimeter waves or microwaves, up through the visible region and extending to the vacuum UV and even X-ray regions. Lasers come in a variety of forms, using many different laser materials, many different atomic systems, and many different kinds of excitation techniques or pumping techniques [9,10]. The beams of radiation that lasers emit or amplify have remarkable properties of directionality, spectral purity, and intensity. These properties have already led to an enormous variety of applications, and others, which undoubtedly have yet to be discovered and developed.

The essential elements of a laser device, as shown in Figure 9.4, are

i. A *laser medium* (or *cavity*) consisting of an appropriate collection of atoms, molecules, ions, or in some instances, a semiconducting crystal. The cavity is used to reflect light from the lasing medium back into itself.

ii. A *pumping process* excites these atoms (molecules, etc.) into higher quantum-mechanical energy levels. Excitation could be electrically pumped, optically pumped, excimer gas pumped, nuclear-pumped method [11], electron-beam pumped, or energy from another laser. A minimum pumping power, P_{min}, is required to maintain the necessary population inversion to keep the laser process going, defined as

$$P_{min} = \frac{N_m}{t_c} E_p \tag{9.49}$$

where t_c and N_m depict the average lifetime and number of modes, respectively. For clarity, population inversion means a transposition in the relative numbers of atoms, molecules, ions, etc. occupying particular energy levels. For instance, suppose there are N_1 and N_2 atoms at the lower and upper levels, respectively. For laser action to occur, pumping must

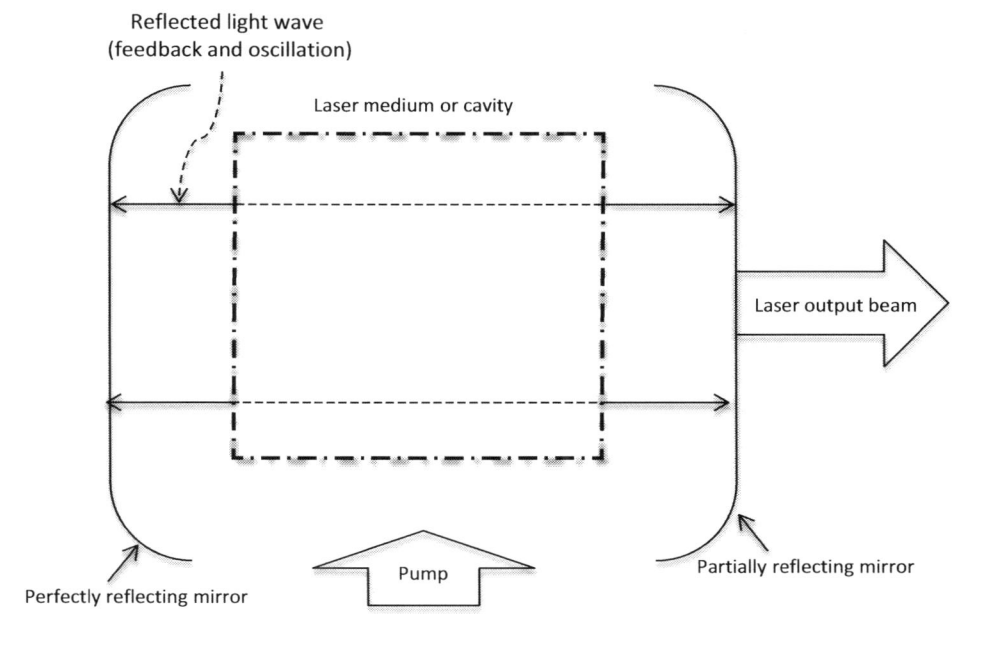

FIGURE 9.4 Elements of a typical laser oscillator.

ensure that $N_2 > N_1$, that is, producing more excited atoms into some higher quantum energy level than are in some lower energy level in the laser medium. This process is known as *population inversion*, Δ, defined by

$$\Delta = N_2 - N_1 \tag{9.50}$$

iii. Suitable *optical feedback elements* that allow a beam of radiation to either pass once through the laser medium (as in a *laser amplifier*) or bounce back and forth relatedly through the laser medium (as in a *laser oscillator*). (More is said about the design of basic amplifiers and oscillators and their characteristics in Chapter 4.)

In principle, we can turn an amplifier into an oscillator by providing *positive feedback*, which can be realized using a pair of plane or concave spherical mirrors. In masers, the active medium is placed into a microwave cavity [12]. Technically, a maser oscillator requires a source of excited atoms or molecules and a resonator to store their radiation. The excitation must force more atoms or molecules into the upper energy level (or energy state) than in the lower, in order for amplification by stimulated emission to predominate over absorption. There are application areas where maser is used, including satellite communication, air-to-air communication, radio technology, amplifiers, and oscillators in microwave, where low noise factor is the most important.

For clarification, there are two related but more familiar phenomena involving the interplay between matter and radiation, which are *spontaneous emission* and *absorption*.

Spontaneous emission happens when an atom in an excited state emits a photon and, in so doing, moves to a lower energy state; as such, the energy difference between the two levels is carried away by the photon. It is a spontaneous process that occurs in the absence of a radiation field and happens without external stimuli. As an example, *spontaneous emission* is something we calculate from the quantum mechanics of an atom, or in the case of continuum radiation something we calculate from the application of electromagnetic theory.

Absorption happens when an atom absorbs a photon and in doing so, moves to a more excited state (i.e. causing a transition from level 1 to level 2); as such, an electron is moved to a higher energy level and the energy difference between the two levels is equal to the energy input from the photon.

Both *spontaneous emission* and *absorption* are thermal processes. In thermal equilibrium, the number of atoms in different energy levels obeys the Boltzmann distribution equation. The Boltzmann distribution law states that the probability of finding an atom or molecule in a particular energy state varies exponentially when the translational energy is divided by its thermal energy, E_{th} (given by Equation 9.48).

The most important notion in quantum electronics is the probability for an electron in an atom or a molecule to make quantum transition from one level to another. In the simplest model of quantum electronics, matter is assumed to consist of separate non-interacting motionless atoms or molecules in external electromagnetic field. The behavior of material particles, in any given external fields, is described by the time-dependent Schrödinger wave equation [3]:

$$-\frac{\hbar^2}{2m}\frac{\partial^2 \psi}{\partial^2 x} + V(x,t)\psi = i\hbar\frac{\partial \psi}{\partial t} \tag{9.51}$$

where the wavefunction $\psi(x, t)$—abbreviated as ψ in Equation (9.51)—is moving in the presence of a potential $V(x, t)$ in the positive x direction at time t, and m is the mass of particle. The potential $V(x, t)$ is assumed mono-dimensional but can be multidimensional, which in that case becomes $V(\mathbf{r}, t)$ instead of $V(x, t)$ with a set of coordinates $\mathbf{r} = \{\mathbf{r}_1, \mathbf{r}_2, \ldots\}$; h-bar ($\hbar$) is the quantization of angular momentum, as defined previously. It is in the setting up and solving Equation (9.51) and then analyzing the physical contents of its solutions that form the basis of wave mechanics.

A keen observer would say that Schrödinger wave equation stems from or looks like the wave equation that describes waves on a stretched string—e.g. the harmonic wavefunction—but it has solutions that represent waves propagating through space. Naturally, the harmonic wavefunction for a free particle of energy E and momentum p is expressed by

$$\psi(x,t) = Ae^{-i(px-Et)/\hbar} \tag{9.52}$$

is a solution of Schrödinger equation with, as appropriate for a free particle, $V(x) = 0$; and, where the precisely determined momentum

$$p = \frac{h}{\lambda} \tag{9.53}$$

Depending on the nature of the potential $V(x)$ experienced by the particle, Equation (9.52) can have distinctly non-wave-like solutions.

Whilst, in general, the solutions to the time-dependent Schrödinger wave equation describe the dynamical behavior of the particle similar to that of Newton's equation[1] of motion—$F = ma$—that describes the dynamics of a particle in classical physics: but with a difference. The major difference is: by solving Newton's equation, one can determine the position of a particle as a function of time; whereas, by solving Schrödinger's equation, what we get is a wavefunction $\psi(x,t)$, which tells us how the probability of finding the particle in space varies as a function of time. It is worth noting that Schrödinger equation also supports the idea of superposition.

The application of the Schrödinger equation extends to the electron transport description in studying *ultimate* performance of subnanometer channel of metal-oxide-semiconductor field-effect transistors (MOSFETs) [13], and in reaching the ultimate scaling limits of a field-effect transistor (FET) structure [14]. More is said about FET and MOSFET in Chapter 3, Section 3.2, and Chapter 5, respectively.

9.4 QUANTUM GATES AND CIRCUITS

A quantum circuit is a sequence of quantum gates. The signals (qubits) may be static while the gates are dynamic. The circuit has a fixed "width" corresponding to the number of qubits being processed. Like in the classical logic design, quantum logic design also attempts to find circuit structures for needed operations that are functionally correct and independent of physical technology, and, wherever possible, that use the minimum number of qubits or gates.

In the classical logic circuits, behavior is governed implicitly by classical physics: no restrictions on copying or measuring signal; signal states are simple bit vectors (e.g. $X = 01010111$), and operations are defined by Boolean algebra. A Boolean gate is said to be reversible if it has the same number of inputs as outputs, and its mapping from input strings to output strings is a bijection. In addition, small well-defined sets of universal gate types (e.g. AND, OR, NOT, NOR, NAND) exist and circuits use fast, scalable, and macroscopic technologies such as transistor-based integrated circuits (e.g. FET, MOSFET, CMOS—as explained in Chapter 6).

The behavior of quantum circuits is governed by quantum mechanics, where signal states are qubit vectors, operations are defined by linear algebra and represented by unitary matrices: for instance, gates and circuits must be logically reversible—meaning the number of output lines must equal to the number of input lines, and cloning (*fanout*) is not allowed—meaning that logic states cannot be copied. Note *fanout* of a gate is the number of gates driven by that gate, i.e. the maximum number of gates (or load) that can exist without impairing the normal operation of the gate.

---

[1] This equation is dubbed Newton's second *law of motion* in classical mechanics, where F is the force acting on a body of mass m and acceleration a of its centre of mass.

(See examples of *fanout* for bipolar junction transistor (BJT): Figure 3.19 in Section 3.4.1 of Chapter 3; and CMOS: Figure 6.11 in Section 6.1.1.1 of Chapter 6). By disallowing *fanout* might seem, on face value, to contradict the universality of the Toffoli and/or Fredkin gates, which are universal only with respect to classical states: it actually does not contradict. (More is said of Toffoli and Fredkin gates later in this chapter.) A typical quantum circuits' inputs are usually computational basis states; $|+\rangle, |-\rangle$. Graphically, quantum circuit diagrams are drawn with time going from left to right, with the quantum gates crossing one or more "wires" (qubits) where appropriate. A quantum circuit has unique features: they are acyclic where loops (feedback) as well as *fan-in* and *fanout* are not allowed. *Fan-in* (equivalent to OR) is not reversible or unitary.

Unitary matrices are used to represent quantum logic gates because unitary operations are the only linear maps that preserve norm. Mathematically, matrix M is unitary if $M^\dagger M = I$ where I is identity matrix; i.e. $\begin{bmatrix} 1 & 1 \\ 1 & 1 \end{bmatrix}$, and $M^\dagger$ is the transpose conjugate matrix of M; i.e. in terms of matrices, $M^\dagger = (M^*)^T$ where * denotes complex conjugation and superscript T denotes transposition. As an example, suppose matrix M is defined by

$$M = \begin{bmatrix} 1+j & 1-j \\ -1 & 1 \end{bmatrix} \tag{9.54a}$$

then

$$M^\dagger = \begin{bmatrix} 1+j & 1-j \\ -1 & 1 \end{bmatrix}^\dagger = \begin{bmatrix} 1-j & -1 \\ 1+j & 1 \end{bmatrix} \tag{9.54b}$$

Since in quantum circuits, we have a set of n qubit systems whose dynamic development is made of a sequence of unitary operators, each of which acts on one or more of the qubits, we can graphically represent the qubits as a set of horizontal lines, and the various stages of their development as a series of boxes (or where necessary by other symbols) showing the structure of operations, as depicted in Figure 9.5. For example, as seen in Figure 9.5, box A is applied to the first qubit and C to the third qubit; then B is applied to the first and second qubits; and finally, D is applied to the second and third qubits. Unlike the digital circuit design structure, time runs from left to right in this case.

The individual steps of the development—the elementary operations on one, two, or a small number of qubits—are usually called *quantum logic gates* (or simply *quantum gates*), by analogy to the basic elements of digital circuit designs (discussed in Chapters 3–6). Quantum gates are logically reversible, like classical reversible gates, but they differ markedly on their universality properties.

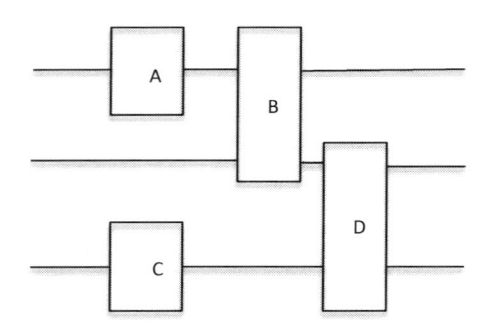

FIGURE 9.5   A 3-qubit quantum circuit.

Logic reversibility is the ability to reconstruct the input from the output of a gate function. For clarity, the reversibility or otherwise issues are explained using the primitive logic gates already discussed in Chapter 2, Section 2.2. For instance, *buffer* and *NOT* gates are reversible because buffer gate's output is the same as its input; and the NOT gate's output is its own inverse. Whereas other "standard" gates—OR, AND, XOR, NOR, and NAND—are irreversible since they do not have an equal number of inputs and outputs, each has only one output: one of their inputs could be considered to have been effectively erased in the process due to change in entropy, causing irretrievable loss of information. As such, quantum gates should not always be thought of in the same way we picture classical gates.

Generically, a quantum logic gate can have any finite number of input qubits, but in practice, we are interested in gates that have a small number of input qubits, which to date is elementary for quantum computation. Small quantum circuits have been demonstrated in the laboratory using various physical technologies. The next subsections explain what constitute quantum logic gates (Section 9.4.1) and how circuits are performed for simple quantum gates (Section 9.4.2).

9.4.1 Elementary Quantum Logic Gates

As earlier demonstrated in the previous chapters, logic gates are the building blocks for processing information. By arranging gates in a circuit, engineers are able to create something akin to a flowchart that enables computers to carry out many kinds of logical operations and perform the kinds of tasks that computers can do. Following a similar analogy, the brain of a quantum computer will consist of chains of logic gates, as quantum logic gates harness quantum phenomena. The next subsections develop simple quantum logic gates.

9.4.1.1 *Buffer* and *NOT* Gates

Buffer and *NOT* gates are examples of single qubit gates.

Buffer gate: Buffer gate is simple: the output state $|+\rangle$ or $|-\rangle$ corresponds to the input state $|+\rangle$ or $|-\rangle$, as shown in Figure 9.6. The graphical representation of Buffer gate is shown in Figure 9.6c symbolized by the • sign.

NOT gates: A *NOT gate's* states are interchangeable; that is, $|0\rangle \rightarrow |1\rangle$ and $|1\rangle \rightarrow |0\rangle$, as shown in Figure 9.7c: a condensed representation of the NOT gate, symbolized by $\oplus$.

We can develop quantum NOT gates from the *Pauli matrices X* and *Z*—expressed by Equation (9.12), and note that Pauli matrices σ_x, σ_y, and σ_z are simply labeled matrices *X*, *Y*, and *Z*, respectively—as well as from the Walsh–Hadamard transform. The developments are discussed as follows:

a. From Pauli-X matrix (or σ_x or X):

$$[X]\begin{bmatrix} |0\rangle \\ |1\rangle \end{bmatrix} = [\sigma_x]\begin{bmatrix} |0\rangle \\ |1\rangle \end{bmatrix} = \begin{bmatrix} 0 & 1 \\ 1 & 0 \end{bmatrix}\begin{bmatrix} |0\rangle \\ |1\rangle \end{bmatrix} = \begin{bmatrix} |1\rangle \\ |0\rangle \end{bmatrix} \tag{9.55}$$

denoted at in Figure 9.8a.

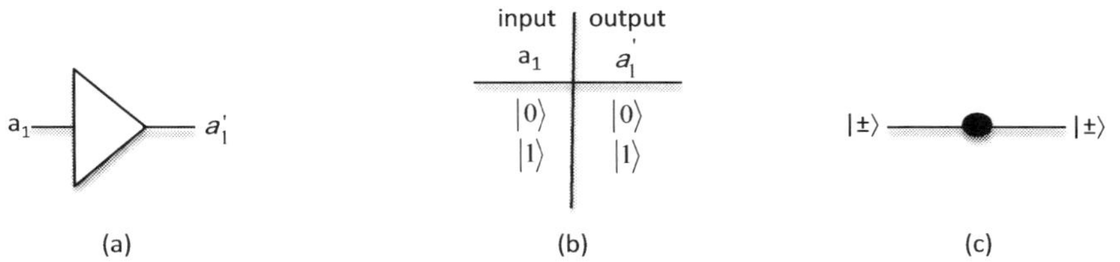

input	output
a_1	a_1'
$|0\rangle$	$|0\rangle$
$|1\rangle$	$|1\rangle$

(a) (b) (c)

FIGURE 9.6 Quantum Buffer gate: (a) classical Buffer gate, (b) truth table, and (c) graphic representation of a Buffer.

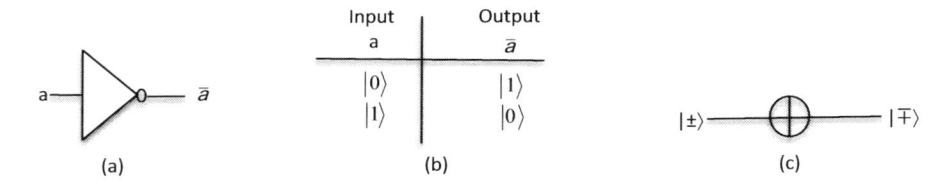

FIGURE 9.7   Quantum NOT gate: (a) classical NOT gate, (b) truth table, and (c) quantum NOT representation.

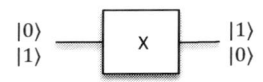

FIGURE 9.8a   Pauli-X gate representation.

The action of NOT gate, in this case, is to take the state $|0\rangle$ and replace it by the state corresponding to the first column of matrix X, while the state $|1\rangle$ is replaced by the state corresponding to the second column of matrix X. This symmetry is depicted by the *truth table* (Figure 9.7b), or as a condensed graphic representation of the NOT gate (Figure 9.7c).

b. From Pauli-Z matrix, (σ_z or Z), a NOT-gate can also be developed:

$$[Z]\begin{bmatrix} |0\rangle \\ |1\rangle \end{bmatrix} = [\sigma_z]\begin{bmatrix} |0\rangle \\ |1\rangle \end{bmatrix} = \begin{bmatrix} 1 & 0 \\ 0 & -1 \end{bmatrix}\begin{bmatrix} |0\rangle \\ |1\rangle \end{bmatrix} = \begin{bmatrix} |0\rangle \\ |-1\rangle \end{bmatrix} \tag{9.56}$$

In Z gate, as shown in Figure 9.8b, the state $|0\rangle$ is unchanged while the state $|1\rangle$ is flipped to give $|-1\rangle$. It is a special case of a phase shift by π. Z gate is sometimes called the phase-flip gate.

If we generalize this by defining the *phase-shift gate* $\Phi_z(\alpha)$ with a matrix representation:

$$\Phi_z(\alpha) = \begin{bmatrix} 1 & 0 \\ 0 & e^{i\alpha} \end{bmatrix} \tag{9.57}$$

then $\sigma_z = \Phi_z(\pi)$. (Note that $e^{i\alpha} = \cos\alpha + i\sin\alpha$. If $\alpha = \pi$, $e^{i\alpha} = -1$.)

A representation of a *phase-shift gate* is shown in Figure 9.8c by ϕ.

For completeness, Pauli-Y gate representation is shown in Figure 9.8d.

c. From Walsh–Hadamard transform (also called the H gate). The H gate is an important special transform that maps qubit basis states $|0\rangle$ and $|1\rangle$ into two superposition states with

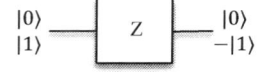

FIGURE 9.8b   Pauli-Z gate representation.

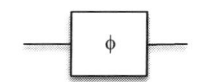

FIGURE 9.8c   Phase-shift gate representation.

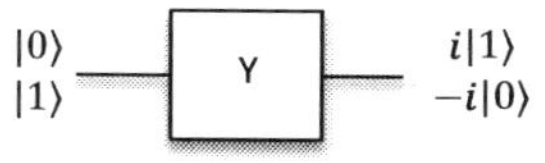

FIGURE 9.8d Pauli-Y gate representation.

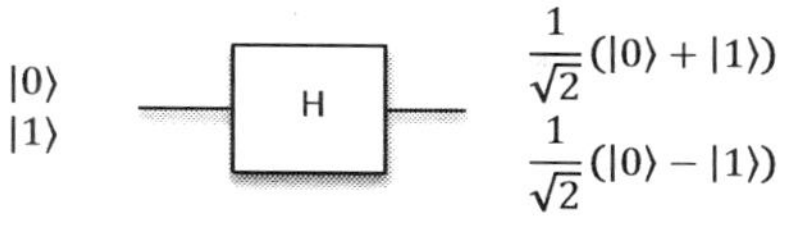

FIGURE 9.9 Hadamard gate representation.

equal weight of the computational basis states $|0\rangle$ and $|1\rangle$. Technically, the H gate creates a superposition state out of a normal 0 or 1, with a 2×2 unitary matrix:

$$H = \frac{1}{\sqrt{2}} \begin{bmatrix} 1 & 1 \\ 1 & -1 \end{bmatrix} \begin{bmatrix} |0\rangle \\ |1\rangle \end{bmatrix} \tag{9.58}$$

which is like the "square-root of NOT" gate (i.e. $\sqrt{NOT}$ gate). The H gate has no classical analogy. This equation (Equation 9.58) transforms the basis states into mixed states; i.e. $\big(|0\rangle + |1\rangle\big)$ and $\big(|0\rangle - |1\rangle\big)$, as shown in Figure 9.9.

Note that, unlike the classical case, the qubits that enter a quantum circuit cannot be assumed to have individual states; in general, the bits may be in a *joint* state. The question that follows is: How then can we describe their joint state?

Suppose we have two qubits $|x\rangle = \big(\alpha_0 |0\rangle + \alpha_1 |1\rangle\big)$ and $|y\rangle = \big(\beta_0 |0\rangle + \beta_1 |1\rangle\big)$, where $a_0, a_1, \beta_0, \beta_1$ are coefficients. The first guess would be using multiplication of some sort, that is, using the *tensor product* notation $\otimes$:

$$|x\rangle \otimes |y\rangle = \big(\alpha_0 |0\rangle + \alpha_1 |1\rangle\big) \otimes \big(\beta_0 |0\rangle + \beta_1 |1\rangle\big)$$

$$= \alpha_0 \beta_0 \underbrace{|0\rangle \otimes |0\rangle}_{|00\rangle} + \alpha_0 \beta_1 \underbrace{|0\rangle \otimes |1\rangle}_{|01\rangle} + \alpha_1 \beta_0 \underbrace{|1\rangle \otimes |0\rangle}_{|10\rangle} + \alpha_1 \beta_1 \underbrace{|1\rangle \otimes |1\rangle}_{|11\rangle} \tag{9.59}$$

Subsequently, the multiplication $|x\rangle \otimes |y\rangle$ looks exactly the same as a linear combination of the four basis states $|00\rangle, |01\rangle, |10\rangle, |11\rangle$.

We can generalize for the case of a quantum state $|\Psi\rangle$ for n single modes $|\psi\rangle_j$, where $j = 1, 2, \cdots, n$. The nodes $|\psi\rangle_j$ can be independent from each other or entangled between each other. So, the quantum state $|\Psi\rangle$ is described as the *tensor product* of the states $|\psi\rangle_j$:

$$|\Psi\rangle = \underset{j}{\otimes} |\psi\rangle_j = |\psi\rangle_1 \otimes |\psi\rangle_2 \otimes |\psi\rangle_3 \cdots \otimes |\psi\rangle_{n-1} \otimes |\psi\rangle_n \tag{9.60}$$

Question 9.1: The reader might ask: with these input mixed states; (i) what will the measurement be? and (ii) will the states have the same amplitude?

Answer:

i. If measured, as represented by Figure 9.10, the result of the observed (or *measured*) output will be unknown: an entangled output. However, since the inputs have *distinct* states; as such, they will behave differently in the way they evolve. For instance, the state $|0\rangle$ is unchanged for both cases, but the state $|1\rangle$ is flipped to give $|-1\rangle$ in the case $\big(|0\rangle - |1\rangle\big)$.

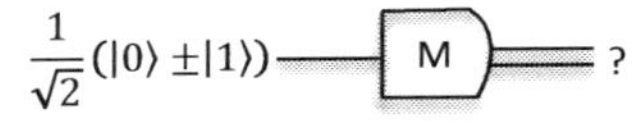

$$\frac{1}{\sqrt{2}}(|0\rangle \pm |1\rangle) \longrightarrow \boxed{\text{M}} \Longrightarrow ?$$

FIGURE 9.10   Example of quantum measurement, M.

This demonstrates that the probabilistic "state" of the register M at some intermediate time in the circuit's execution reflects only the uncertainty that we, the observers, have about the register's values.

ii. Yes, the states will have the same probability. Drawing from Equation (9.58), for example, we can write each state's vector thus

$$\frac{1}{\sqrt{2}} \begin{bmatrix} 1 \\ 1 \end{bmatrix} \quad \text{for state} \quad (|0\rangle + |1\rangle);$$

$$\frac{1}{\sqrt{2}} \begin{bmatrix} 1 \\ -1 \end{bmatrix} \quad \text{for state} \quad (|0\rangle - |1\rangle);$$

and, when measured in the computational basis, gives either outcome with amplitude $\left| \frac{1}{\sqrt{2}} \right|^2 = \frac{1}{2}.$

9.4.1.2   *XOR Gates* (CNOT or C-NOT Gates)

A simple, reversible gate, with two input bits and two output bits, is called the *controlled-NOT* (denoted as CNOT or C-NOT) gate, drawn as in Figure 9.11. The CNOT is sometimes called XOR, since it performs an *exclusive*-OR operation on the two-input qubits and writes the output to the second qubit. For example, the first input qubit x is always passed through directly; the second qubit (or target qubit) y gets NOT applied to it if and only if the input "control" qubit x is $|1\rangle$. (The decision to negate or not negate the second qubit is controlled by the value of the first qubit x: hence, the name *controlled-NOT*). By convention, the dot symbol attached to input–output gate with a vertical line means creating a "controlled" gate.

We can make exchange from the CNOTs. Exchange generates the "SWAP" operation.

SWAP gate: A SWAP gate simply interchanges the states of the qubits it is handed—a more complicated example of a reversible logic gate. Swapping the states of qubits allows you to permute the information stored in a set of qubits. The circuit representations of SWAP gate, for 2-qubit input, are shown in Figure 9.12.

Figure 9.12a appears suggestive that some operations occur by crossing wires. But many implementations of quantum circuits do not have physical wires as such. Hence, to avoid misconception, a SWAP operation is achieved as the result of a sequence of applied fields. For example, by using a sequence of three CNOT gates that swaps the states of two qubits, a simple SWAP operation can be constructed, thus $SWAP_{xy} = CNOT_{xy} CNOT_{yx} CNOT_{xy}$; i.e. $(x, y) \rightarrow (x, x \oplus y) \rightarrow (y, x \oplus y) \rightarrow (y, x) -$ the states of the two qubits have been swapped, as shown in Figure 9.13.

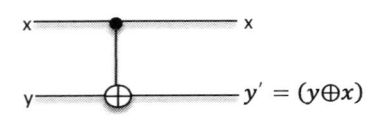

FIGURE 9.11   CNOT (or XOR) gate.

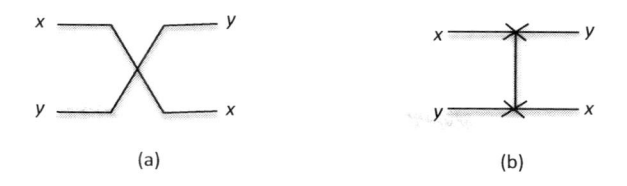

FIGURE 9.12 SWAP gate circuit representations: (a) classical representation and (b) quantum circuit representation.

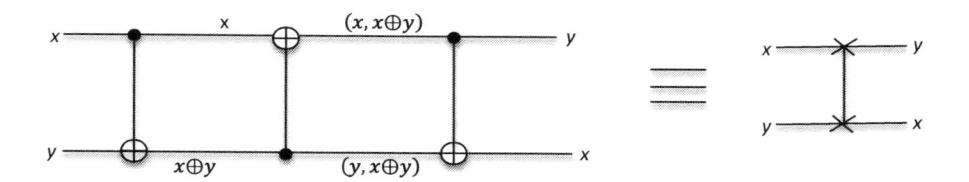

FIGURE 9.13 A SWAP gate formed from three back-to-back CNOT gates.

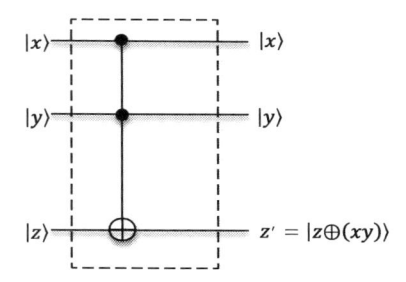

FIGURE 9.14 A Toffoli gate.

9.4.1.3 OR, AND, and NAND Gates

We now describe a small but important universal 3-bit reversible gate, called Toffoli gate [15], which is also called *controlled*-CNOT gate (abbreviated as CCNOT or CC-NOT). An arrangement of CCNOT gate is shown in Figure 9.14. The first two input qubits (x, y) to the CCNOT gate are passed through directly, and the third input qubit z is negated if and only if "control" input-qubits x and y are both $|1\rangle$. In other word, the values of the first two-input qubits control whether the third input qubit is flipped. The Toffoli gate is its own inverse.

The Toffoli gate is especially interesting because, depending on the input, the gate is extremely handy in the sense that it can be arranged to perform logical OR, AND, NAND, and fanout operations. For example

a. *NAND gate*: When the third qubit $|z\rangle$ is fixed to be $|1\rangle$, the Toffoli gate writes the NAND of the first two qubits on the third qubit, as shown in Figure 9.15. In this arrangement, the first two qubits $|x, y\rangle$ represent the input of the NAND gate. The third qubit, prepared as standard state $|1\rangle$, is sometimes called an *ancilla* state (also called "ancillary" state). An ancilla can be interpreted physically as an extra quantum system that is assumed as having a state space with the required properties. It allows extra working space during the computation. Simply, an input that is "hardwired" to the constant qubit—$|1\rangle$ in this case—for the purpose of assisting the computation.

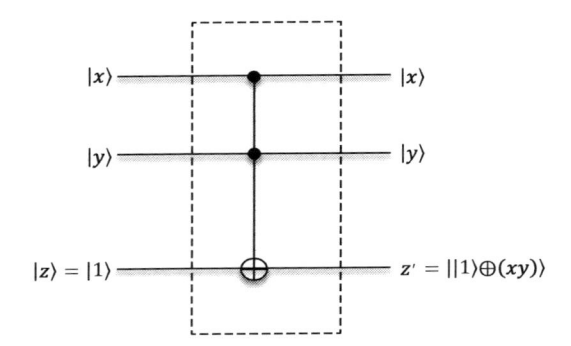

FIGURE 9.15 A *NAND gate* from Toffoli gate.

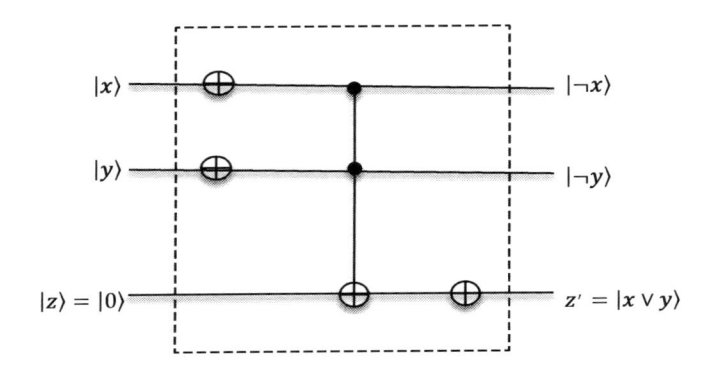

FIGURE 9.16 An *OR gate* from Toffoli gate. Note: Symbols $\vee$ and $\neg$ denote logic OR and logic NOT operations, respectively.

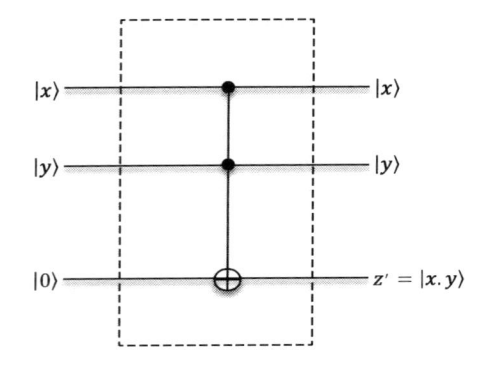

FIGURE 9.17 An *AND gate* from Toffoli gate.

Other gates are developed from the generic Toffoli gate as follows:

b. OR gate (Figure 9.16).

c. AND gate (Figure 9.17).

d. Fanout of a gate is the number of gates driven by that gate, i.e. the maximum number of gates (or load) that can exist without impairing the normal operation of the gate—examples of classical fanout gates are shown in Figure 3.19, Chapter 3, Section 3.4.1 for

BJT and Figure 6.11, Chapter 6, Section 6.1.1.1 for CMOS. In the case of quantum electronics, *fanout* of a gate is arranged with Toffoli gate, in this case in two ways, as shown in Figure 9.18. In the first arrangement, Figure 9.18a, the second qubit *y* is the input to the fanout while the other two qubits (*x, z*) become the standard *ancilla* states, and the output from fanout appearing on the second and third qubits. In case of Figure 9.18b, the fanout's input is the first qubit *x* while the ancilla states are (*y, z*) and its output appearing on the first and third qubits.

By following discussions for generating NAND (Figure 9.15) and Fanout (Figure 9.18a) using both $|1\rangle$ ancilla and $|0\rangle$ ancilla, respectively, we can also use CCNOT gates to generate $|0\rangle$'s from $|1\rangle$ ancillas—states interchangeable, i.e. $|1\rangle \rightarrow |0\rangle$—thus (Figure 9.19).

Another well-known universal, reversible gate is the Fredkin (or controlled-SWAP) gate, shown in Figure 9.20. It has three-input bits (a_1, a_2, a_3) and three-output bits (a_1', a_2', a_3'). The bit a_1 is a *control* bit whose value does not change by the action of the gate, meaning $a_1 = a_1'$. When the control bit is set to 0 (i.e. $a_1 = 0$), the values of a_2 and a_3 are unchanged (i.e. $a_2 = a_2'$ and $a_3 = a_3'$). However, if the control bit is set to 1 (i.e. $a_1 = 1$), the values of a_2 and a_3 are swapped; that is, $a_3' = a_2$ and $a_2' = a_3$. Technically, the Fredkin gate is a *multiplexer*.

The beauty of the Fredkin gate is that it is easy to recover the original inputs a_1, a_2, a_3: all we need is to apply another Fredkin gate to a_1', a_2', a_3'.

The Fredkin gate can be configured to simulate elementary gates such as *AND, NOT,* and *Crossover*—a primitive routing function—as shown in Figure 9.21a–c, respectively. A closer look at Figure 9.21b shows that it also performs *fanout* function since it produces two copies of *x* at the output.

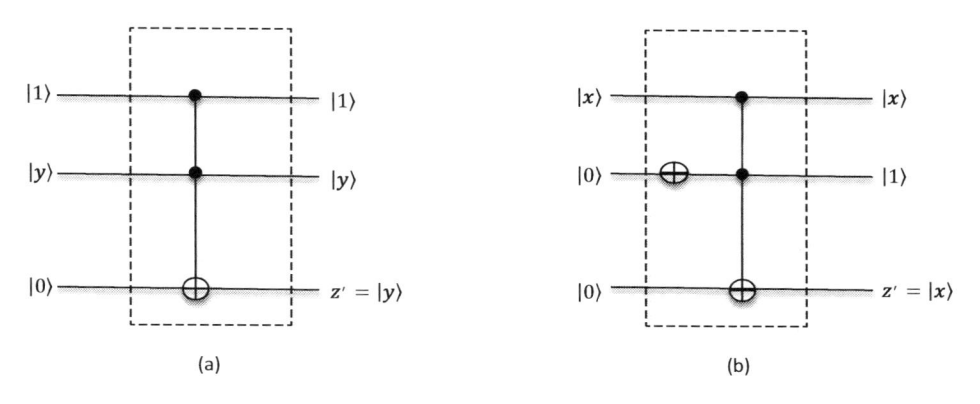

(a)   (b)

FIGURE 9.18   Arrangements of *Fanout*: (a) a *Fanout* with the Toffoli gate and (b) another arrangement of *Fanout* with the Toffoli gate.

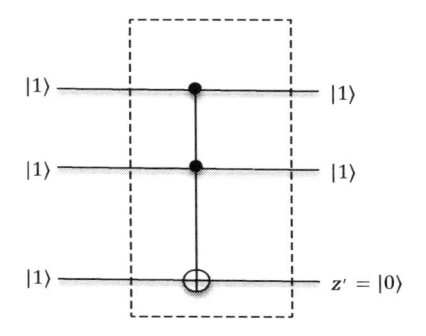

FIGURE 9.19   Using CCNOT gates to generate $|0\rangle$'s from $|1\rangle$ ancillas.

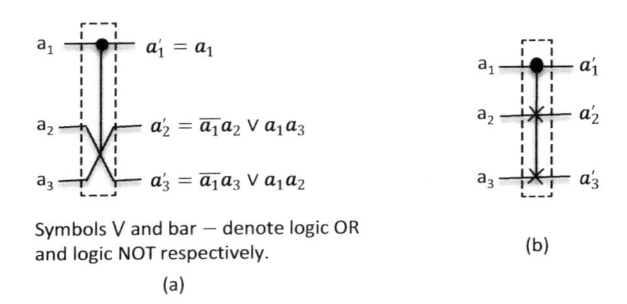

Symbols V and bar — denote logic OR and logic NOT respectively.

(a)

(b)

FIGURE 9.20 Fredkin (controlled-Swap) gate circuit: (a) controlled-Swap gate and (b) quantum circuit representation.

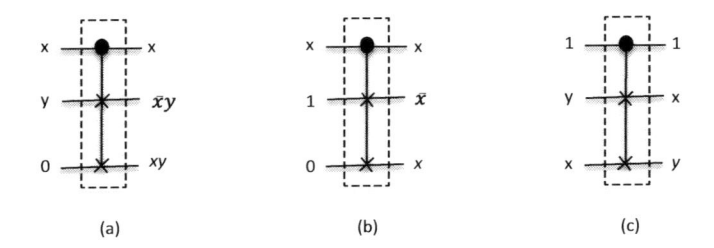

(a)   (b)   (c)

FIGURE 9.21 Derivation of elementary gates from Fredkin gate: (a) AND gate, (b) NOT gate, and (c) crossover gate.

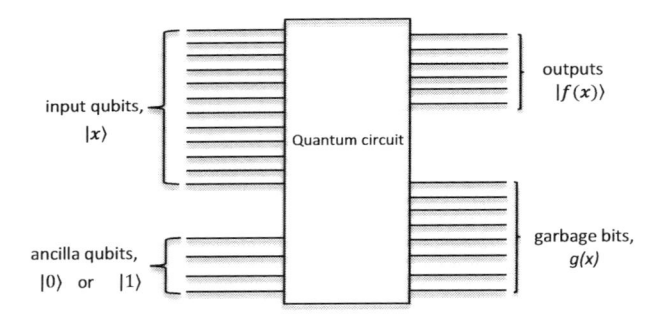

FIGURE 9.22 A representation of a quantum circuit containing ancilla qubits and garbage bits.

These configured elementary gates from the Toffoli and Fredkin processes can be cascaded to create any classical circuit whatsoever. It should be noted that each of these configurations and circuits requires the use of extra "ancilla" qubits (or simply *ancillae*) as part of input states, and the output may contain "garbage" bits, $g(x)$. A *garbage* is an output that is not used as input to other gate, or not considered to be as a primary output. Garbages are unwanted and may not impact the computational functions of the circuit, as shown in Figure 9.22.

Often, the input ancilla qubits act like an input that is "hardwired" to qubits $|0\rangle$, or $|1\rangle$, depending on what functions the designated circuits are performing to facilitate things. The ancilla and garbage bits are only important to make the computation reversible. For some quantum circuits to effectively emulate classical (Boolean) circuits, they preserve input and ancilla qubits—in this case, storing entangled states, thereby enabling tasks that would not normally be possible with local operations and classical communication. It is anticipated that quantum computers will use ancilla bits for QEC. (More is said about *Quantum errors and processing* in Section 9.4.3.)

9.4.2 Quantum Circuits

A quantum circuit is a compact representation of a computational system, consisting of quantum gates and quantum bits (qubits) as input. As noted in Chapter 2, Section 3 (*Logic circuits design*), the goal of electronic circuits' design is to build hardware that will compute given function(s). We can build quantum circuits of any complexity out of these simple quantum gates (barest essentials). Efficient design of quantum circuits to performing arithmetic operation and other axillary functions needs special attention in quantum systems and, as well as in computing, and should be fault tolerant because reversible requirement makes quantum circuit design more difficult than classical circuit design. As such, the number of qubits is an important metric in a quantum circuit design and must be kept minimal. Few examples of quantum circuits derived from quantum gates are described as follows.

9.4.2.1 Adders

Conventional binary addition is not reversible. As such, we need a new *adder* construction to fulfil the reversibility concept of quantum systems.

Quantum half adder: Reversible quantum half adder (HA) can be implemented using two reversible gates: a CCNOT (Toffoli gate) and one NOT gate, as shown in Figure 9.23, where the adder formulating expressions are

$$S = |x \oplus y\rangle \tag{9.61a}$$

$$C_{out} = |z\rangle \oplus (|x\rangle|y\rangle) = |xy\rangle \tag{9.61b}$$

$$G = |x\rangle \tag{9.61c}$$

G is the garbage output, in this case.

Two cascaded HAs may be used to build a *quantum full adder* (FA), as shown in Figure 9.24, where their formulating expressions are

$$S = |c_0\rangle \oplus |x\rangle \oplus |y\rangle \tag{9.62a}$$

$$C_{out} = |z\rangle \oplus \{|c_0\rangle \oplus |x\rangle|y\rangle\} = |c_0\rangle(|x\rangle \oplus |y\rangle) \oplus |x\rangle|y\rangle \tag{9.62b}$$

$$G_1 = |c_0\rangle \tag{9.62c}$$

$$G_2 = |x\rangle \oplus |c_0\rangle \tag{9.62d}$$

G_1 and G_2 are output garbages, in this case.

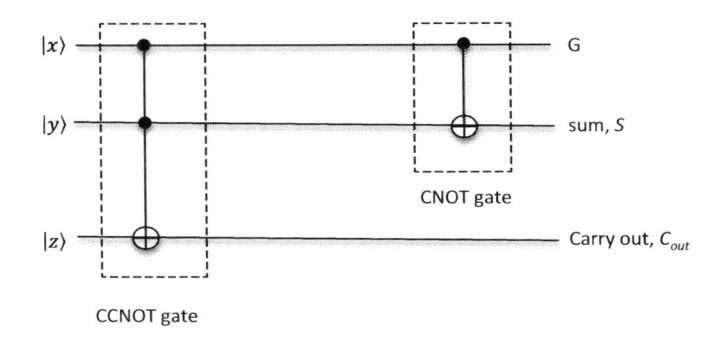

FIGURE 9.23 Quantum HA circuit.

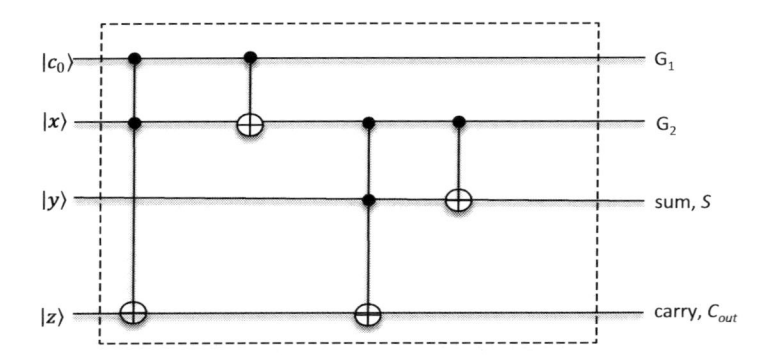

FIGURE 9.24 Quantum FA circuit.

Quantum ripple-carry adder: By defining recursively the *carry string* for x and y, we let $c_0 = 0$ and $MAJ(x_i, y_i, c_i)$ for $i \geq 0$; where $MAJ(x_i, y_i, c_i) = x_i y_i \oplus x_i c_i \oplus y_i c_i$; that is, Equation (2.24) in Chapter 2. The sum S_i {Equation (2.23) in Chapter 2} is for $i < n$, and $S_n = C_n$. The $MAJ(x_i, y_i, c_i)$ gate computes the majority of three bits in place and can be constructed using first two CNOTs followed by one Toffoli (as demonstrated by Figure 9.25a). The sum component (Equation 2.23), which is not majorly and adds—tagged $UMA(x_i, y_i, c_i)$—can be constructed similarly with three CNOTs (as seen in Figure 9.25b). Note that an isolated XOR symbol $\oplus$ in Figure 9.25b means a quantum NOT-gate. Of course, the UMA gate can be constructed conceptually simpler with two CNOTs (as shown in Figure 9.25c). However, to reduce the depth of the basic circuit, we use Figure 9.25b in addition to Figure 9.25a. Consequently, we can build a *ripple-carry* adder by stringing together

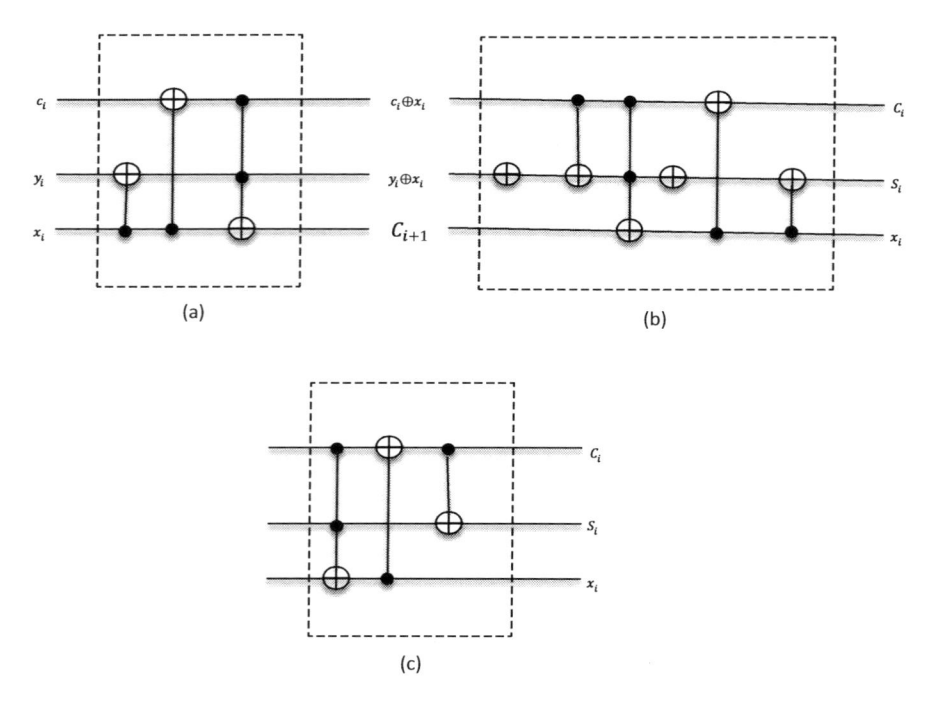

FIGURE 9.25 Implementation of *addition function* using CNOTs and Toffoli gates when initial carry bit, c_o, is zero (i.e. $c_o = 0$): (a) $MAJ^{-1}(x_i, y_i, c_i)$ gate, (b) $UMA(x_i, y_i, c_i)$ gate with three CNOTs, and (c) $UMA(x_i, y_i, c_i)$ gate with two CNOTs.

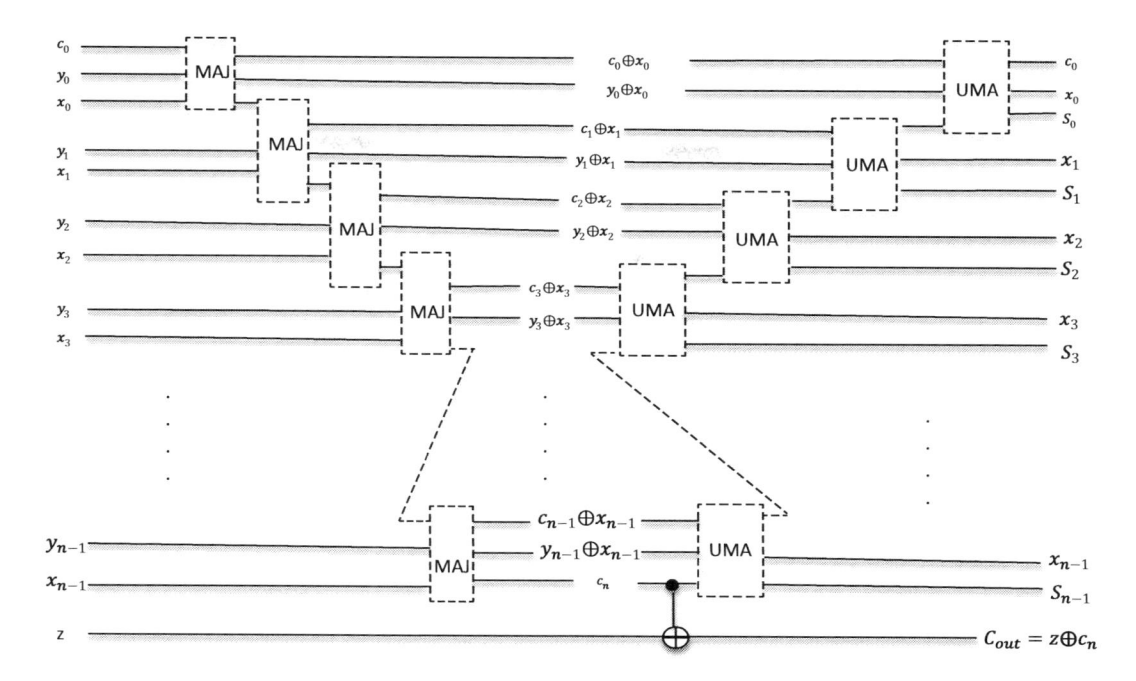

FIGURE 9.26   A ripple-carry adder with one ancilla z.

MAJ (Figure 9.25a) and UMA (Figure 9.25b) gates. An example of this stringing process is that demonstrated by Cuccaro et al. [16] as shown in Figure 9.26.

A ripple-carry adder (as seen in Figure 9.26) is built by stringing together MAJ and UMA gates starting with the low-order bits of the input and working upwardly to the high-order bits for $0 \leq i \leq n-1$. As such, the circuit requires only one ancilla z, which is set to "0."

9.4.2.2   Quantum Comparator Circuits

As demonstrated in [17], *comparators* $x < y$ are similar to subtractors—one subtracts x from y (that is, $x - y$) and checks $x - y < 0$. This is done by modifying the Cuccaro et al. (2005) adder circuit built by stringing together MAJ and UMA, as shown in Figure 9.25, to perform comparison by leaving their data inputs unchanged and producing a 1-bit result as the most significant carry-bit of subtraction. Unlike the adder where UMA blocks follow the MAJ blocks, for the *comparator* one uses "inverse MAJ" blocks (as shown in Figure 9.27) instead of UMA blocks used in the *adders*. Note that for the first MAJ block $c_i = 0$.

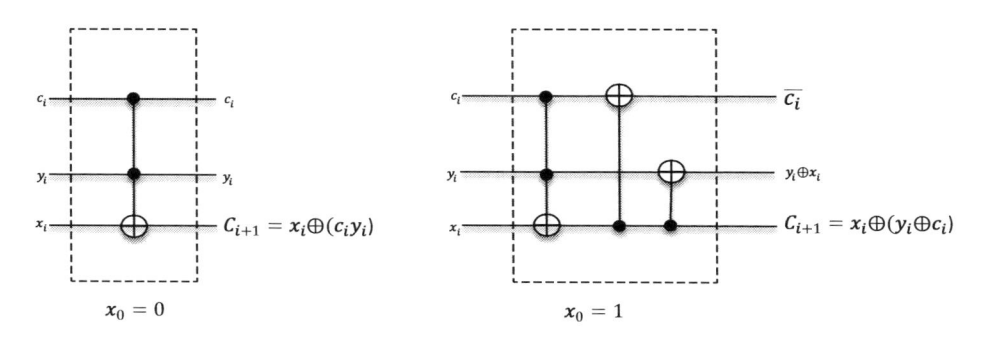

FIGURE 9.27 $MAJ^{-1}(x_i, y_i, c_i)$ gate for $0 < i < n$.

9.4.2.3   Quantum Multiplier Circuits

Like the multiplication algorithm detailed in Chapter 2, Section 2.3, Figure 2.40, to developing *Binary Multiplier Circuits*, similar algorithm can be applied to construct quantum multiplier circuits with some modifications; that is, avoiding unnecessary garbage ancillae. The common multiplication method is the "add and shift" algorithm. An example of a 4-bit multiplicand by 4-bit multiplier is shown in Table 9.2.

As observed in Table 9.2, the multiplication process has a regular structure, and before summing operation, each successive partial product is *shifted* one position to the left relative to the preceding partial product. The sum (addition) is performed with normal propagate FA. It can be seen that the addition is done serially as well as in parallel, thus adding to operational delay and area deficiency. As both multiplicand and multiplier may be *positive* or *negative*, a modified Baugh and Wooley 2's complement number system has been reconstituted, as shown in Table. 9.3, by treating both positive and negative numbers uniformly [18,19]. Following the process shown in Table. 9.3, a *quantum multiplier circuit* is implemented as in Figure 9.28.

TABLE 9.2

Two 4-bit Multiplication Process

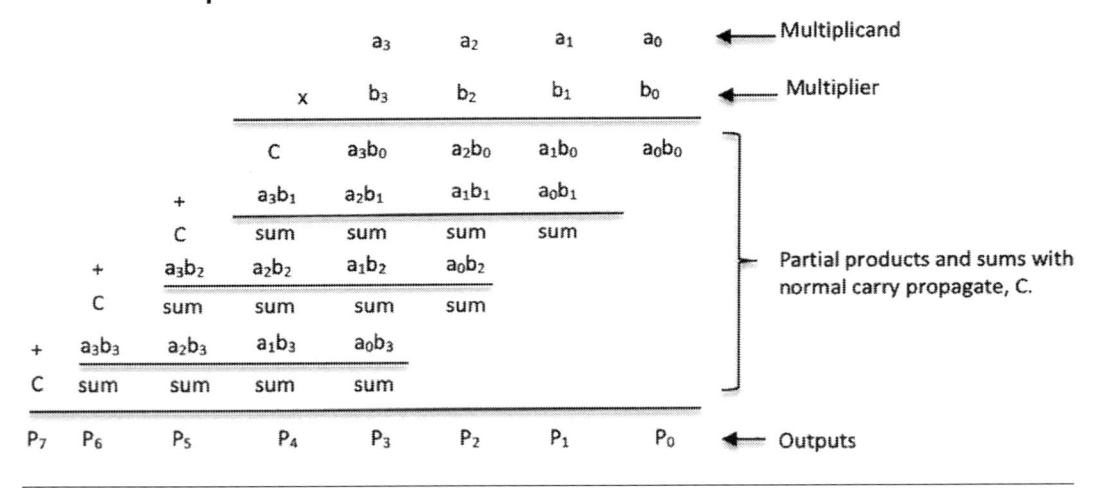

TABLE 9.3

An Illustration of the Baugh–Wooley 2's Complement Number System's Multiplier

					a_3	a_2	a_1	a_0
				x	b_3	b_2	b_1	b_0
				1	$\overline{a_3b_0}$	a_2b_0	a_1b_0	a_0b_0
			$\overline{a_3b_1}$		a_2b_1	a_1b_1	a_0b_1	
		$\overline{a_3b_2}$		a_2b_2	a_1b_2	a_0b_2		
	1	a_3b_3	$\overline{a_2b_3}$	$\overline{a_1b_3}$	$\overline{a_0b_3}$			
P_7	P_6	P_5	P_4	P_3	P_2	P_1	P_0	

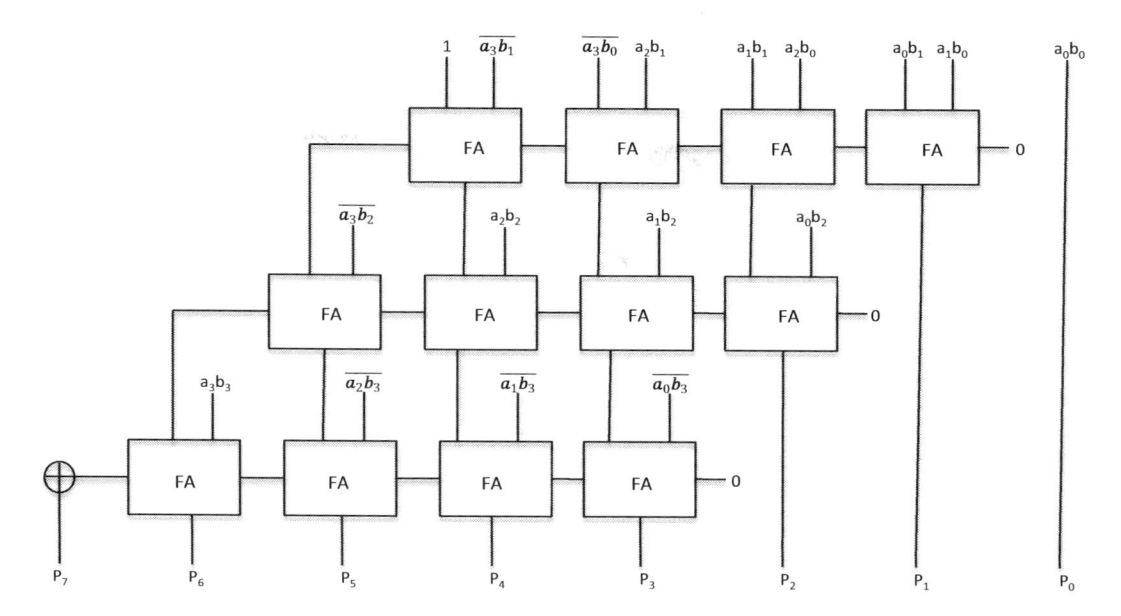

FIGURE 9.28   A quantum multiplier circuit, where FA implies *full adder.*

An emerging QCA multiplier that focused on serial–parallel multiplier technique is being taunted as having wire delay minimization due to its simple structure [20]. More is said about QCA in Section 9.5. In summary, efficient design of quantum circuits to performing arithmetic operation and other axillary functions needs special attention in quantum systems and computing and should be fault tolerant. Existing quantum hardware is limited in terms of the number of available qubits.

9.4.3   Quantum Error-Correction Scheme

Like error correction being central to classical information theory, QEC is similarly foundational in quantum information theory. Both classical information theory and quantum information theory are concerned with the fundamental problem of error propagation in communication channel, as well as in information storage, in the presence of noise. Due to the uncertainty of its physical quantity, quantum noise appears as shot noise—quantum-limited intensity noise—caused by current fluctuations due to discreteness of the electrical charge. (*Shot noise* has been discussed in Chapter 6, Section 6.1.) Error-correction techniques are themselves susceptible to noise. For instance, when we perform multiple-qubit gates during a quantum computation, any existing errors in the system can propagate to other qubits, even if the gate itself is perfect. As such, we need to avoid, or at least limit or reduce error propagation in communication process.

9.4.3.1   But How Do We Describe a Quantum Error?

The most general type of errors in the state of a qubit can be described as follows. Suppose a single qubit $|\psi\rangle$ in the state $\alpha|0\rangle + \beta|1\rangle$ undergoes some random unitary transformation, or decoheres, by becoming entangled with the environment it transits through. We define the state of the environment before the interaction with the qubit as $|E\rangle$. Basically, the qubit will undergo some unitary transformation in the combined system of qubit and environment. The most general unitary transformation on system and environment can be described as [21]

$$U = |\psi\rangle \otimes |E\rangle \tag{9.63a}$$

that is,

$$U: \quad |0\rangle \otimes |E\rangle \quad |0\rangle \otimes |E_{00}\rangle + |1\rangle \otimes |E_{01}\rangle \tag{9.63b}$$

$$|1\rangle \otimes |E\rangle \quad |0\rangle \otimes |E_{10}\rangle + |1\rangle \otimes |E_{11}\rangle \tag{9.63c}$$

In this case, $|E_{ij}\rangle$ represents not necessarily orthogonal or normalized states of the environment, and the unitary U entangles the qubit with the environment. Potentially, this entanglement will lead to decoherence of the information stored in the qubit. If the qubit is afflicted by an error, it evolves to

$$(\alpha|0\rangle + \beta|1\rangle) \otimes |E\rangle = \alpha(|0\rangle \otimes |E_{00}\rangle) + |1\rangle \otimes |E_{01}\rangle + \beta(|0\rangle \otimes |E_{10}\rangle + |1\rangle \otimes |E_{11}\rangle)$$

$$= (\alpha|0\rangle + \beta|1\rangle) \otimes \frac{1}{2}(|E_{00}\rangle + |E_{11}\rangle) \quad \text{identity}$$

$$+ (\alpha|0\rangle - \beta|1\rangle) \otimes \frac{1}{2}(|E_{00}\rangle - |E_{11}\rangle) \quad \text{phase flip}$$

$$+ (\alpha|1\rangle + \beta|0\rangle) \otimes \frac{1}{2}(|E_{01}\rangle - |E_{10}\rangle) \quad \text{bit-flip}$$

$$+ (\alpha|1\rangle + \beta|0\rangle) \otimes \frac{1}{2}(|E_{01}\rangle - |E_{10}\rangle) \quad \text{bit and phase flip} \tag{9.64}$$

where, in this case, the coefficients α and β may not necessarily be numbers, they can be states that are orthogonal to both $|0\rangle$ and $|1\rangle$. Intuitively, we can infer that one of the four things happens to the qubit: nothing (identity), a bit flip, a phase flip, or a combination of bit flip and phase flip. In some literature and books, the four errors acting on a qubit are presented as so-called Pauli group, thus

$$\underbrace{I = \begin{pmatrix} 1 & 0 \\ 0 & 1 \end{pmatrix}}_{\text{identity}} \quad \underbrace{\sigma_x = \begin{pmatrix} 1 & 0 \\ 0 & -1 \end{pmatrix}}_{\text{phase flip}} \quad \underbrace{\sigma_z = \begin{pmatrix} 0 & 1 \\ 1 & 0 \end{pmatrix}}_{\text{bit flip}} \quad \underbrace{\sigma_x\sigma_z = \begin{pmatrix} 0 & -1 \\ 1 & 0 \end{pmatrix}}_{\text{bit and phase flip}}$$

Naturally, when an error affects one qubit, it is equally likely to be σ_x, σ_z, or $\sigma_x\sigma_z$ error. Note that the error analysis discussed thus far is primarily due to decoherence, which is somewhat simplified. However, we have neglected errors due to imperfections in the quantum gates, in preparation of the initial states, and even in the measurement process [21]. All these operations can be faulty. Also, the single qubit considered may be part of a larger quantum state of several qubits. It might be entangled with other qubits that are unaffected by errors.

But how do we construct a QEC scheme?

Figure 9.29 shows the general schematic of a QEC process.

To achieve reliable transmission in the presence of noise, similar to classical ideas of error-correcting codes, the errors in the quantum domain may be corrected by employing QEC codes [22]. Due to issue of quantum entanglement, analogous classical approach might be inadequate for constructing an error-correcting scheme. As noted in [23,24], a quantum error-correcting scheme can take advantage of entanglement in two ways:

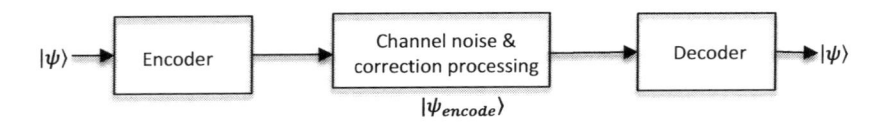

FIGURE 9.29   QEC process.

a. Entangle one qubit carrying information with $(n-1)$ qubits, and create an n-qubit quantum codeword that is more resilient to errors.

b. Entangle the n qubits of this quantum codeword with ancilla qubits and construct the error syndrome without measuring the qubits of the codeword and thus without altering the state.

A simplified approach to encoding and decoding is discussed as follows. Suppose a single qubit $|\psi\rangle$ is encoded and deal with bit flip errors. To encode $|\psi\rangle$, we use a 3-bit repetition code on the basis states like

$$|0\rangle \rightarrow |000\rangle \quad |1\rangle \rightarrow |111\rangle$$

such that,

$$\alpha|0\rangle + \beta|1\rangle \quad \mapsto \quad \alpha|000\rangle + \beta|111\rangle$$

The extra qubit is like an ancilla qubit (redundancy). Following Shor [23], we encode a qubit using nine qubits in the following way:

$$|0\rangle_{encode} = \frac{1}{2\sqrt{2}}(|000\rangle + |111\rangle)(|000\rangle + |111\rangle)(|000\rangle + |111\rangle) \tag{9.65a}$$

$$|1\rangle_{encode} = \frac{1}{2\sqrt{2}}(|000\rangle - |111\rangle)(|000\rangle - |111\rangle)(|000\rangle - |111\rangle) \tag{9.65b}$$

Suppose the first qubit is flipped, then both $|100\rangle$ and $|011\rangle$ have the same parity 1 on the first two qubits. If no qubit is flipped and the code word is still in the state of $\alpha|000\rangle + \beta|111\rangle$, their parity will be 0 for both $|000\rangle$ and $|111\rangle$. If the error is a *linear* combination of identity and bit flip, similar to Equation (9.64), then the measurement will collapse the state into either identity or bit flip. The three parities give a complete information about the location of the bit flip error, and thereby constitute what is called the error *syndrome* measurement. The syndrome measurement does not acquire any information about the encoded superposition, and hence it does not destroy it [21]. Depending on the outcome of the syndrome measurement, we can correct the error by applying bit flip to the appropriate qubit. In principle, the encoding is simply repeating the bit to be transmitted three times, and decoding is just taking the most frequently appearing bit.

Can this encoding and correction process where each of the blocks of three qubits encoded with a repetition code be applicable to phase flip errors?

The answer is yes. Suppose one of the first three qubits, with phase flip error, acts as

$$|0_{encode}\rangle \rightarrow \frac{1}{2\sqrt{2}}(|000\rangle - |111\rangle)(|000\rangle + |111\rangle)(|000\rangle + |111\rangle) \tag{9.66a}$$

$$|1_{encode}\rangle \rightarrow \frac{1}{2\sqrt{2}}(|000\rangle + |111\rangle)(|000\rangle - |111\rangle)(|000\rangle - |111\rangle) \tag{9.66b}$$

What we need, therefore, is to detect this *phase flip* without measuring the information in the state. We achieve this by following the same process developed for the bit flip and measure the parity of the phases on each pair of two of the three blocks.

Theoretically, if each of the qubits undergoes some error at rate ε, then we can assume that the 9-bit encoding scheme with a repetition is applicable to $\sigma_x\sigma_z$ (bit and phase flip) error and can protect against bit flip, phase flip, and a combination of both (when both bit and phase flip errors are detected).

Contrary to the usually conceived assumption that the encoding and decoding circuits are fault-tolerant, the encoders (transmitters) or decoders (receivers) could be erroneous (corrupted during the process of communication). As such, the environmental effect (noise) imposed would need to be appropriately adjusted by using special entanglement distillation protocols to clean up the erroneous pairs and/or errors (or decoherence) in the quantum channel. A quantum channel is any medium that allows light to propagate through. Entanglement distillation is a process of extracting a group of strongly entangled states from a larger group of weakly entangled states.

As noted in [4], protocols based upon entanglement distillation offer performance superior to more conventional QEC techniques in enabling noise free communication of qubits. Could this assertion hold for both pure and mixed states? We are still in the early stage of understanding quantum behavior. How then do we secure exchanges during quantum communication session? This brings us to the issue of *quantum cryptography*.

9.4.4 QUANTUM CRYPTOGRAPHY

Quantum cryptography—another name for Quantum key distribution (QKD)—constitutes an approach for the distribution of cryptographic keys. In contrast to classical key distribution schemes, QKD security is based on the laws of quantum physics. In classical transmission system, the security of the information being transmitted is ensured by using cryptographic protocol: encrypting the information being sent from site A to site B and decrypting the encrypted information at the other end (site B), while preventing a malevolent third-party eavesdropping [26]. The classical cryptographic functions—encryption, decryption, and key distribution or management—have been discussed in detail in [26], Chapter 1.

The assertion that a quantum state contains hidden information that is not accessible to measurement—that is, the indistinguishability of non-orthogonal quantum states, plays a key role in quantum cryptography. As earlier explained, in Section 9.4.1.1 *Question* 9.1, given the state $|0\rangle$, the measurement will yield 0 with probability 1. However, when we measure $\frac{1}{\sqrt{2}}(|0\rangle + |1\rangle)$, the measurement will yield 0 with probability 1/2 and 1 with probability 1/2. Thus, while a measurement result of 1 implies that state must have been $\frac{1}{\sqrt{2}}(|0\rangle + |1\rangle)$, since it could not have been $|0\rangle$, we cannot infer anything about the identity of the quantum state from a measurement result of 0. To quantitatively apply quantum information to cryptography depends how well to develop measures quantifying how well non-orthogonal quantum states may be distinguished. One of these challenges is dealing with encryption—the security method where information is encoded in such a way that only authorized parties can access or decrypt the encoded information with the correct encryption key or keys. It is arguable that encryption itself does not prevent interference; it only denies the intelligible content to eavesdropper. However, the fragility of quantum information makes it likely that some form of QEC will be required in any practical quantum information processing, including the distribution of the cryptographic keys.

9.4.4.1 How Do We Exchange Cryptographic Keys?

The first part of a quantum key exchange is the distribution and measurement of quantum objects—the so-called raw key exchange—where both parties obtain a list of qubits, which could contain errors [27]. In this case, an error-correction step is performed to remove the errors from the raw key using either low-density parity check (LDPC) [28], cascade [29], or polar codes [30]. The LDPC and polar codes are non-interactive decoding schemes while cascade is interactive.

The second part of a quantum key exchange is postprocessing, where pure classical algorithms can be applied to the raw key [27]. The security of the privacy amplification step in QKD postprocessing is particularly important because privacy amplification is the final step in the postprocessing and no subsequent step can make up for leakage(s) in this step. Privacy amplification is a process that allows two parties to distil a secret key from a common random variable about which an eavesdropper has partial information [31]; a key agreement protocol.

The violation of a Bell inequality is a fundamental manifestation of quantum systems; it also benchmarks the performance of these systems when involved in certain quantum protocols. Current study demonstrated that a violation of Bell inequality provides a good benchmark on how a quantum protocol will perform [32]. By extension, it could be argued that the issue of quantum security, particularly the aspect of key distribution and management schemes in dealing with key generation, key storage, and key maintenance ensuring the key life-cycle remains secure, should not be overlooked. Also, importantly, is how to destroy the key after the expiration of its useful life.

9.5 FABRICATION OF QUANTUM GATES

Single qubit gates, like *buffer* and *NOT* gates, are relatively easy to fabricate. For example, atoms are used as qubits and their states are controlled with laser light pulses of carefully selected frequency, intensity, and duration. Any prescribed superposition of two selected atomic states can be prepared this way [33].

At the quantum level, fabricating (or packing) of transistors would require them to shrink down to the atomic length scales. At noted in Chapter 1, problems arise when the size of a transistor shrinks towards the ultimate limit due to quantum mechanical effects—the very nature of the electronic motion—as device size reduces so its wavelength changes, which in turn necessitates additional quantized energy. This will profoundly influence the operation of devices as their dimensions shrink beyond certain limits, as well as posing integration interfacing problem to other optical or electronic devices. Despite, it is envisaged that quantum electronic circuits with superconducting interference devices may be readily integrated with existing microelectronics. Whilst couplings between a single atom and a single microwave photon in a superconducting circuit are too weak for practical coherent interfaces, there is a strong inclination that quantum interfaces between solid state qubits, atomic qubits, and light may ensure direct quantum connection between electrical and atomic metrology standards [34].

The next subsections describe two different fabrication techniques for producing superconducting qubits devices: *tunnel junctions* and QCA.

9.5.1 TUNNEL JUNCTIONS

Different nanofabricated superconducting circuits techniques, based on Josephson tunnel junction [35], have achieved a degree of quantum coherence that allow the manipulation of their quantum states to build quantum circuits. Figure 9.30 depicts a Josephson tunnel junction: in its simplest form, it consists of two superconducting electrodes (SE1, SE2) separated by a barrier, *BR*. The barrier can be an insulating tunnel barrier, thin normal metal, etc. If the barrier is thin enough, the macroscopic wavefunctions of both electrodes overlap and form a weakly coupled system. Josephson tunnel junctions can be fabricated using electron-beam lithography (EBL)—already discussed in Chapter 1, Section 1.6—and shadow evaporation. Multiqubit gates depend on the relative frequencies of the qubits. As such, to reliably build multiqubit devices therefore requires a careful fabrication of Josephson junctions in order to precisely set their critical currents [36].

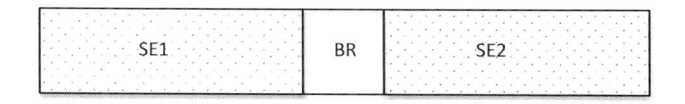

FIGURE 9.30 A simple Josephson junction consisting of superconducting electrodes (SE1 and SE2) interposed with a thin insulating barrier (BR).

Another fabrication technique to building these qubits is *scanning tunneling microscope* (STM). STM is an instrument for imaging surfaces at the atomic level, based on the concept of quantum tunneling. Tunneling is a functioning concept that describes the behavior of a quantum object (with a very small mass, such as the electron) when it hits a barrier. If the barrier is thick, the object bounces off. However, if the barrier is thin enough, the object may sometimes get through. The thinner the barrier, the more likely the object passes. A metal is made up of quantum atoms and electrons. If we approach a very thin tip electrically powered, it may tear the electrons from the metal by tunnel effect. By measuring the electric current passing through the tip, we can reconstruct where the atoms are. This is the principle of STM. Using STM concept, the University of New South Wales, Sydney Australia, proved for the first time and fabricated a working quantum transistor consisting of a single atom placed precisely in a silicon crystal [37]. Figure 9.31 illustrates the operation of a single atom transistor. A voltage is applied across the phosphorus electrodes, which induces a current that forms electrical leads for a single phosphorus atom positioned precisely in the center of the silicon crystal.

The reader might wonder: how does a single atom transistor work compared to a traditional transistor?

As discussed and shown in Chapter 3, a typical semiconductor-device has three terminals: namely, *source*, *drain*, and *gate*. The gate controls the electron flow from source to drain. Whereas, a single-atom transistor allows the electrons to pass from source to drain. The single-atom transistor works as an atomic switch, or an atomic relay, where the switchable atom opens and closes the gap between two tiny electrodes called source and drain.

9.5.2 QUANTUM-DOT CELLULAR AUTOMATA

9.5.2.1 QCA Basics

A quantum dot is a nanoparticle made of any semiconductor material, for example, silicon (Si), cadmium selenide (CdSe), cadmium sulphide (CdS), or indium arsenide (InAs). As noted in Chapter 8, Section 8.4, Tougaw and Lent [38] proposed the technology of QCA given that CMOS technology has reached its scaling limit due to quantum mechanical effects. QCA is based essentially on a cell. Each cell consists of four quantum-dots (simply dots) placed in the corner of a square populated with two excess (mobile) electrons that are allowed to tunnel between the dots. The basic QCA cell is shown in Figure 9.32. The dot is basically a region of space with energy barriers surrounding it. These barriers are large and high enough so that the charge within it is quantized to a multiple elementary charge. As a result of the electrostatic interactions (repulsion) between the charges,

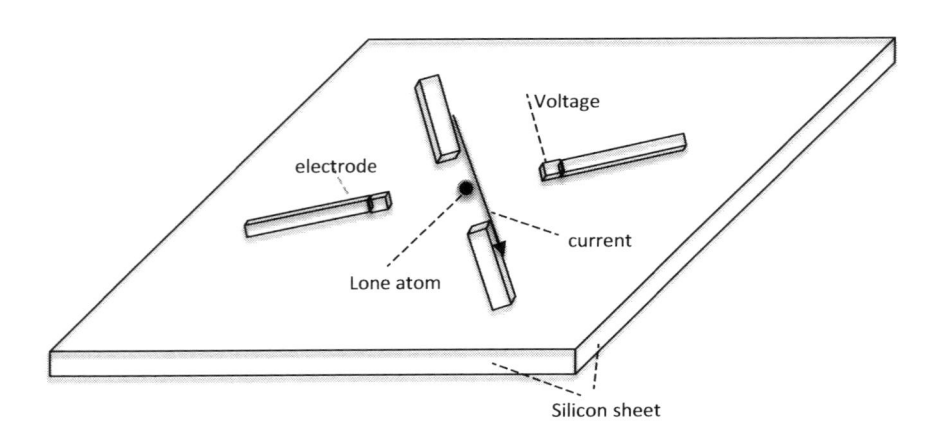

FIGURE 9.31 A single-atom transistor.

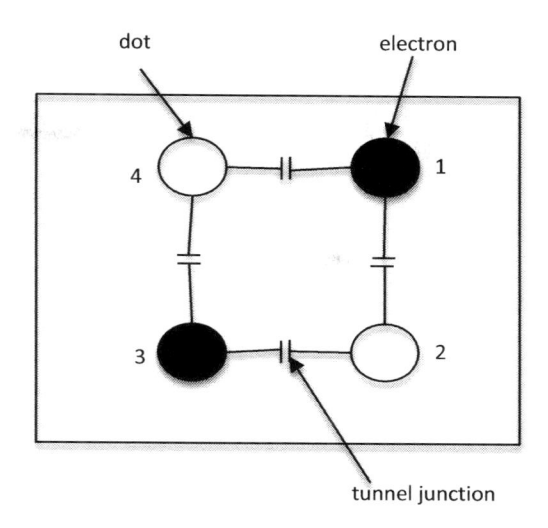

FIGURE 9.32 QCA cell and tunnel junction.

the two excess electrons in this cell are forced to dots at diagonal sites of the cell. As annotated in Figure 9.32, the numbering of the dots in the cell goes clockwise starting from the top right. There are only two stable configurations in a cell, as there are only two diagonals in a square. Accordingly, the two distinct configurations in each cell would have polarizations that are energetically equivalent ground states of the cell. These polarizations −1 and +1 are used to represent the logic states 0 and 1, respectively.

A polarization P in a cell, which measures charge distributed among the four dots, is defined as [38]

$$P = \frac{(\rho_1 + \rho_3) - (\rho_2 + \rho_4)}{\rho_1 + \rho_2 + \rho_3 + \rho_4} \tag{9.67}$$

where ρ_i is the probability of the presence of an electronic charge in each dot of a four-dot cell.

Once polarized, a QCA cell can be in any one of these two possible states depending on the polarization of the charges in the cell. Because of Coulombic interactions, the two most likely polarization states are those defined as $P = +1$ and $P = −1$ when observed, as shown in Figure 9.33, without the symbolic tunnel junctions. Note that these two states are called "most likely" and not the only two polarization states that are because of the small (almost negligible) likelihood of existence of an erroneous state in a continuum of states until it is observed (or measured). Due to the potential barriers between cells, electron tunneling out of one cell to another and/or between the neighboring cells is not possible due to high inter-cell potential. For completeness, the dots in each cell can be arranged in two ways: 90° and 45° format, as seen in Figure 9.33a and b, respectively. However, the 90° format of dot representation in each cell is used in the subsequent developments without the symbolic tunnel junction wires.

Quantum dots can be any charge containers, with discrete electrical energy states (there may be more than two states, but only two are used), sometimes called artificial atoms [39]. Of course, some molecules have well-defined energy states and, therefore, are suitable for supporting the operation of QCA systems. Molecular QCA emanates from building QCA devices out of a single molecule in which each molecular QCA cell consists of a pair of identical allyl groups. Small metal pieces can also behave as quantum dots, if the energy states an electron can occupy are distinguishable, instead of the usual energy band. This means that the difference between two consecutive energy states must be well above the thermal energy (as given by Equation (9.48)).

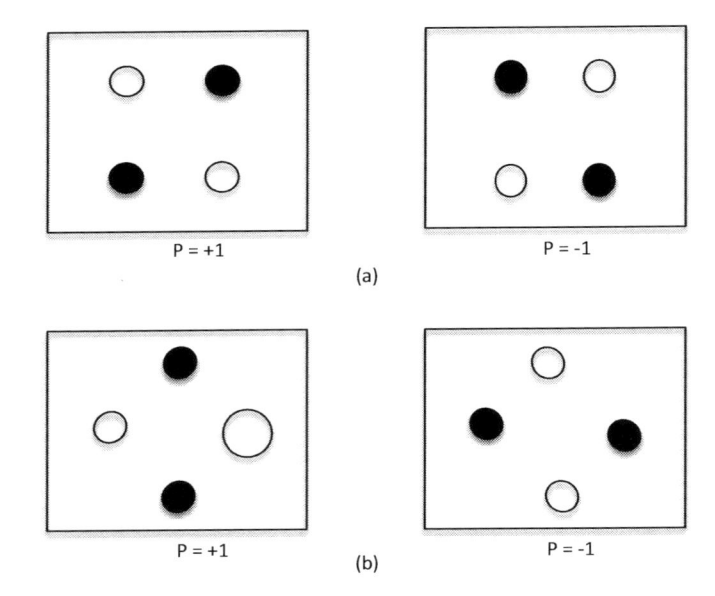

FIGURE 9.33 Two possible polarizations of QCA cells: (a) four dots using 90° arrangement cells and (b) four dots using 45° arrangement cells.

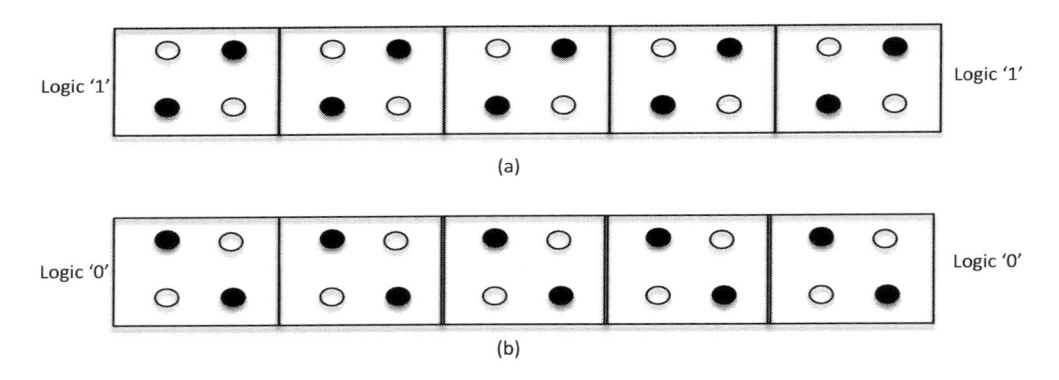

FIGURE 9.34 Logic "1" and "0" pipeline wiring: (a) $P = +1$ and (b) $P = -1$.

9.5.2.2 QCA Gates

QCA gates can be formulated based on the cells' arrangement. Given that all cells are assigned specific polarization (i.e. -1, or $+1$) depending on the input cell (or driver cell), by arranging the cells side-by-side any logic information can be transferred. For example, if two of the input cells are low (i.e. with polarization $P = -1$), then the output cell is low. If two of the input cells are high (i.e. with polarization $P = +1$), then the output cell is high. As seen in Figure 9.34, a line of cells is formed where the input cell's polarization is kept fixed, which determines the state of the whole array. Adjacent QCA cells interact in an attempt to settle to a ground state determined by the current state of the inputs. The polarization of each cell tends to align with that of its neighbors. Importantly, the polarization of the input cell is propagated down the wire, as a result of the system attempting to settle to a ground state. Any cell along the wire that are anti-polarized to the input cell would be at a higher level and would soon settle to the correct ground state. Comparing Figure 9.34a with Figure 9.34b demonstrates that flipping the polarity of the input cell results in the flipping of all other cells. The arrangements in Figure 9.34 can be considered as playing the function of a wire, or a buffer gate, or binary interconnection, where the adjacent cells interact in an attempt to settle to a ground state determined by the current state of the input(s). The polarization of each cell is propagated down the wire.

Fanout, an important part of implementing logically significant QCA arrays as it provides a mechanism for signal splitting, can be formulated as shown in Figure 9.35, where a signal applied to a single input cell is amplified by that cell and sent to two output cells. For an information to propagate through a fanout gate, it is important to ensure that the input cell's charge configuration dictates that of the two output cells.

An array of physically interacting cells can be used to realize Boolean logic functions. As observed in Chapters 2–6 in digital gates developments, for any functional binary expression, the output is the majority of the inputs. Suppose $MAJ(A,B,C) = (ABC) + (A+B+C) + \overline{A}$, with three inputs A, B, and C, each of the constituents is a logic gate with its output represented by F. These constituent QCA gates can be arranged as follows.

AND and *OR gates*: Logic gates AND ($F = ABC$) and OR ($F = A + B + C$) implementations are possible by keeping one of the inputs of *MAJ* to fixed polarization −1 and +1, as shown in Figure 9.36a and b, respectively. Note that the middle cell is called the device cell, which must have the polarization of the majority of the inputs. The right-hand side cell is the output cell which must have the same polarization as device (also called driver) cell.

Inverter gate: By placing cells in a diagonal position, the polarizations of these cells will be reversed. Two standard cells in a diagonal orientation are geometrically similar to two rotated cells in a horizontal orientation. For this reason, standard cells in a diagonal orientation tend to align in opposite polarization directions as in the inverter chain [40]. Based on this characteristic, two arrangements of QCA inverter ($A \rightarrow \overline{A}$) can be constructed, as shown in Figure 9.37, where the input A splits into two parallel wires and is inverted at the point of convergence ($F = \overline{A}$).

In essence, the configuration of adjacent cells is determined by minimizing Coulomb repulsion in each gate's implementation.

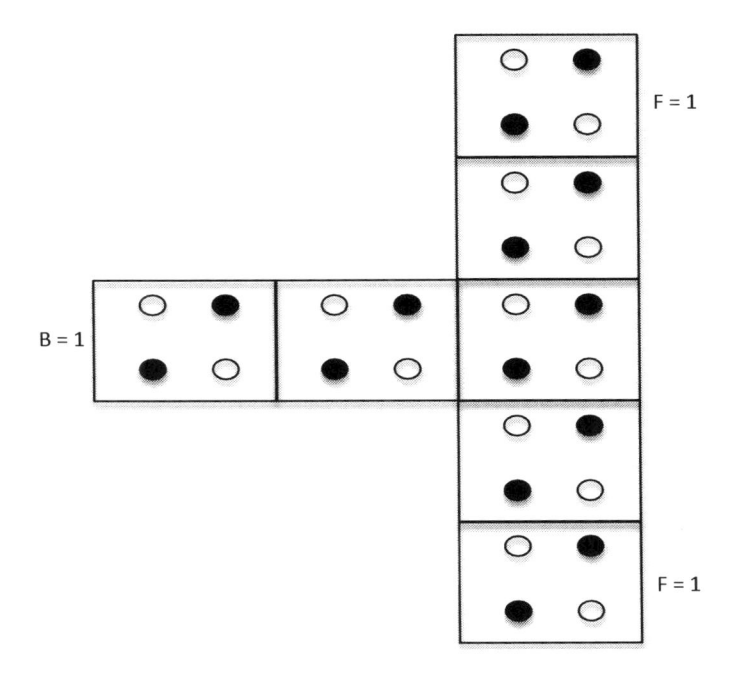

FIGURE 9.35   Implementation of a fanout.

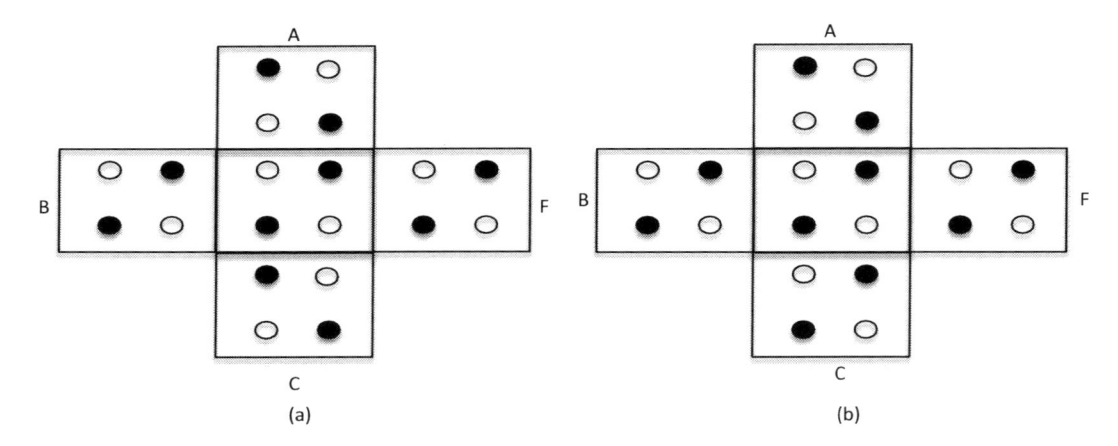

FIGURE 9.36 Implementation of AND and OR gates: (a) AND gate and (b) OR gate.

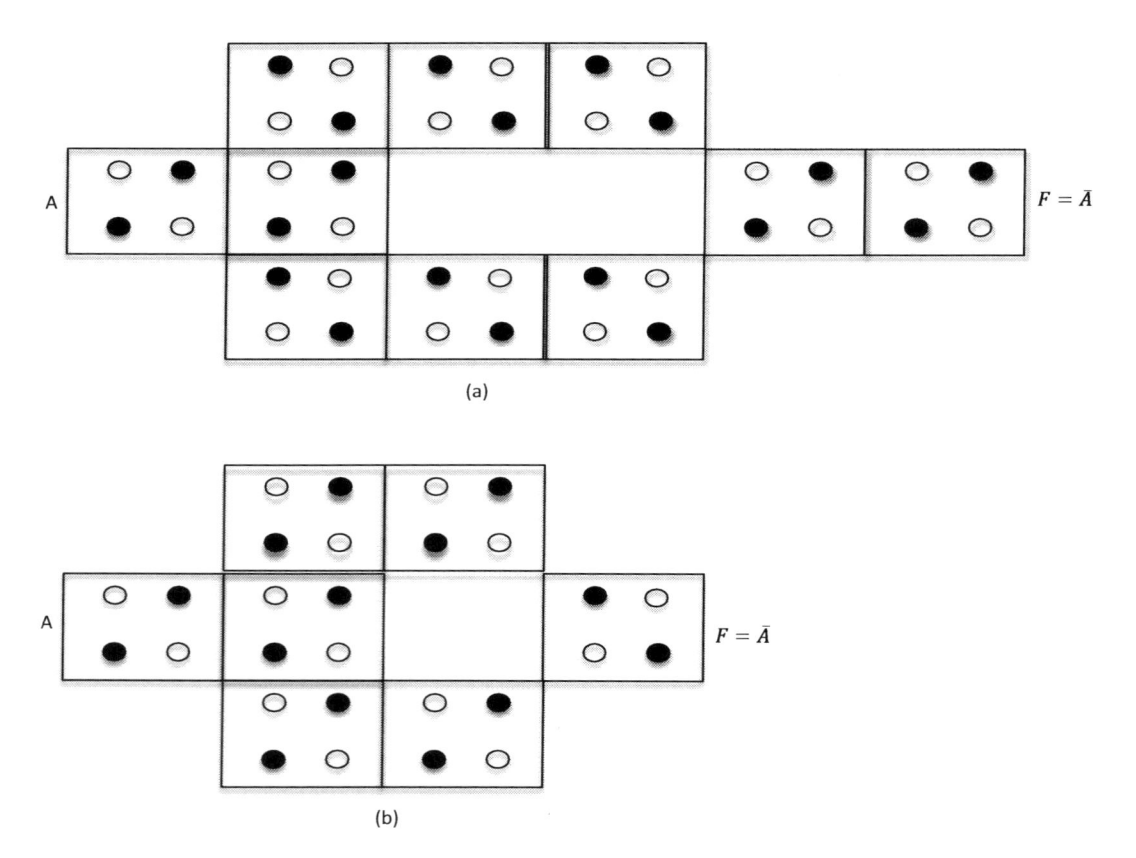

FIGURE 9.37 Two arrangements of an inverter: (a) an inverter arrangement and (b) another inverter arrangement.

9.5.2.3 QCA Circuits

QCA circuits—a network of QCA gates—can be developed with one or multiple layers of cells cross-overing. As observed thus far, the above simple gate design comprises of an array of QCA cells of different polarizations. In order to propagate (or transfer) intended information through

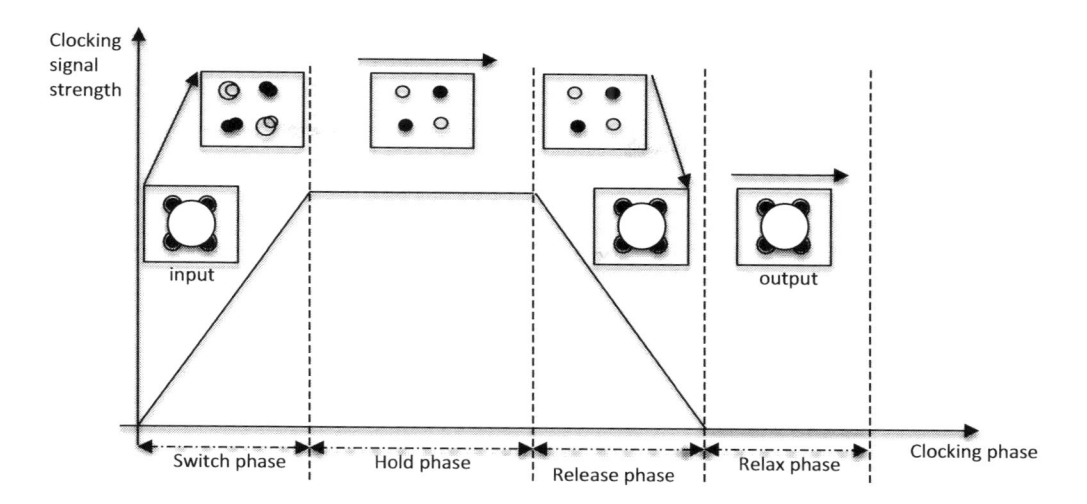

FIGURE 9.38   QCA clocking periodic phases.

QCA circuits without randomly adjusting the cells, an accurate timing process that guarantees delivery of input data at their intended outputs needs to be incorporated. In case of QCA, clocking is a sort of cyclical manipulation of inter-dot tunnel barriers. Timing in QCA is obtained by clocking in four distinct periodic phases: switch, hold, release, and relax [41–43], as depicted in Figure 9.38. Each phase is described as follows. In the

- *Switch phase*: the potential barriers are raised from low to high and the cells become polarized according to the polarization of their driver. It is in this clock phase that actual computation occurs. At the end of this phase, the barriers remain high and suppress any kind of tunneling;
- *Hold phase*: the potential barriers are held at a high value preserving the switch-phase-polarized state. The cell now acts as an input to the next stage;
- *Release phase*: the potential barriers are lowered and the cells are allowed to relax to an unpolarized state; and
- *Relax phase*: the potential barriers remain lowered, the cells remain at unpolarized neutral state, and the cells are ready to switch again and repolarize.

After these four phases, the information is transferred and propagated in QCA circuits by keeping the ground-state polarization at all time [44]. Therefore, QCA clocking controls information flow, and, hopefully, with appropriate power to run the circuits. Each QCA clock phase should consist of at least two cells to maintain its influence during the next clock phase [45–47].

Like in synchronous digital circuits where a clock signal coordinates or initiates the operations to be performed; theoretically, QCA circuits would also require clock sending out regularly spaced signals when to change states. The unique clocking scheme and wire-level pipelining in QCA present serious timing issues. Consequently, QCA circuit designs would encounter timing challenges that make it difficult to assign proper clocking zones and design feedback [46]. Besides, physical realization of QCA circuits are, at this stage, still largely hypothetical because the engineering required to construct quantum electronic components is very difficult.

9.5.2.4   Fabrication of QCA Circuits
Theoretically, QCA circuits can be fabricated using two methods, namely: electrostatic interaction-based QCA method and magnetic QCA method. The electrostatic interaction-based method

comprises of metal islands, semiconductors, nanomagnetics, and molecular structures. Aluminum islands with aluminum-oxide tunneling junctions are implemented on a silicon wafer to create metal-island QCA cells [48]. In addition, the semiconductor QCA type can be implemented using gallium arsenide (GaAs)/gallium aluminum arsenide (AlGaAs) heterostructure material with high mobility electrons using EBL [49,50]. The EBL has been described in Chapter 1, Section 1.6.

In the molecular QCA, free electrons are induced to switch between ferrocene groups (that act like quantum-dots) and cobalt group (e.g. graphene) in the cell's center that provides a bridging ligand acting like a tunneling path [51].

Moreover, nanomagnetic QCA can be fabricated using EBL [52,53]. More research studies are active in the use of QCA technology.

9.6   SUMMARY

The chapter develops the fundamental principles of quantum computation with the development of quantum computational toolkit including reversible quantum gates, and establishes the basic building blocks for quantum circuits. Existing quantum hardware is limited in terms of the number of available qubits. In general, efficient design of quantum circuits to performing arithmetic operation and other axillary functions needs special attention in quantum systems (and computing) and should be fault tolerant.

Also, in this chapter, an overview of the protocols required to ensure uninfected transmitted information is received while transiting through the quantum communication channels is discussed. However, noise imposed on the quantum communication channels would need to be appropriately adjusted by using QEC techniques to clean up the erroneous pairs and/or errors (or decoherence) in the quantum channel. It is argued that there is a connection between distillable entanglement and QEC. As a result, entanglement distillation protocols can be used as a type of error correction for quantum communications channels between two authorized parties. There would be variable algorithms to be developed that would enable secured information traverse the quantum communication channels.

Moreover, the principle behind the QCA technology was discussed. In the QCA architecture, information is transferred between neighboring cells by mutual interaction from cell to cell, leading to the formulation of QCA logic gates from an array of cells of different polarizations. QCA circuits can be developed with one or multiple layers and inputs by an appropriate arrangement of cells. The unique clocking scheme and wire-level pipelining in QCA present serious timing issues. Consequently, QCA circuit designs would encounter timing challenges that make it difficult to assign proper clocking zones and design feedback. More research studies are active in the use of QCA technology.

PROBLEMS

9.1.   How can you estimate the bond length between different elements in molecules?

9.2.   Is quantum entanglement a physical phenomena?

9.3.   When is a quantum state not collapsed?

9.4.   What notation is used for vector spaces in quantum mechanics?

9.5.   What happens when adjacent QCA cells interact?

9.6.   If the size of the nanoparticle grains is manipulated, what effect will it have on the property changes in the bulk material?

9.7.   Is carbon nanotube suitable for FET manufacturing?

9.8.   Is tunneling criterion the only obstacle to factor in when considering manufacturing electronic devices?

9.9.   What does Coulomb blockade mean?

REFERENCES

1. Levitov, L.S., Lee, H.-W. and Lesovik, G. (1996). Electron counting statistics and coherent states of electric current. *Journal of Mathematical Physics*, 37(10), 4845.
2. Bell, J.S. (1964). On the Einstein-Podolsky-Rosen paradox. *Physics*, 1, 195–200.
3. Schrödinger, E. (1935). Discussion of probability relations between separated systems. *Mathematical Proceedings of the Cambridge Philosophical Society*, 31(4), 555–563.
4. Nielsen, M.A. and Chuang, I.L. (2010). *Quantum Computation and Quantum Information*, 10th edition. Cambridge: Cambridge University Press.
5. Pauli, W. (1927). Zur Quantenmechanik des magnetischen Elektrons. *Zeitschrift für Physik*, 43, 601.
6. Cooper, M.D. (2006). Electric dipole moment of the neutron. *Los Alamos Science*, 30, 208–210.
7. Dirac, P.A.M. (1926). The Compton effect in wave mechanics. *Proceedings of the Cambridge Philosophical Society*, 23, 500–507.
8. Susskind, L. and Friedman, A. (2014). *Quantum Mechanics: The Theoretical Minimum*. London: Penguin Random House.
9. Greiner, W. (2000). *Quantum Mechanics: An Introduction*. Berlin: Springer.
10. Quimby, R.S. (2006). *Photonics and Lasers: An Introduction*. Worcester, MA: Wiley.
11. Prelas, M. (2016). *Nuclear-Pumped Lasers*. New York: Springer.
12. Klyshko, D.N., Chekhova, M. and Kulik, S. (2011). *Physical Foundations of Quantum Electronics*. Singapore: World Scientific Publishing.
13. Sverdlov, V.A., Walls, T.J. and Likharev, K.K. (2003). Nanoscale silicon MOSFETs: A theoretical study. *IEEE Transactions on Electron Devices*, 50(9), 1926–1933.
14. Walls, T.J. (2008). Theoretical analysis of prospective nanoelectronic devices. *PhD Thesis*, Stony Brook University, New York.
15. Toffoli, T. (1980). Reversible computing. Technical Memo MIT/LCS/TM-151, MIT Lab for Computer Science.
16. Cuccaro, S.A., Draper, T.G., Kutin, S.A. and Petrie, D. (2005). A new quantum ripple-carry addition circuit. 8th *Workshop on Quantum Information Processing*, Cambridge, 1–9.
17. Markov, I.L. and Saeedi, M. (2012). Constant-optimized quantum circuits for modular multiplication and exponentiation. *Quantum Information and Computation*, 12 (5–6), 361–394.
18. Baugh, C.R. and Wooley, B.A. (1973). A two's complement parallel array multiplication algorithm. *IEEE Transactions on Computers*, 22, 1045–1047.
19. Sridharan, K. and Vikramkumar, P. (2015). *Design of Arithmetic Circuits in Quantum Dot Cellular Automata Nanotechnology*. Basel: Springer.
20. Cho, H. and Swartzlander, E.E. (2009). Adder and multiplier design in quantum-dot cellular automata. *IEEE Transactions on Computers*, 58(6), 721–727.
21. Kempe, J. (2005). Approaches to quantum error correction. *Seminaire Poincare*, 1, 65–93.
22. Botsinis, P., Babar, Z., Alanis, D., Chandra, D., Nguyen, H., Ng, S.X. and Hanzo, L. (2016). Quantum error correction protects quantum search algorithms against decoherence. *Scientific Reports*, 6(38095), 1–13. doi: 10.1038/srep38095.
23. Shor, P.W. (1995). Scheme for reducing decoherence in quantum memory, *Physical Review A*, 52, 2493–2496.
24. Shor, P.W. (1997). Polynomial-time algorithms for prime factorization and discrete logarithms on a quantum computer. *SIAM Journal of Computing*, 26(5), 1484–1509.
25. Marinescu, D.C. and Marinescu, G.M. (2012). Quantum error-correcting codes. In *Classical and Quantum Information*. MA: Academic Press, 455–562.
26. Kolawole, M.O. (2013). *Satellite Communication Engineering*. Boca Raton, FL: CRC Press.
27. Nikiforov, O., Sauer, A., Schickel, J., Weber, A., Alber, G., Mantel, H. and Walther, T. (2018). Side-channel analysis of privacy amplification in postprocessing software for a quantum key distribution system. Technische Universität Darmstadt, Technical Report TUD-CS-2018-0024.
28. Gallager, R. (1962). Low-density parity-check codes. *IRE Transactions on Information Theory*, 8(1), 21–28.
29. Brassard, G. and Salvail, L. (1993). Secret-key reconciliation by public discussion. *Proceedings of the 12th Workshop on the Theory and Application of Cryptographic Techniques (EUROCRYPT)*, Berlin, 410–423.
30. Jouguet, P. and Kunz-Jacques, S. (2014). High performance error correction for quantum key distribution using polar codes. *Quantum Information and Computation*, 14(3–4), 329–338.

31. Bennett, C.H., Brassard, G. Crepeau, C. and Maurer, U.M. (1995). Generalized privacy amplification. *IEEE Transactions on Information Theory*, 41(6), 1915–1923.

32. Moreau, P-A., Toninell, E., Gregory, T., Aspden, R.S., Morris, P.A. and Padgett, M.J. (2019). Imaging Bell-type nonlocal behaviour. *Science Advances*, 5(7), eaaw2563.

33. Ekert, A., Hayden, P. and Inamori, H. (2000). *Basic Concepts in Quantum Computation*. Los Alamos, NM. e-print arXiv: quant-ph/0011013.

34. Kielpinski, D., Kafri, D., Woolley, M., Milburn, G. and Taylor, J. (2012). Quantum interface between an electrical circuit and a single atom. *Physical Review Letters*, 108, 130504.

35. Göppl, M.V. (2009). Engineering quantum electronic chips—Realization and characterization of circuit quantum electrodynamics systems. *PhD Thesis*, Diplom-Physiker (University), TU München.

36. Rosenblatt, S., Hertzberg, J., Brink, M., Chow, J., Gambetta, J., Leng, Z., Houck, A., Nelson, J.J., Plourde, B., Wu, X., Lake, R., Shainline, J., Pappas, D., Patel, U. and McDermott, R. (2017). Variability metrics in Josephson junction fabrication for quantum computing circuits. *American Physical Society (APS) Meeting*, 62(4). http://meetings.aps.org/link/BAPS.2017.MAR.Y46.2.

37. Fuechsle, M., Miwa, J.A., Mahapatra, S., Ryu, H., Lee, S., Warschkow, O., Hollenberg, L.C.L., Klimeck, G. and Simmons, M.Y. (2012). A single-atom transistor. *Nature Nanotechnology*. doi: 10.1038/nnano.2012.21.

38. Tougaw, P.D. and Lent, C.S. (1996). Dynamic behavior of quantum cellular automata. *Journal of Applied Physics*, 80, 4722–4736.

39. Sasamal, T.N., Singh, A.K. and Mohan, A. (2019). *Quantum-Dot Cellular Automata Based Digital Logic Circuits: A Design Perspective*. Singapore: Springer.

40. Tougaw, P.D. and Lent, C.S. (1994). Logical devices implemented using quantum cellular automata. *Journal of Applied Physics*, 75(3), 1818–1825.

41. Lent, C.S. and Tougaw, P.D. (1997). Device architecture for computing with quantum dots. *Proceedings of the IEEE*, 85(4), 541–557.

42. Frost, S.E., Dysart, T.J., Kogge, P.M. and Lent, C.S. (2004). Carbon nanotubes for quantum-dot cellular automata clocking. *4th IEEE Conference on Nanotechnology*, Munich, 171–173.

43. Niu, M. (ed) (2018). *Advanced Electronic Circuits—Principles, Architecture and Applications on Emerging Technologies*. London: IntechOpen.

44. Mehta, U. and Dhare, V. (2015). Defect characterization and testing of QCA devices and circuits: A survey. *19th International Symposium on VLSI Design and Test (VDAT)*, Ahmedabad, Vol. 1, 1–2.

45. Kim, K., Wu, K. and Karri, R. (2007). The robust QCA adder designs using composable QCA building blocks. *IEEE Transactions on Computer—Aided Design of Integrated Circuits and Systems*, 1, 176–183.

46. Liu, W., Swatzlander, E.E. and O'Neill (2013). *Design of Semiconductor QCA Systems*. London: Artech House.

47. Sheikhfaal, S., Navi, K., Angizi, S. and Navin, A.H. (2015). Designing high speed sequential circuits by quantum-dot cellular automata: Memory cell and counter study. *Quantum Matter*, 4, 190–197.

48. Orlov, A., Amlani, A., Bernstein, G., Lent, C. and Snider, G. (1997). Realization of a functional cell for quantum-dot cellular automata. *Science*, 277, 928–930.

49. Smith, C., Gardelis, S., Rushforth, A., Crook, R., Cooper, J. and Ritchie, D. (2003). Realization of quantum-dot cellular automata using semiconductor quantum dots. *Superlattices Microstructures*, 34, 195–203.

50. Wang, Z. and Liu, F. (2011). Nanopatterned graphene quantum dots as building blocks for quantum cellular automata. *Nanoscale*, 3, 4201–4205.

51. Pulimeno, A., Graziano, M., Sanginario, A., Cauda, V., Demarchi, D. and Piccinini, G. (2013). Bis-ferrocene molecular QCA wire: Ab initio simulations of fabrication driven fault tolerance. *IEEE Trans Nanotechnology*, 12, 498–507.

52. Cowburn, R. and Welland, M. (2000). Room temperature magnetic quantum cellular automata. *Science*, 287, 1466–1468.

53. Orlov, A., Imre, A., Csaba, G., Ji, L., Porod, W. and Bernstein, G.H. (2008). More research studies in the QCA area are active. *Journal of Nanoelectronics and Optoelectronics*, 3, 1–14.

Appendix A
Notations

The symbols have been chosen carefully as possible to prevent confusion. In a few instances, the same symbol has been used. When this occurs, a clear distinction is made in the meaning of each, and where it is used in the text is indicated.

A_c:	MOS (*metal oxide semiconductor*) capacitor area.
A_I:	Current gain
A_v:	Voltage gain
A_{tc}:	transconductance gain
A_{tr}:	transresistance gain
c:	speed of light in vacuum ($\approx 3 \times 10^8$ m s^{-1})
C:	capacitance
C_{fringe}:	substrate fringing capacitance
C_i:	binary adder functional *carry in*
C_{i+1}:	binary adder functional *carry out*
C_{ox}:	thin-oxide field capacitance
C_{pp}:	parallel-plate capacitance
C_W:	interconnect wire's capacitance
e:	electronic charge ($\approx 1.602 \times 10^{-19}$C) in Chapter 9
DNL:	differential nonlinearity error
d_g:	nanoparticle grain size
d_s:	separation distance
E_B:	nuclear binding energy
E_d:	energy dissipated per bit
E_g:	bandgap
E_p:	photon energy (also called *quantum energy*)
E_{th}:	thermal energy
E_{Tdy}:	total energy dissipated
f:	frequency of photon or electromagnetic radiation
F:	output of logic gate functional expression (Chapters 2–6 and 9—Section 9.5.2), or the force acting on a body of mass from Newton's second *law of motion* in classical mechanics (Chapter 9, Section 9.3)
g_m:	transconductance
G_{ion}:	gain factor due to ionization
i_b:	base current (small letter "i" denotes "variable" current)
i_c:	collector current (small letter "i" denotes "variable" current)
i_e:	emitter current (small letter "i" denotes "variable" current)
I:	current through a diode or identity matrix (Chapter 9).
$I(t)$:	elementary current pulse
$I_{C(\text{ion})}$:	current due to the onset of ionization
I_D:	drain current
I_G:	gate current

I_s:	saturation current
INL:	integral nonlinearity error
K:	amplifier gain (Chapter 4)
k:	Boltzmann's constant ($\approx 1.3805 \times 10^{-23}$ JK^{-1})
k_1:	Process-dependent (Rayleigh) constant
k_n:	NMOS aspect ratio
k_p:	PMOS aspect ratio
k_y:	nanomaterial strengthening coefficient
L:	polysilicon integrated circuit (IC) sheet length and width
L_D:	Debye length
L_{drawn}:	transistor drawn length
L_{eff}:	effective channel length
L_N:	total angular momenta
L_{poly}:	polysilicon gate length after etches
L_s:	spin angular momentum
MAJ(x, y, z):	majority circuit of a Boolean expression
m_n:	mass of a neutron
m_p:	mass of a proton
M_a:	actual mass of an atom
NA:	numerical aperture of the lens system
N_n:	number of neutrons
N_m:	number of modes
n_i:	intrinsic carrier density
P:	output power of the amplifier (Chapter 4) or polarization in a cell (Chapter 9)
P_{dy}:	dynamic power dissipated by the logic gate
P_{min}:	minimum pumping power to keep the laser process going
Q:	surface charge
Q_B:	Total substrate charge
Q_{IS}:	Equivalent interface charge
Q_G:	metal gate charge
Q_N:	Total inversion or channel charge
R_Δ:	resistance between the conductors
R_n:	atomic radius
R_{res}:	resolution of the optical lithographic system
R_s:	sheet resistance
R_W:	wire (or line) resistance
$S(f)$:	power spectral density of a noise waveform
S_i:	binary adder functional sum
$S_{tni}(f)$:	thermal noise spectral density
S_x, S_y, S_z:	the components of spin operator
$S_{tn-m}(f)$:	MOS transistor thermal noise spectral density when operating in saturation mode
$S_{tnv}(f)$:	thermal noise spectral density of resistive conductor
SNR:	signal-to-noise ratio
t:	time
t_c:	average lifetime required to maintain the necessary population inversion
t_{ox}:	silicon-oxide wafer thickness
T:	absolute temperature
T_s:	sampling time of measurement
U:	unitary transformation

V:	potential or voltage
V_T:	thermal voltage
V_{BR}:	*breakdown voltage* (Chapter 3)
V_{droop}:	droop rate
V_G:	gate voltage
V_{GS}:	gate-source voltage
v_{fn-v}^2:	spectral density of flicker noise
v_{in}:	input voltage
V_m:	CMOS inverter switching threshold
V_o:	output voltage
V_{ref}:	reference voltage
V_{sp}:	inverter switching point voltage
V_{th}:	device voltage threshold
W:	polysilicon integrated circuit (IC) sheet width
W_{eff}:	effective channel width
X_C:	capacitive reactance
Z:	gate output functional expression (Chapter 2)
Z_C:	impedance
α:	plate area (Chapter 1), *forward common-base current gain* (Chapter 3), or coefficient (Chapter 9)
β:	forward current gain (Chapter 3), feedback network gain (Chapter 4), or coefficient (Chapter 9)
δx_n:	quantization noise
ε_0:	permittivity of vacuum or free space, or simply electric constant
ε_{ox}:	silicon-oxide wafer permittivity
ε_s:	body substrate's permittivity
γ:	*gamma factor* (Chapters 1 and 9); *early-effect* factor (Chapter 4); *body-effect* factor (Chapters 5 and 6)
ζ:	transconductance parameter
h:	Planck's constant ($\approx 6.62608 \times 10^{-34}\ Js$)
$\hbar$:	quantization of angular momentum
λ:	wavelength
Δ:	correction factor
ΔL_{diff}:	channel length offset
Δx_n:	conversion error
τ_D:	total delay through the distributed RC circuit
σ_o:	material yield strength
Σ_x:	phase flip error
Σ_z:	bit flip error
$\Sigma_x\sigma_z$:	bit and phase flip error
σ_x:	Pauli-X matrix
σ_y:	Pauli-Y matrix
σ_z:	Pauli-Z matrix
q:	electronic charge ($\approx 1.602 \times 10^{-19}$C) in Chapter 1
Φ_b:	semiconductor *work function*
μ:	mobility (Chapter 1)
μ_N:	magnetic dipole moment
μ_0:	permeability of free space ($= 4\pi \times 10^{-7}$ H m^{-1})
x_{ref}:	conversion scaling factor

$\psi(x, t)$:	wavefunction
$\lvert E \rangle$:	state of the environment noise before interaction with the qubit
$\lvert \psi \rangle$:	qubit
$\lvert \psi \rangle$:	vector space
$\lvert + \rangle$ or $\lvert - \rangle$:	quantum state
$\lVert v \rVert$:	norm of $\lvert v \rangle$

Appendix B
Glossary of Terms

Absorption: A phenomenon involving the interplay between matter and radiation: it happens when an atom absorbs a photon.

Adder: A digital logic circuit that performs addition of numbers.

Amplifier: A device that responds to a small input signal (voltage, current, or power) and delivers a larger output signal that contains the essential waveform features of the input signal.

Arithmetic Logic Unit (ALU): A combinational logic built from circuits where the clock governs its register behavior which times the process.

Astable circuit: A circuit that allows oscillation between the two states with no external triggering to produce the change in state.

Backgate: The phenomenon known as coupling effect between front and back gates. Also called the "body effect."

Band-pass filter: A filter that passes some particular band of frequencies and rejects all frequencies outside the range.

Bandgap: The energy range (or band) in a solid where no electron states can exist. The energy gap between the bands depends on the strength of bonds between atoms. For example, carbon forms very strong C-C bonds so it has a very large bandgap.

Barkhausen criterion: The necessary condition for a stable output sine wave where the loop gain must be unity at a specific frequency for which its phase is zero degree.

Bistable: A one-bit memory that has two stable internal states.

Bode magnitude: One of the two "Bode plots" plotted as a function of frequency.

Buffer: A single-input device that has unity gain.

Carbon nanotubes: The carbon structures in which each carbon forms three bonds with other carbon atoms but forms tubes rather than balls. They have excellent charge transport properties. Notably, they are not the same as carbon fibers. Carbon nanotubes can be divided into two groups, namely:

> **Single-walled carbon nanotube (SWCNT):** A hollow and cylindrically shaped one-dimensional graphitic carbon allotrope that appears as a rolled-up sheet of monolayer graphene.

> **Multiwalled carbon nanotube (MWCNT):** It is composed of a number of SWCNTs nested inside one another and can be thought of as a rolled-up sheet of multilayer graphene.

Central processing unit (CPU): The part of a computer in which operations are controlled and executed.

Colpitts oscillator: Another type of oscillator, whose circuit layout is technically identical to the Pierce oscillator except for the location of the ground point.

Converter: A device that receives one type of signal and outputs another type of signal.

Comparator: A combinational circuit that compares two binary numbers and determines their relationship.

Coulombic effects: The charging effects and resonant tunneling in quantum dots, or the intercell interaction between electrons of the neighboring cells, or simply the effect of the interactions between electrons, or coulomb forces acting between electrons, or inelastic scattering in the moving reference frames.

Crystal oscillator: Another type of oscillator based on the property of piezoelectricity exhibited by some crystals and ceramic materials.

Cutoff region: The region where the emitter–base junction and the base–collector junctions are reverse biased.

Debye length: The spatial extension of a perturbation inside a semiconductor.

Decoder: A decoder is an inverse operation of an encoder.

DeMorgan's law: A law that describes how mathematical statements and concepts are related through their opposites.

Depletion mode: A channel exists even with zero gate voltage.

Digital signal processing (DSP): A process of performing a wide variety of signal processing operations.

Duality: The circumstance when the current or voltage in one circuit behaves in a similar manner as the voltage or current, respectively, in another circuit.

Electrical conductivity: A measure of a material's ability to allow the flow of electrons through it.

Encoder: A device that converts information from one code format to another with the aim of shrinking information size for speed, secrecy, or security purpose.

Fanout: The number of gates (or load) that can exist without impairing the normal operation of the gate.

Flip-flop: A *synchronous* basic logic gate that samples its inputs and changes its inputs only at times determined by a clocking signal.

High-pass filter: A filter that passes high frequencies and blocks low frequencies.

Karnaugh map: Another sort of truth table which is constructed so that all adjacent entries on the table represent reducible minterm pairs.

Kirchhoff's Current Law: It states that the sum of currents entering and leaving a node is zero.

Laplace Transform: An integral transform that converts functions with a real dependent variable (such as time) into functions with a complex dependent variable (such as frequency, often represented by s).

Laser: A laser is a device that generates or amplifies light. It stands for light amplification by stimulated emission of radiation.

Latch: An *asynchronous* basic logic gate that watches all of its inputs continuously and changes its outputs at any time, independent of a clocking signal.

Linear time-invariant (LTI): When a system satisfies both the additive and the scaling criteria and its behavior does not change over time.

Logic gate: A circuit that implements a Boolean function and performs a logical operation on one or more binary inputs and produces a single binary output. Examples of logic gate include AND, NAND, OR, XOR, NOT, NOR, and XNOR.

Least significant bit (LSB): The lowest bit in binary.

Low-pass filter: A filter that passes low frequencies and rejects high frequencies.

Maser: A maser produces and amplifies electromagnetic radiation mainly in the microwave region of the spectrum. It stands for microwave amplification by stimulated emission of radiation.

Maxterm: A *sum* (OR) term that contains each of the n variables as factors in either complemented or uncomplemented form.

Mextram: A scalable model that describes the various intrinsic and extrinsic regions of bipolar transistors.

Miller Effect: It basically explains how an impedance sitting across an amplifier is effectively converted into a smaller impedance at the input and a roughly equivalent one at the output.

Minterm: A *product* (AND) term that contains each of the n variables (literals) as factors in either complemented form or uncomplemented form.

Monostable circuit: A circuit that allows its output to change its state for a period, after a trigger, and then returns to the base state.

Most significant bit (MSB): The highest bit in binary.

Multivibrator: A two-stage amplifier operating in two modes or states.

Nano: A unit prefix used primarily with the metric system, which denotes a factor of 10^{-9}.

Nanoelectronics: It generally refers to semiconductor or microelectronic devices that have been shrunk to nanoscale.

Nanofabrication: A manufacturing process that involves making devices at the smallest dimensions.

Nanomaterials: Materials that have particles or constituents of nanoscale dimensions and can be metals, ceramics, polymeric materials, or composite materials.

Nanotechnology: The way discoveries made at the nanoscale are put to work that involves the manipulation of matter at the atomic level.

Noise: Unwanted influences caused by random motion of electrons in a conductor. Types include (Chapter 6):

Flicker noise: Present whenever a direct current (dc) flows in a discontinuous material.

Shot noise: Results from the discrete movement of charge across a potential barrier.

Thermal noise: Caused by the thermal agitation of charge carriers (electrons or holes) in a conductor.

Quantization noise (also called* quantization error*): A rounding error between the analog input voltage to the analog-to-digital conversion (ADC) and the output digitized value. It is, however, strongly dependent on and, in general, correlated with the input signal, and hence cannot be precisely modeled as signal-independent additive noise (Chapter 7).

Switching noise: The noise present on the supply rails resulting from the switching of large currents in the presence of parasitic resistances and inductors (Chapters 1, 4, 6, and 7).

Ohmic region: The region where MOSFET behaves like a resistor.

Orthogonal frequency-division multiplexing system: A modulation technique used in many new and emerging broadband technologies including Internet and digital audio and video broadcasting.

Oscillator: A circuit or device that generates a given waveform at a constant peak amplitude and specific frequency and maintains this waveform within certain limits of amplitude and frequency.

Pass transistors: A chain of transmission gates.

Pierce oscillator: Another type of oscillator.

Photonic accelerator: A processor that can process time-serial data either in an analog or digital fashion on a real-time basis.

Polarization: A measure of charge distributed among the four dots in a cell.

Polycide: A method of patterning the silicide (a compound that has silicon with more electropositive elements) on the polysilicon gate electrode.

Quantum dot: A nanoparticle made of any semiconductor material, for example, silicon (Si), cadmium selenide (CdSe), cadmium sulfide (CdS), or indium arsenide (InAs). Technically, quantum dots are any charge containers with discrete electrical energy states.

Quantum-dot cellular automata (QCA): A concept that works on the intercell interaction mechanism where Coulombic effects (see definition above) dominate over tunneling (see definition below).

Quiescent point (Q-point): The switching threshold, which represents the dc bias, or starting point, for the transistor circuit and that the variations about this point carry the information (ac signals) in the circuit.

Rayleigh Criterion: The minimum required criterion used to calculate the resolution of photolithography equipment.

Sample-and-hold circuit: A circuit that samples (or captures) its input voltage and holds that voltage at its output until the next sample is taken.

Schmitt trigger: A comparator circuit with hysteresis implemented by applying positive feedback to the non-inverting input of a differential or comparator amplifier. It has different transition voltages, depending on whether the input signal is changing from high to low or low to high.

Semiconductor: A unique material usually a solid chemical element—such as silicon (Si), or compound—such as gallium arsenide (GaAs) or indium antimonide (InSb), which can conduct electricity or not under certain conditions.

Sequential circuit: A logic circuit whose output depends not only on the present value of its input signals but also on the sequence of past inputs.

Signal: Any time-varying, or spatial-varying, quantity.

Silicide: A method of self-aligning a silicide layer to all exposed silicon regions.

Spontaneous emission: A phenomenon involving the interplay between matter and radiation: it happens when an atom in an excited state emits a photon.

Stability factor: An indicator of the degree of change in operating point due to variation in temperature that is used to compare different biasing circuits' performances.

State: A snapshot of the memory.

Switch: A device that connects the output of one circuit to the input of another.

Threshold: An agreed voltage or current level.

***Transfer function*:** An important characteristic that embodies the dynamic or control behavior of a circuit or system, determined from the ratio of the output signal to the input signal. It is also called *system function*.

Transistor: A solid-state electronic device that is used to control the flow of electricity in electronic equipment, which usually comprises at least three electrodes.

Tunneling: A functioning concept that describes the behavior of a quantum object when it hits a barrier.

Wien bridge circuit oscillator: A two-stage RC coupled amplifier circuit; an oscillator susceptible to distortion, which might make it unstable.

Work function: The energy difference between the Fermi level and the vacuum level. In solid-state physics, it is the minimum thermodynamic work needed to remove an electron from a solid to a point in the vacuum immediately outside the solid surface.

Index